Biomedical Signal Processing

This book presents the theoretical basis and applications of biomedical signal analysis and processing. Initially, the nature of the most common biomedical signals, such as electroencephalography, electromyography, electrocardiography and others, is described. The theoretical basis of linear signal processing is summarized, with continuous and discrete representation, linear filters and convolutions, Fourier and Wavelets transforms. Machine learning concepts are also presented, from classic methods to deep neural networks. Finally, several applications in neuroscience are presented and discussed, involving diagnosis and therapy, in addition to other applications.

Features:

- Explains signal processing of neuroscience applications using modern data science techniques.
- Provides comprehensible review on biomedical signals nature and acquisition aspects.
- Focusses on selected applications of neurosciences, cardiovascular and muscle-related biomedical areas.
- Includes computational intelligence, machine learning and biomedical signal processing and analysis.
- Reviews theoretical basis of deep learning and state-of-the-art biomedical signal processing and analysis.

This book is aimed at researchers, graduate students in biomedical signal processing, signal processing, electrical engineering, neuroscience and computer science.

Biomedical Signal and Image Processing

Series Editor: Ganesh R. Naik

Adelaide Institute for Sleep Health, Flinders University, Adelaide, Australia

Biomedical signal and image processing involve the analysis of physiological measurements to provide useful information upon which clinicians can make decisions. Working with traditional bio-measurement tools, the signals and images can be computed by software to provide physicians with real-time data and greater insights to aid in clinical assessments. These challenges motivate the further study of biomedical signal and image analysis, and this book series intends to report the new results of these efforts. Topic coverage aims to include Neurological signal processing, Pathological speech signal analysis, Electromyography (EMG) signal processing, EEG, ECG, BCI and all associated analytical techniques. This book series is intended for both biomedical Engineers and clinicians (researchers and graduate students) who wish to get novel research ideas and some training in theory and recent advances of Clinical, Neural engineering and Biomedical signal/image processing applications.

Biomedical Signal Processing
A Modern Approach
Edited by Ganesh R. Naik and Wellington Pinheiro dos Santos

For more information about this series, please visit: www.routledge.com/Biomedical-Signal-and-Image-Processing/book-series/BSIP

Biomedical Signal Processing

A Modern Approach

Edited by Ganesh R. Naik and
Wellington Pinheiro dos Santos

CRC Press
Taylor & Francis Group
Boca Raton London New York

CRC Press is an imprint of the
Taylor & Francis Group, an **informa** business

Contents

About the Editors viii
List of Contributors ix
Preface x

SECTION 1 *Physiological Signal Processing— Challenges*

Chapter 1 Signal Processing for Understanding Physiological
Mechanisms for Information Decoding 3

P. Geethanjali

Chapter 2 Automated Recognition of Alzheimer's Dementia:
A Review of Recent Developments in the Context of
Interspeech ADReSS Challenges 43

*Muhammad Shehram Shah Syed, Zafi Sherhan Syed,
Margaret Lech and Elena Pirogova*

Chapter 3 Electrogastrogram Signal Processing: Techniques and
Challenges with Application for Simulator Sickness
Assessment 62

Nadica Miljković, Nenad B. Popović, and Jaka Sodnik

Chapter 4 Impact of Cognitive Demand on the Voice Responses of
Parkinson's Disease and Healthy Cohorts 90

Rekha Viswanathan and Sridhar P. Arjunan

SECTION 2 *EEG—ECG Signal Processing*

Chapter 5 Electroencephalography and Epileptic Discharge
Identification 115

*Mohd Syakir Fathillah, Theeban Raj Shivaraja,
Khalida Azudin and Kalaivani Chellappan*

Chapter 6   A Novel End-to-End Secure System for Automatic
                Classification of Cardiac Arrhythmia 139

                Narendra K. C., Pradyumna G. R. and Roopa B. Hegde

Chapter 7   Machine Learning for Detection and Classification of
                Motor Imagery in Electroencephalographic Signals 149

                *Juliana C. Gomes, Vanessa Marques, Caio de Brito, Yasmin
                Nascimento, Gabriel Miranda, Nathália Córdula, Camila
                Fragoso, Arianne Torcarte, Maíra A. Santana, Giselle Moreno
                and Wellington Pinheiro dos Santos*

Chapter 8   Emotion Recognition from Electroencephalographic and
                Peripheral Physiological Signals Using Artificial Intelligence
                with Explicit Features 172

                *Maíra A. Santana, Juliana C. Gomes, Arianne S. Torcate,
                Flávio S. Fonseca, Amanda Suarez, Gabriel M. Souza, Giselle
                M. M. Moreno and Wellington Pinheiro dos Santos*

Chapter 9   Identification of Emotion Parameters in Music to Modulate
                Human Affective States: Towards Emotional Biofeedback as
                a Therapy Support 198

                *Maíra A. Santana, Ingrid B. Nunes, Flávio S. Fonseca, Arianne
                S. Torcate, Amanda Suarez, Vanessa Marques, Nathália
                Córdula, Juliana C. Gomes, Giselle M. M. Moreno and
                Wellington Pinheiro dos Santos*

SECTION 3   Gait—Balance Signal Processing

Chapter 10   Updated ICA Weight Matrix for Lower Limb Myoelectric
                Classification 225

                Ganesh R. Naik

Chapter 11   Cortical Correlates of Unilateral Transfemoral Amputees
                during a Balance Control Task with Vibrotactile Feedback 235

                Aayushi Khajuria, Upinderpal Singh and Deepak Joshi

Chapter 12   Assessing the Impact of Body Mass Index on Gait Symmetry
                of Able-Bodied Adults Using Pressure-Sensitive Insole 246

                *Maria Rashid, Asim Waris, Syed Omer Gilani, Faddy
                Al-Najjar, Amit N. Pujari and Imran Khan Niazi*

Chapter 13 Analysis of Lower Limb Muscle Activities during Walking
and Jogging at Different Speeds 258

Ganesh R. Naik

SECTION 4   Wearables—Sensors Signal Processing

Chapter 14 Biosensors in Optical Devices for Sensing and Signal
Processing Applications 271

Shwetha M. and Ganesh R. Naik

Index 282

About the Editors

Ganesh R. Naik, ranked as top 2% of researchers in Biomedical Engineering (Stanford University Research), is a leading expert in biomedical engineering and signal processing. He received his PhD degree in electronics engineering, specializing in biomedical engineering and signal processing, from RMIT University, Melbourne, Australia, in December 2009. He held a Postdoctoral Research Fellow position at MARCS Institute, Western Sydney University (WSU) from July 2017 to July 2020 and worked on a CRC project for sleep. During his tenure at WSU, he has developed several novel algorithms for wearables related to sleep projects. Before that, he held a Chancellor's Post-Doctoral Research Fellowship position in the Centre for Health Technologies, University of Technology Sydney (UTS), between February 2013 and June 2017. As an early mid-career researcher, he has edited 12 books and authored around 150 papers in peer-reviewed journals and conferences. Ganesh serves as an associate editor for IEEE ACCESS, Frontiers in Neurorobotics and two journals (*Circuits, Systems, and Signal Processing* and *Australasian Physical & Engineering Sciences in Medicine*). He is a Baden–Württemberg Scholarship recipient from Berufsakademie, Stuttgart, Germany (2006–2007). In 2010, he was awarded an ISSI overseas fellowship from Skilled Institute Victoria, Australia.

Wellington Pinheiro dos Santos holds a degree in electrical and electronic engineering (2001) and a master's degree in electrical engineering (2003) from the Federal University of Pernambuco, and a PhD in electrical engineering from the Federal University of Campina Grande (2009). He is currently an associate professor (exclusive dedication) at the Department of Biomedical Engineering at the Center for Technology and Geosciences/School of Engineering of Pernambuco, Federal University of Pernambuco, working in the undergraduate program in biomedical engineering and the graduate program in biomedical engineering. He is also a member of the graduate program in computer engineering at Escola Politécnica de Pernambuco, Universidade de Pernambuco, since 2009. He has experience in the field of computer science, with an emphasis on graphics processing, working mainly in the following areas: digital image processing, pattern recognition, computer vision, evolutionary computing, numerical optimization methods, computational intelligence, image formation techniques, virtual reality, game design and applications of Computing and Engineering in Medicine and Biology. He is a member of the Brazilian Society of Biomedical Engineering (SBEB), of the Brazilian Society of Computational Intelligence (SBIC, ex-SBRN) and of the International Federation of Medical and Biological Engineering (IFMBE).

Contributors

Sridhar P. Arjunan
SRM University, Chennai, TN, India

Kalaivani Chellappan
Universiti Kebangsaan, Malaysia

Wellington Pinheiro dos Santos
Federal University of Pernambuco, Recife, Brazil

P. Geethanjali
VTU University, Vellore, TN, India

Roopa B. Hegde
NMAM Institute of Technology, Nitte, Karnataka, India

Aayushi Khajuria
IIT Delhi, Delhi, India

Shwetha M.
Presidency University, Bangalore, India

Nadica Miljković
University of Belgrade

Ganesh R. Naik
Flinders University, Adelaide, Australia

Elena Pirogova
RMIT University, Melbourne, Australia

Maria Rashid
National University of Sciences and Technology, Islamabad, Pakistan

Preface

This book presents recent advances in biomedical signal processing applications. The recent advances in this field yield the potentials to improve the accuracy and reliability of medical diagnoses, and their integration in the everyday personal electronic devices will re-ignite human-computer interaction enabling a more intimate relationship with electronic devices that will help to close the loop of personalized medicine empowering patients of their own health focusing on the real lifestyle and daily routine.

Our editorial goal is to provide a forum for researchers to exchange ideas and foster a better understanding of the "state of the art" in this matter; therefore, this book is intended for biomedical signal processing experts and biomedical, computer science and electronics engineers (researchers and graduate students) who wish to get novel research ideas and some training in biomedical signal processing applications. Additionally, the research results previously scattered in many scientific articles worldwide are collected methodically and presented in the book in a unified form.

The book is organized into four sections. The first section is devoted to: *Physiological signal processing—Challenges.* In this section, we have collected four chapters with several novel contributions. The second section focuses on the various applications of *EEG and ECG singal processing* and contains five chapters. The third section covers *Gait—Balance signal processing* and has four chapters in it. The final section, *Wearables—Sensors Signal Processing*, covers one chapter.

We want to thank the authors for their excellent submissions (chapters) to this book and their significant contributions to the review process, which have helped ensure this publication's high quality. Without their contributions, it would not have been possible for the book to come successfully into existence.

Ganesh R. Naik and Wellington Pinheiro dos Santos
May 2022

Section 1

Physiological Signal Processing—Challenges

1 Signal Processing for Understanding Physiological Mechanisms for Information Decoding

P. Geethanjali

1.1 INTRODUCTION

The key factor to implement the translation of the user's bioelectric signals in a pattern recognition approach into communication or control commands in a typical HCI paradigm is the feature extraction and classification. The choice of feature extraction may affect the feature classification and vice-versa. A typical signal processing block diagram in the human computer interface application is shown in Figure 1.1. The block diagram shows HCI architecture with signal acquisition, conditioning, processing and control modules. This chapter discusses various elements of signal processing modules used by several researchers in order to implement HCI.

The primary stage of continuous data translation via computer is the data segmentation/windowing. The size of segmentation is not known; however, windowing of data depends on the hardware properties and its application as well. Further, the elimination of interference from the signals is essential to improve the translation of information from the signals through a chain of stages. In some cases, the signal to noise ratio, a measure of usefulness of physiologic information in bioelectric signal, is poor especially in very low magnitude signals like EEG. Therefore, data segmentation may be followed by digital filtering and or preprocessing, to clean the signals by removing noise from various sources such as motion artifacts, artifacts from other undesirable physiological sources, power line interference, etc.

A common stage of signal identification is the extraction of m-dimensional features to identify the relevant information from the n-dimensional data. The m-dimensional feature space may be reduced using feature selection/reduction techniques to reduce computation burden and eliminate redundant or irrelevant data. It is also possible to reduce the number of channels as well, depending on the significance of contribution of relevant information. The final stage of signal processing is the classification. This stage predicts intention, in order to communicate and or drive the devices in the external world for human computer interface applications from the processed data of the previous stage.

DOI: 10.1201/9781003201137-2

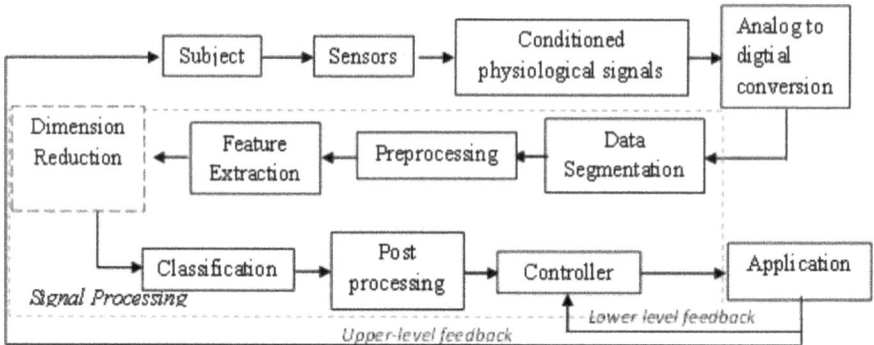

FIGURE 1.1 General block diagram of stages of signal processing in the human computer interface application.

In pattern recognition, there are some signal processing approaches that play a different role at different stages of application. For example, principal component analysis may be used for de-noising the signals in preprocessing, reducing the dimension through transformation in feature reduction and as a classifier as well. The linear discriminant analysis technique may be used as a feature transformation technique or as a classifier. Therefore, it is necessary to understand about the different stages and various methods applied at each stage. The subsequent section of the chapter discusses various elements of pattern recognition along with different techniques which are attempted by various researchers.

1.2   DATA SEGMENTATION AND WINDOWING

Longer time data introduce computational complexity in the analysis; therefore, it is necessary to segment the data prior to feature extraction. A data segment is a time slot for acquiring raw data considered for further processing. In general, each segment is assumed to be stationary and it is appropriate to extract the feature from a segment of data. The major problem is frequency of feature extraction from data segment, which depends on the real-time applications and hardware capabilities. In real-time HCI applications, a feature should be extracted over a data segment/window spanning over a time period (past to present instant) to identify the intention of the subject. The data segment could be overlapped window–a current data segment overlaps with the small portion of previous segment–or non-overlapped window–disjoint adjacent data segments. However, there is some consideration for window length in feature extraction, such as processing hardware, application, psychophysiological inference time period, variations in bias and variance, etc. Typically the window length should be small enough to minimize the misinterpretation in the estimation. However, the smaller window segment makes the translation difficult during transient state and increases the bias as well as variance of the feature, and subsequently the classification performance is reduced. On the other hand, the longer window affects physiological responses in case of stimuli based physiological information, introduces a time delay in real-time applications, etc. Therefore, there is a trade-off between classification accuracy and speed of computation.

1.3   SIGNAL PREPROCESSING

Bioelectric signals of interest may include physiological signals that are not associated/generated by the user and interference from various external sources. Signal preprocessing aims at cancelling/reducing the unwanted signals to improve signal quality by improving the so-called signal-to-noise ratio (SNR) to make the subsequent processing operations without losing relevant information. A low value of SNR signifies the difficulty in detecting the buried relevant patterns and a large SNR, on the other hand, simplifies the detection.

The effective preprocessing methods improve the signal-to-noise ratio and spatial resolution. Since the analog filtering is not always adequate, the quality of recorded bioelectric signal is reduced due to irrelevant signals from the hardware and the non-associated physiological signals from the subject. Further, analog filters also distort the signal in terms of phase and make the circuit bulky. Signal preprocessing can be done using various forms such as referencing, filtering, blind source separation techniques, etc. In referencing technique, the SNR can be improved by appropriate selection of reference electrode or linear combination of electrodes by appropriate weighing to measure electrode potential at desired location. There are different referencing methods available in the literature and discussed in the spatial filtering section. In order to eliminate signals that lie outside the range of signal of interest, digital filtering namely impulse response (FIR) filter and infinite impulse response (IIR) filter may suffice. Another powerful technique to extract source signal from the observed signals is the blind source separation (BSS), where the term blind refers to unknown source signal as well as unknown mixing (hybrid) system i.e., the coefficients of mixer. BSS technique estimates various sources based on certain considerations such as statistical independence, sparseness, space-time independence and smoothness. The most common methods of signal preprocessing, i.e., spatial filtering, temporal filtering and BSS applications in ICA are discussed briefly in the following subsection.

1.3.1   SPATIAL FILTERING

There are several ways to enhance the signal quality in spatial filtering. In referencing, spatial filters are designed as a weighted sum of observed signals like fixed/data-independent spatial filters common average reference (CAR), Laplacian reference (Wolpaw et al. 2002) etc., data driven spatial filters Principle component analysis (PCA) (Cheng et al. 2002), independent component analysis (ICA) (Jung et al. 2000), Common spatial patterns (CSP) (Blankertz et al. 2006) canonical correlation analysis (CCA), etc. The prior mathematical transformations combined with spatial filtering e.g., linear transformation with component analysis, techniques is beneficial to improve the quality of the signal. But, the direct association is lost between the spatial filtered component of the channel to the signal due to linear transformation. The next section discusses referencing, unsupervised data-driven PCA and ICA, supervised common spatial pattern.

1.3.1.1   Referencing Methods

Referencing methods enhance the signal quality using average of recorded signal. A common average reference (CAR) spatial filter transforms the signal space to component space as given next.

(i) Computing average of all recording channels with a common reference
(ii) Adjusting each recorded channel point by subtracting average from the preceding step eliminates the effect of reference electrode

This tends to enhance the signal if the impact of noise is common at all channels, e.g. power line interference and not effective in reduction of artifacts from different non-interested physiological sources. Therefore, it is essential to use other techniques like PCA, ICA, etc.

Laplacian reference is another method of transformation, similar to CAR, varied in the computation of average, which could be a small Laplacian reference or large Laplacian reference. In the small Laplacian reference, the average of recording from nearest electrodes in radial is considered in adjusting each channel point. In large Laplacian reference, the average of recording from next to immediate nearest radial electrodes is considered. Therefore, the Laplacian reference results vary with the spacing of electrodes considered in adjusting the point in each channel. This filter emphasizes localized activity.

1.3.2 Principal Component Analysis (PCA)

PCA and ICA transform the signal space to components using the linear transformation technique. In component analysis, set of spatial signals is considered to be a linear combination of unknown signals to decorrelate. In PCA, the transformation is constrained to be an orthogonal linear transformation, so that uncorrelated components, that is the expected value of the components must satisfy $E\{x_1 x_2\} = E\{x_1\}E\{x_2\}$, account for the maximum amplitude variance of the set of spatial signals. This section discusses PCA and the following section discusses the ICA.

PCA determines the transformation matrix $W = [w_1, w_2, \ldots w_n]$, which transform the n-dimensional observational signals $X \in R$ into a new signal Y having the same dimensionality i.e., $Y = W^T X$. The resulting signals ordered principal components in decreasing order of variance. That is, the first principal component in the filtered signal Y, associated with the linear combination of the signals in X, contains the highest amplitude variance and the amplitude variance in the successive components decreases. The decomposed principal components are orthogonal to each other (resulting in new signal Y) and are uncorrelated with each other. The dimensionality of data or channel can be reduced by considering first few principal components, which capture most of the variance and discarding the remaining components. That is, a typical high-dimensional input channel/data is reduced to a smaller number of output channels/data and called channel reduction/feature reduction.

Mathematically, the PCA transformation matrix can be computed from the correlation matrix of observed data which is influenced by temporal features of data as given next.

(i) Compute the covariance matrix from the normalized observed signal.
(ii) Compute eigen values and eigen vectors of the covariance matrix.
(iii) Arrange the eigenvector corresponding to highest to lowest eigenvalue reflecting the highest to the lowest variation of amplitude variance.
(iv) Construct the transformation matrix with the arranged eigenvectors.

One disadvantage of PCA is, the largest amplitude variance may not be correlated with the signals of interest. For instance, if all of the source signals are corrupted with a very large level of power line interference, it is likely that the principal PCA component would be power line signals rather than of signals of interest. Further, the principal components computation is based on second order statistical properties of the probability distribution function and assumption that the mean of observed data is zero. Therefore, PCA is suitable for observed signals with a Gaussian distributed source signal. In order to cancel noise, higher order statistics, i.e., non-linear function are essential like in ICA and kernel PCA. Kernel PCA represents nonlinear PCA and uses higher-order statistics to cancel noise signals considering the orthogonal constraint.

1.3.3 INDEPENDENT COMPONENT ANALYSIS (ICA)

Although transformed data from PCA are uncorrelated, they are not necessarily statistically independent due to the assumption that mean of the observation is zero. ICA is based on the assumption of statistical independence, i.e., the joint probability distribution function of signals s_1 and s_2 must satisfy the following:

$$P(s_1, s_2) = P(s_1) P(s_2)$$

Mathematically, ICA is computed with the following assumptions to effectively separate the sources $S = [s_1, s_2, \ldots \ldots s_n]$, from the observed mixture of source and noise $X = [x_1, x_2, \ldots \ldots x_n]$, that is, $X = AS$ where A is the mixing matrix.

(i) A stricter statistical independence is a constraint on the statistical relationship than uncorrelatedness. A zero-mean ($E[s_i] = 0$) also essential in sources and observed data, refers to unit variance i.e., $E[s_i{}^2] = 1$ and uncorrelatedness of observed signals. The process of transforming observed data with zero mean is referred to as whitening or sphering. The process of whitening reduces the estimation of parameters in ICA. Centering simplifies whitening by subtracting the mean of observation with each point.

(ii) Assuming that the number of observed signals (n) is greater than or equal to the source signals (m), ICA determines weight/demixing matrix W of dimension $n \times n$ as $Y = WX$. In this, the data are correlated.

(iii) The source signals can be separated, when the number of source signals with a Gaussian distribution is not greater than 1.

(iv) Similar to PCA, noise from various sources should be small. Otherwise, with BSS technique, noise can be discarded by considering noise as the signal of interest.

(v) A priori knowledge on the probability distribution of the source signals is necessary.

There are plenty of ICA algorithms available in the literature such as information maximization (Infomax) approach, FastICA, etc. Each algorithm uses different approaches to obtain independent components. However, before applying ICA, it is vital to preprocess the signals before using ICA. The preprocessing consists of centering and whitening of data as discussed in assumption (i). It is common to apply PCA for whitening prior to ICA to discard irrelevant data/channel for uncorrelation.

1.3.4 COMMON SPATIAL PATTERNS (CSP)

CSP is an approach closely related to PCA, except that CSP is supervised and requires class label information to be incorporated when determining transformation matrix. The transformation matrix, in contrast to PCA, maximizes the variance of signals in one condition and minimizes variance for the other, and vice versa. Also, by inverting the CSP transformation matrix W, it is possible to visualize the actual spatial patterns of CSP components. The first transformed signal of *WX*, reflects most of the variance for condition 1 and least for condition 2. Similarly, the last component of *WX*, reflects most of the variance for condition 2 and least for condition 1. In this way, the CSP algorithms are suitable for discriminating between two classes or conditions. As with PCA and ICA, the relevant CSP components for identifying the condition/classes are retained and can be fed directly to the classifier without further processing i.e., feature extraction and reduction.

1.3.4.1 Mathematical Description of the Pattern Recognition Task

The description and implementation of signal processing algorithms using suitable mathematical notation are useful to comprehend the pattern recognition of signals. A pre-processed/raw data of single channel is a discrete function and may be represented as a vector $d = \{d_1, d_2, d_3, \ldots \ldots \ldots d_L\}^T$ of length L. Thus, discrete signal space may be denoted as $d \in D \subseteq \Re^L$. A pattern (feature vector) from a channel may be said to consist of h number of variables and denoted by a vector $\{f_1, f_2, f_3, \ldots \ldots \ldots f_h\}^T$. The feature vector extracted from each channel of a single segment (window size) is concatenated into a single feature vector expressed as $V = \{f^{(1)}_1, f^{(1)}_2, f^{(1)}_3, \ldots \ldots \ldots f^{(1)}_h, f^{(2)}_1, f^{(2)}_2, f^{(2)}_3, \ldots \ldots \ldots f^{(2)}_h, \ldots \ldots \ldots f^{(j)}_1, f^{(j)}_2, f^{(j)}_3, \ldots \ldots \ldots f^{(j)}_h\}$ = feature-channel set = $\{v_1, v_2, v_3, \ldots \ldots \ldots v_w\}$, where, superscript $1, 2, \ldots j$ represents channel number. From the preceding, feature space is defined as $v_i \in V$ such that total number of features, product of number of channels (J) and number of features extracted from a channel (H), $W \leq L$. The extracted number of features may be reduced in dimension using the feature/channel selection or feature transformation/reduction algorithm. Therefore, feature vectors considered for classification $\Omega = \{\omega_1, \omega_2, \ldots \omega_E \ldots \ldots \omega_I\}$ associated with I number of classes is denoted as $Y = \{y_1, y_2, \ldots \ldots y_I\} = \{1, 2, \ldots \ldots I\}$, where $y_j \in Y$. It means that j denotes the index of the class of the input. From the preceding consideration the output space is defined as $\Omega \times Y$ which is set of all pairs of feature vector and the corresponding class labels (ω, y).

The classifiers assume all information available in input features to classify the class to which it belongs to. To identify the class, the classifiers use the classification function c_i where i represents the class number, consisting of features considered for classification. The objective of pattern recognition is to identify an observed data to one of the classes. If the estimated class matches with the actual class, the result of classifier is correct, otherwise it is not.

1.4 FEATURE EXTRACTION

Directly processing raw data is computationally too high, has large amount of noise and is difficult to manage. It would be better if irrelevant or redundant attributes are discarded and the relevant and important one is isolated. The useful features reflecting the characteristics of the data/describing the data have to be extracted from the preprocessed/raw data under consideration, to make it computationally less expensive. In order to implement an HCI based on pattern recognition method, features must be extracted from the data. For instance, amplitudes and frequencies are some of the few features in HCIs for characterizing the data. This section represents frequently used features and the formulae to compute the features. The signal features represent the data in higher-level from low-level and need to be translated into a specific output. The feature extraction process provides a feature vector consisting of computed feature values from the acquired data. This process is often the most complex and computationally expensive in the multichannel processing. The feature extraction process can have a significant influence on the identification of the user's intent from the data. Feature vector includes diverse features, namely time domain features, frequency domain features, time-scale features from the raw data. The extracted feature can independently represent the original data, and no means are available to measure the relevance of extracted feature in the pattern recognition task. However, researchers found that composite features consisting of diverse features achieve robust performance. The condensed representation of features in a feature vector is called a pattern containing the relevant information from the raw/preprocessed data. A few examples of time domain features used for pattern recognition application by various researchers are (Krusienski et al. 2007; Huang and Chen 1999; Hudgins et al. 1993; Zardoshti-Kermani et al. 1995; Park and Lee 1998; Du and Vuskovic 2004; Boostani and Moradi 2003) given next.

1.4.1 Time Domain Feature Extraction

To date, time domain feature extraction has been found computationally simple, due to the fact that these features do not require transformation in comparison to other methods of feature extraction. Widely used time-domain features are discussed next especially in the EMG signal classification (Tkach et al. 2010).

Mean Absolute Value (MAV): This is the mean absolute value of the measure of the average of the absolute value of data of length L and is defined by equation (1.1). This is used as an onset index in EMG based pattern recognition approach.

$$MAV = \frac{1}{L}\sum_{i=1}^{L} \mid d_i \mid \qquad (1.1)$$

Where,
d_i is the ith sample in a window segment

Zero Crossings (ZC): This parameter measures the frequency by counting the number of times the waveform crosses zero in time domain. A threshold (ε) must be included to discard noise zero crossings in the calculation. Given two consecutive data d_i and d_{i+1}, the zero crossing count is incremented if,

$$\{d_i > 0 \, and \, d_{i+1} < 0\} \, or \, \{d_i < 0 \, and \, d_{i+1} > 0\} \, and \, |d_i - d_{i+1}| \geq \varepsilon \qquad (1.2)$$

Where, ε is the threshold

Slope Sign Changes (SSC): In this feature extraction, the number of slope sign changes in the signal is considered, another way to measure frequency. Similarly, as with the aforementioned case, a suitable threshold must be included to discard slope sign changes due to noise. Given three consecutive samples, d_{i-1}, d_i and d_{i+1}, the slope sign change count is incremented if,

$$\left(\{d_i > d_{i-1} \, and \, d_i > d_{i+1}\} \, or \, \{d_i < d_{i-1} \, and \, d_i < d_{i+1}\}\right) and$$
$$\left(|d_i - d_{i+1}| \geq \varepsilon \, or \, |d_i - d_{i-1}| \geq \varepsilon\right) \qquad (1.3)$$

Waveform Length (WL): In this, feature provides information on the waveform complexity, which is simply a cumulative length of the waveform over the segment of length L. This feature is a measure of the waveform amplitude, frequency and duration within a single parameter as defined by equation (1.4).

$$WL = \sum_{i=1}^{L} |d_i - d_{i+1}| \qquad (1.4)$$

Mean Absolute Value Slope (MAVS): This is simply the difference between the MAVs in adjacent segments, i and $i + 1$, as defined by equation (1.5).

$$MAVS = MAC_{i+1} - MAV_i \qquad (1.5)$$

Variance (VAR): In this, the feature is extracted by averaging square values of data, describes the power density measurement of the data using the equation (1.6).

$$VAR = \frac{1}{L-1} \sum_{i=1}^{L} d_i^2 \qquad (1.6)$$

Root Mean Square (RMS): This is another power index feature, measure power of the signal. The RMS represents the subject's fatigue and force in case of EMG. It is given by equation (1.7).

$$RMS = \sqrt{\frac{1}{L} \sum_{i=1}^{L} d_i^2} \qquad (1.7)$$

Willison Amplitude (WAMP): This feature indicates the contraction level of muscles in EMG and is measured by counting the number of changes of the EMG signal amplitude in two adjoining data that exceeds a pre-defined threshold (ε). It is expressed by equation (1.8). This is also a measure of frequency similar to ZC.

$$WAMP = \sum_{i=1}^{L-1}\left|d_i - d_{i-1}\right| > \varepsilon \qquad (1.8)$$

Histogram: This feature is also another version of ZC and WAMP in measurement of frequency. It is a measure of number of data over a range of voltage in an "*X*" equally divided segments/bin. The recommended number of segments/bin size is 9.

Integrated EMG (IEMG): This feature is used to find the activity of the muscle from absolute value of EMG over a window segment. The mathematical formula for (IEMG) is expressed by equation (1.9).

$$IEMG = \sum_{i=1}^{L}\left|d_i\right| \qquad (1.9)$$

V-order (VO): This feature is nonlinear, used in EMG for estimating the contraction force produced by the muscle. The mathematical formula for (VO) is expressed as a function of EMG signal generation (g_i).

$$g_i = \left(\beta f_i^{\varsigma}\right)c_i \qquad (1.10)$$

Where, β and z are constants
c_i is the class of ergodic Gaussian process

$$VO = \left(\frac{1}{N}\sum_{i=1}^{N}g_i^{\,v}\right)^{1/v} \qquad (1.11)$$

Log detector (LD): Similar to v-order feature, this feature also estimates the contraction force produced by the muscle. This can be calculated using logarithm given by (1.12).

$$LD = e^{\left(\frac{1}{N}\sum_{i=1}^{N}\log|d_i|\right)} \qquad (1.12)$$

Peak picking: The simplest and straight forward method of feature extraction is peak picking. It identifies the minimum and maximum value of data in a data segment/window.

Correlation and Template Matching: The similarity of data may be used as a feature. The correlation will be high if the template closely matches with segment of data and decreases depending on the level of matching. This is one of the time domain features considered in BCI.

Derivation of Time Series: A time series is a chronological sequence of observations of amplitude of the EMG data. There are various models to represent a time series. A model that depends only on the previous outputs of the system is called Auto regressive (AR) model. Moving average (MA) model depends only on present values of the system. Further, autoregressive-moving-average model (ARMA) depends on both inputs and outputs of the system.

AR model is a common technique used for representing the data in time series in a linear second-moment stationary model. AR model involves determination of coefficients using statistics such as variance for modelling the time series data. The AR model is expressed by equation (1.13).

$$a_k = -\sum_{i=1}^{Q} \lambda_i a_{k-i} + \delta_k \qquad (1.13)$$

Where,
a_k is the estimated data
λ_i are the AR-coefficients
δ_k is the estimation error
Q is the order of the model (number of coefficients)

The AR model depends only on the previous outputs of the system and is approximated to finite order Q. The order of an AR model represents the amount of information necessary to predict, an estimated data with minimum variance that can be represented by the AR-coefficients. The AR model is simple with fast computation.

Cepstral/Cepstrum Analysis is based on non-linear modelling and models the data in stochastic time series. It is defined as the inverse of Fourier transform of the logarithm of power spectrum magnitude of the data. The Cepstral coefficients are expressed by

$$CC_1 = -C_1 \qquad (1.14)$$

$$CC_q = -C_q - \sum_{p=1}^{q-1}\left(1 - \frac{p}{q}\right)C_{k-p}C_q; \quad 1 < p < Q \qquad (1.15)$$

Where,
CC_q is the Cepstral coefficients
Q is the order of Cepstral model

In addition to the aforementioned features, there are other time domain features such as multiple hamming windows, multiple trapezoidal windows, maximum amplitude, average amplitude change, etc. However, time-domain extraction is based on the assumption that the signals are stationary and statistical properties of signal do not change over time. The appropriate selection of features leads to good pattern recognition depending on the nature of the signal and application. Therefore, researchers have investigated frequency domain approaches to improve the recognition accuracy.

1.4.2 FREQUENCY DOMAIN FEATURE EXTRACTION

The feature extraction in frequency domain is useful in studying the variation of the signals accurately. The frequency domain feature extraction process has a wide variety of methods, namely time-based, space-based and time-space methods. The time-based method includes band-pass filtering, Fourier-based spectral analysis, parametric methods such as autoregressive spectral methods and use of wavelets.

Space-based methods include Laplacian filters, principal components, independent components and common spatial patterns. Time-space methods include component analysis in time and space, multivariate autoregressive models and coherence. In general, fast Fourier transforms (FFT) is used to transform the data from time domain to frequency domain for frequency analysis, filtering, power spectral analysis, etc. The FFT is an efficient method of frequency domain transformation in the discrete time domain. In some cases, the coefficients obtained during time domain to frequency domain is reduced/transformed to input to the classifier. The coefficients may be transformed using one of the techniques discussed in feature reduction methods. Frequency/spectral analysis is studied using power spectral density (PSD). The frequency domain characteristics are extracted by measuring statistics over the PSD. The other methods of frequency domain feature extraction from power spectral analysis include band power, AR modeling.

Band Power: In this technique, band pass filter isolates a frequency content of interest in a specific frequency range. The filtered signals are squared to obtain power samples. The resultant is smoothed through averaging over a time or low pass filtering to reduce variability due to peaks. Logarithmic transform makes the response more Gaussian. However, the smoothened signal is delayed slightly and this method may not be suitable for analysis of multiple frequency content. The analysis of multiple frequency components is possible in FFT.

1.4.2.1 Fourier Analysis

Much of signal processing technology requires filtering, modulation and spectral analysis. This time-frequency analysis is rooted in Fourier analysis. The transformation constitutes decomposing of the signal into individual sinusoidal components represented as coefficients. The analog periodic signals are expressed as the sum of amplitude scaled and phase shifted sinusoids in Fourier series. The Fourier transform is to decompose any arbitrary aperiodic analog signals.

In order to model bioelectric signals, it is necessary to apply Fourier transform due to the aperiodic nature in frequency domain analysis. A signal $y(t)$ of the finite time segment is expressed in Fourier series. The Fourier series can be represented in trigonometric form, magnitude-phase form and complex exponential form. In exponential form, the magnitude (scale) and phase (shift) of the individual sinusoid at angular frequency (ω) are expressed using equation (1.16–1.17).

$$Y(j\omega) = \int_{-\infty}^{\infty} y(t)e^{-j\omega t}\,dt = \int_{-\infty}^{\infty} y(t)\left[\cos \omega t + j \sin \omega t\right]dt \tag{1.16}$$

$$Y(j\omega) = \int_{-\infty}^{\infty} y(t)\left[\cos \omega t\right]dt + \int_{-\infty}^{\infty} y(t)\left[j \sin \omega t\right]dt = R(\omega) + jI(\omega) = Y(\omega) \tag{1.17}$$

The magnitude and phase of individual sinusoid components are computed as given follows.

$$|Y(\omega)| = \sqrt{R^2(\omega) + I^2(\omega)} \tag{1.18}$$

$$\theta = \arg Y(\omega) = \tan^{-1}(I(\omega)/R(\omega)) \tag{1.19}$$

The plot of coefficients vs frequency is referred to as spectrum to describe the spectral content or the distribution of different frequencies in the signal under consideration. The decomposed signal can be reconstructed using inverse Fourier transform from the magnitude and phase in time domain. The inverse Fourier transform for analog signals is expressed as

$$y(t) = \int_{-\infty}^{\infty} |Y(\omega)| [\cos(\omega t + \theta(\omega))] d\omega \tag{1.20}$$

The continuous Fourier transform cannot be applied for frequency analysis, digital signals due to some differences between continuous time and discrete time signals property. One of the differences between continuous time and discrete time sinusoid component is that discrete time sinusoidal signals are periodic with period 2π. Therefore, frequency domain transformation of discrete time signals can be implemented using discrete time Fourier transform (DTFT). The DFT of a digital sequence $y(n)$ is expressed as given next.

$$y(k) = \sum_{n=0}^{N-1} y(n) e^{-j2\pi kn/N} \ for\, k = 0,1,...N-1 \tag{1.21}$$

From the preceding equation, it is clear that the size of DFT is N. The computation of DFT requires N complex multiplications and N complex additions.

1.4.2.2 Fast Fourier Transform (FFT)

The FFT is an efficient implementation of the discrete Fourier transform, exploiting the periodicity and symmetry properties of discrete time complex exponential to reduce the number of computations. The FFT algorithm uses a base 2 value for improving the computational efficiency and solves the sequence $y(n)$ of length, having a power of two. If the length of discrete sequence is not a power of two, the length is increased by padding zeros. There are two approaches in FFT to represent the discrete time sequence in frequency domain, namely decimation in time and decimation in frequency.

The decimation in time (DIT) takes an N-sample digital signal and divide the N point sequence as $N/2$ point, even indexed and $N/2$ point odd indexed and then breaks $N/2$ DFT into $N/4$ point DFT and continues till 2-point DFT appear. Due to the base of 2, the number of complex multiplication and addition is $Nlog_2 N$ compared to N^2 computation in DFT. The decimation in frequency (DIF) method divides the transformed sequence $X(k)$ (Proakis and Manolakis 2007). The discrete time sequence is reconstructed from frequency domain using inverse discrete time Fourier transform as given next.

$$y(n) = \frac{1}{N} \sum_{k=0}^{N-1} Y(k) e^{j\Omega kn/N} = \sum_{k=0}^{N-1} y(n) e^{j2\pi kn/N} \ for\, n = 0,1,...N-1 \tag{1.22}$$

The FFT will return N complex values that can be converted to magnitude and phase as given in equation to obtain magnitude spectrum and phase spectrum. The FFT spectrum of N-sample sequence of a real-world signal is symmetrical around DC component such that only the positive half of the frequency spectrum is needed to be shown for several sinusoids in two sided spectrum. The frequency bins/data segment from DC to sampling rate/2 is the redundant of the positive frequency bins. Therefore, for an N-sample real signal, the two sided spectrum is converted to single sided by considering spectral frequency bins from DC to sampling rate/2.

The spectrum resolution of FFT depends on the length of the sequence and is typically sample-rate/FFT-points. Zero padding to increase the length of the sequence does not increase the spectral resolution, but it does provide an interpolated spectrum with different bin frequencies. If N-sample signal sequence is non-integral of number of cycles, then the spectral leakage might be encountered and artificial ripples tend to be produced around the peaks of the spectrum. This can be mitigated by synchronous sampling.

The amplitude spectrum is commonly referred to as the power spectrum rather than the amplitude spectrum. The signal power is proportional to squared signal amplitude. Each bin of the FFT magnitude spectrum traces the sinusoidal amplitude of the signal at the corresponding frequency. A simple estimate of the power spectrum can be obtained by simply squaring the single sided amplitude spectrum. A robust power spectral density (PSD) is estimated in periodogram, using a square of the FT of signal divided by the length of the sequence. The Fourier transform method of the PSD estimate may have the disadvantage of spectral leakage, low frequency resolution, large variance, periodicity assumption. Therefore, the PSD estimation from the AR model may mitigate the aforementioned disadvantages. The PSD is estimated using mean frequency and median frequency. In addition to PSD there are other frequency domain features such as mean power, total power, peak frequency, variance, etc. Some of the frequency domain features are (Oskoei and Hu 2008; Biopac Systems, Inc. 2010; Du and Vuskovic 2004) discussed next with their mathematical formula.

Mean Frequency (MNF): The mean frequency represents the muscle force variation in case of EMG and is calculated using the equation

$$MNF = \left. \sum_{m=0}^{N-1} f_m P_m \middle/ \sum_{m=0}^{N-1} P_m \right. \tag{1.23}$$

Where,
f_m is the spectral frequency for m^{th} bin
P_m is the power spectrum for m^{th} bin
N is the length of frequency bin

Median Frequency (MDF): The mean frequency represents 50% of total power.

$$MDF = \frac{1}{2} \sum_{m=0}^{N-1} P_m \tag{1.24}$$

Where,
f_m is the spectral frequency for m^{th} bin
P_m is the power spectrum for m^{th} bin
N is the length of frequency bin

Total Power (TP): The total power is a summation of the power spectrum of the sequence and is given by

$$TP = \sum_{m=0}^{N-1} P_m \tag{1.25}$$

Mean Power (MP): The mean of total power is a summation of the power spectrum of the sequence and is given by

$$MP = \left.\sum_{m=0}^{N-1} P_m \middle/ M \right. \tag{1.26}$$

PSD Estimation From AR Modelling: AR modelling with shorter data segment provides better resolution than the FFT. The signal modelling in AR is, generated using infinite impulse response (IIR) filter by passing white noise, function of delayed output and not delayed input. The power (P) spectrum of the signal is estimated using IIR filter weights using the equation

$$P(\omega) = \frac{|C(0)|^2}{\left|1 + \sum_{k=1}^{q} a_q(k)e^{-jk\omega}\right|} \tag{1.27}$$

Where,
$C(0)$, $a_q(k)$ are IIR filter weights
q is the order of the AR model

The order of AR model depends on the spectral content of the signal. The order of the model influences the spectrum very much. A much lower order AR model produces the blur spectrum and overly higher order may cause peaks in the spectrum artificially. The order of the AR model may be chosen using several criteria. The simplest way to select the AR model is based on mean squared residual error (MSRE) value. The MSRE varies inversely with the order of the model and the estimation method also affects MSRE significantly. The order is chosen such that the rate of decrease of MSRE is insignificant beyond the selected value.

There are several algorithms (Anderson et al. 1998; Pfurtscheller et al. 1998; Burke et al. 2005) such as Burg algorithm, Yule-Walker, Covariance and modified covariance etc., to estimate AR model weights. Each method has its own merits and demerits. The commonly used algorithm for AR model parameter estimation is Burg algorithm, order-recursive least square lattice method, producing stable model. The advantage of this model is high frequency resolution. However, it suffers from the disadvantage of spectral line splitting in the case of high SNR and in case of higher order model with spurious peaks. These problems may be mitigated using a windowing technique or by varying the weights on the basis of squared forward and backward errors.

1.4.3 TIME-FREQUENCY DOMAIN FEATURE EXTRACTION

In frequency analysis, the FT does not provide timely information to identify the occurrence of an event. Also, FT is applicable for stationary signals. But the bioelectric signals are non-stationary in nature. This problem is mitigated with windowing of FT called a short time Fourier transform (STFT), Gabor transforms. This chapter discusses short time Fourier transforms.

Short Time Fourier Transform (STFT): The short time/term Fourier transform resolves the long discrete sequence into small segments with/without overlapped window and or windowed to find DFT. STFT is generalized Gabor transform (a Gaussian windowed) amplitude modulated by allowing general window function. Multiplying the sequence of samples by a short time tapering windowing function such as uniform/rectangular window, Hanning window and Flat top window produce broad transforms i.e. the width of the main lobe varies inversely with window length. The short time Fourier transform is mathematically defined by

$$Y(k,n) = \sum_{m=0}^{N-1} y(m+n)w(m)e^{-j2\pi km/N} \ for \ k = 0,1,...N-1 \qquad (1.28)$$

Where,
$W(m)$ is the window sequence

The product of window segment with sliding data segment possibly reduces the spectral leakage. The convolution of the small window amplitude modulates the sequence and reflects the DFT of the sequence with a lower frequency resolution. Although windowing removes the burst, there is a tradeoff for obtaining a smoother spectrum and spectral resolution.

The squared signal amplitude of the STFT gives the spectrogram of the signal. Spectrogram evince the energy distribution of the signal along the direction of frequency at a given time.

1.4.4 TIME-SCALE DOMAIN FEATURE EXTRACTION

The STFT provides a smoothed spectrum reducing ripples, but it lowers the spectral resolution with the expansion of width of frequency peaks. In addition, STFT has limited precision due to the fixed size of the window and depends on the size of the window. Therefore, STFT is difficult to use to analyze the signals of type containing high and low frequency content or abrupt transition. Wavelet analysis provides time scale analysis to compare the source signals with variable time-domain prototype signal retaining similar shape. The variable time-domain prototype provides all frequencies in the local region and retains the shape with dilation factor. Besides feature extraction, the wavelet analysis found numerous applications in signal processing such as de-noising without much degradation, isolation of muscle activity in EMG, etc. There are different ways to classify the wavelet transforms based on situation and applications. In general, based on the orthogonality of the wavelet function, the transform techniques could be continuous wavelet transform (use non-orthogonal wavelet) and discrete wavelet transform (use orthogonal wavelet).

1.4.4.1 Continuous Wavelet Transform (CWT)

Just as FT, wavelet transform decomposes the signal into wavelets, by dilations and translation from the mother wavelet. It is also considered as a measure of similarity between scaled wavelet $\Psi(t)$ at location (τ) and signal $y(t)$ at different frequencies. Therefore, continuous wavelet transform of a signal $y(t)$ is mathematically defined as

$$\psi(t) = \frac{1}{\sqrt{a}} \psi \left(\frac{t-\tau}{a} \right) \tag{1.29}$$

Where,
a is scaling factor. The scaling factor dilates the wavelet when it is greater than 1 and contracts the wavelet when it is less than 1. The wavelet $\Psi(t)$ is multiplied by a constant $a^{-1/2}$ to normalize energy. The wavelets are normalized in amplitude with the constant a^{-1}
τ is translation factor

Variation of dilation and translation factor produces daughter/baby wavelet, or wavelet atoms. The continuous wavelet transform of signal $y(t)$ is given by equation (1.30)

$$WT_{y(\tau,a)} = \frac{1}{\sqrt{a}} \int_{-\infty}^{\infty} y(t) \psi^* \left(\frac{t-\tau}{a} \right) dt \tag{1.30}$$

Where,
Ψ^* represents complex conjugate of Ψ
$WT_{y(\tau,a)}$ represents wavelet coefficients

The WT of a signal is the correlation between signal and the scaled/dilated wavelets. If the signal is closely related to the wavelet, wavelet coefficient value is large and small for weaker correlation between mother wavelet and the signal under consideration. The translation parameter provides the position of strongest and weakest correlation from the value of WT coefficients. Further, the wavelet transform is localized in time and frequency as well.

WT coefficient determines the frequency content of the signal in a signal train to localize time and frequency information of signal $y(t)$. In other words, the wavelet transform is useful for multiresolution signal analysis by varying scale factor. In multiresolution analysis, the signal is decomposed into multiple frequency bands, to process signals at different frequency bands differently and independently.

A square integrable function is wavelet, if the function has finite energy and satisfies admissible condition. To reconstruct the signal using inverse wavelet transform and wavelet coefficients, the wavelet function must satisfy the admissible condition. Another condition imposed on wavelet is regularity condition, so that, wavelet coefficients decay at the fastest rate for decreasing scale. The properties of wavelet are listed next.

The *admissibility criterion* implies that the Fourier Transform of $\Psi(t)$ vanishes at zero frequency (mean value of wavelets is zero) and is given by

$$\int_{-\infty}^{\infty} \psi(t)dt = 0 \qquad (1.31)$$

The preceding equation implies that wavelet function oscillates about time axis and the mean value in time domain is zero. In frequency domain, the FT of the wavelet function vanishes at zero frequency; therefore the wavelet is the band-pass filter in the frequency domain. The center frequency varies with the scale.

The *regularity criterion* implies that wavelet function smoothness as well as concentration in the frequency domain. This criterion is specified as zero moment property of the wavelet as given next.

$$\int_{-\infty}^{\infty} t^m \psi(t)dt = 0 \qquad for\ m = 0,1,2 \dots \dots N\text{--}1 \qquad (1.32)$$

The zeroth moment is zero as per admissibility criterion and making other moments to zero leads to the rapid decay of wavelet coefficients for smooth function. For fast decay of coefficient, the FT of mother wavelet should possess smoothness and concentrations in the domain of frequency. The choice of wavelet influences the frequency resolution of wavelet transform. There are different types of wavelets and some of the wavelets are listed next.

Haar Wavelet: A bipolar step wavelet is defined by equation

$$\psi(t) = \begin{cases} 1 & if\ 0 \le t \le \frac{1}{2} \\ -1 & if\ \frac{1}{2} \le t \le -1 \\ 0 & otherwise \end{cases} \qquad (1.33)$$

Mexican Hat Wavelet: A second order derivative of Gaussian function is defined by equation

$$\psi(t) = (1 - t^2)\exp\left(\frac{-t^2}{2}\right) \qquad (1.34)$$

Gabor Wavelet: The Gabor function with Gaussian window is defined by equation

$$\psi(t) = \exp(j\omega_o t)\exp\left(\frac{-(t-\tau)^2}{2}\right) \qquad (1.35)$$

FT of Wavelet Transform: FT of wavelet is given by equation (1.36 and 1.37)

$$\Psi(\omega) = \int_{-\infty}^{\infty} \frac{1}{\sqrt{a}} \psi\left(\frac{t-\tau}{a}\right) e^{-j\omega t} dt \qquad (1.36)$$

$$\Psi(\omega) = \sqrt{a}\Psi(a\omega) e^{-j\omega\tau} dt \qquad (1.37)$$

Where,
$\Psi(\omega)$ is FT of the mother wavelet $\Psi(t)$

In the frequency domain the wavelet is normalized by $a^{-1/2}$, scaled by $1/a$ and multiplied by the phase factor $exp(-j\tau\omega)$. In frequency domain, scale is referred to as frequency for the WT. In this domain, wavelet is localized and is zero at zero frequency according to regularity and admissibility condition respectively. Therefore, wavelet is a bandpass filter, intrinsically. The FT of wavelets is referred to as filter and the impulse response of the filter is the scaled wavelet. Therefore, the WT is a wavelet transform filter bank having different scales, *a*. If the scale is small, the wavelet is concentrated in time and wavelet analysis gives a detailed view of the signal. If the scale is large, the wavelet is stretched in time and wavelet analysis gives the global view and takes into account the long-time behavior of the signal. Therefore, WT is considered as multi-resolution band pass filters.

The WT of one dimension function returns two-dimensional data, two-dimensional function returns four-dimensional data and so on. The CWT gives highly correlated data and suffers from the drawback of redundancy due to use of non-orthogonal wavelet basis. In CWT, the time-bandwidth product is square of that of the signal. Further, CWT has an infinite number of wavelets during transformation and causes computation burden, but discrete wavelet transforms (DWT) reduce the computational burden as well as time bandwidth product. But CWT provides a good decomposition for the signals with low SNR.

1.4.4.2   Discrete Wavelet Transform

In DWT, wavelet is continuous function and the scale and the translation factors are discrete. Discrete wavelets are scalable and translatable only at discrete steps. The discrete wavelets are expressed as

$$\psi(t) = a_d^{-i/2}\psi(a_d^{-i}(t - k\tau_d a_d^i))$$
$$= a_d^{-i/2}\psi(a_d^{-i}t - k\tau_d) \qquad (1.38)$$

Where,
a_d^i is discrete scaling factor
$k\tau_d a_d^i$ is discrete translation factor
i,k are integers

The translation parameter of DWT depends on dilation step. Typically, the scaling factor is chosen as 2 and translation factor τ as 1 to have dyadic sampling of time and frequency axis. The DWT is computed at discrete scales and times called sampling in the time-scale space. The sampling interval in time axis is $\tau_d\, a^i_d$ and varies with the scale a^i_d. The varying time-scale in DWT enables it to focus analysis on singularities of the signal with more number of concentrated wavelets of very small scale. The small-scale wavelet analysis is carried out by the small time sampling step and large-scale wavelet analysis is carried out with the large time sampling step. Small time sampling step in wavelet transform provides detailed analysis with few small time translation steps. This helps to analyze the transient signal.

The discrete wavelet behavior is a function of τ and a. The discrete wavelet is close to continuous wavelet for small value of τ and scale $a = 1$. The localization of discrete wavelet is along scale axis logarithmic $log\ a = ilog\ (a_d)$. In small scale wavelet, the translational steps are small for small positive values of i and large for large positive value with large scale wavelets. The localization of time axis for scale factor $a = 2^i$ is generally selected in most of the HCI application. The signal analysis of dyadic wavelet is compared with time, frequency, time-frequency in Figure 1.2.

Wavelet transform allows the representation of signal with various resolutions. The multiresolution signal analysis is used to analyze the multiple frequency bands that exist in the given signal. Pyramid and packet are the two existing approaches to decompose the signals based on subband coding. In both the approaches two-bank filters are used to obtain the multiresolution tree structure.

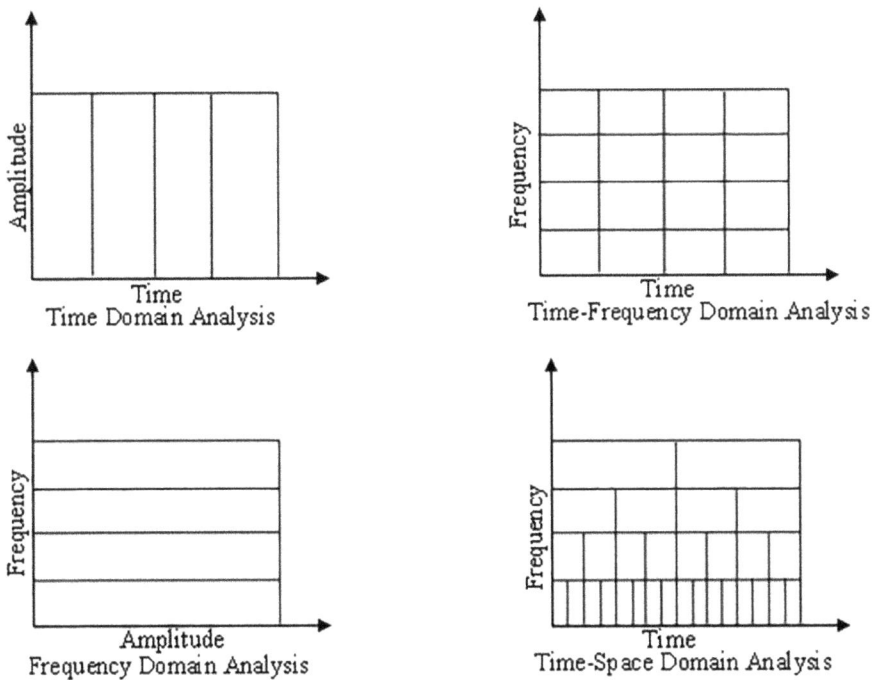

FIGURE 1.2 Analysis of signal in different domain.

In pyramid decomposition, the signal spectrum/original data sequence is divided as the number of independent subbands. The low pass filtering of the input sequence with filter having impulse response $l(n)$, returns the lower resolution approximation of the sequence. The detailed information, lost in low pass filtering, is obtained by high pass filtering the data sequence with a filter having response $h(n)$. The sequence for the first level is obtained by down sampling the output of filters by a factor of 2. The first stage sequence in each subband has half the resolution of the original data sequence. In pyramidal approach, only subbands with lower approximation are decomposed into two subbands and it is shown in Figure 1.3. The pyramidal decomposition is called wavelet transforms. In case of packet approach each subband is further divided into two subbands; therefore, the resultant subbands have the resolution, one fourth of the original signal. The decomposition using packet approach is called wavelet packet transform, which contains most of the information in the signal. The tree structure of the wavelet packet transform is shown in Figure 1.4.

The multiresolution analysis using orthonormal wavelet, specifically dyadic orthonormal wavelet, allows the decomposition of tree at a faster rate. The orthonormal wavelet basis is generated by a basis of scaling function such that two bases are mutually orthogonal at every stage. The scaling function and wavelet function satisfy the regularity and orthonormality property.

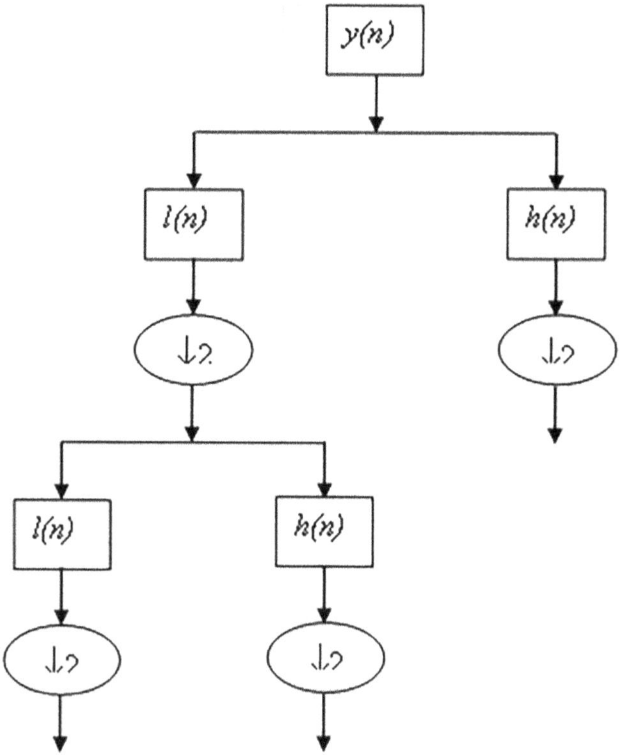

FIGURE 1.3   A wavelet transform tree structure.

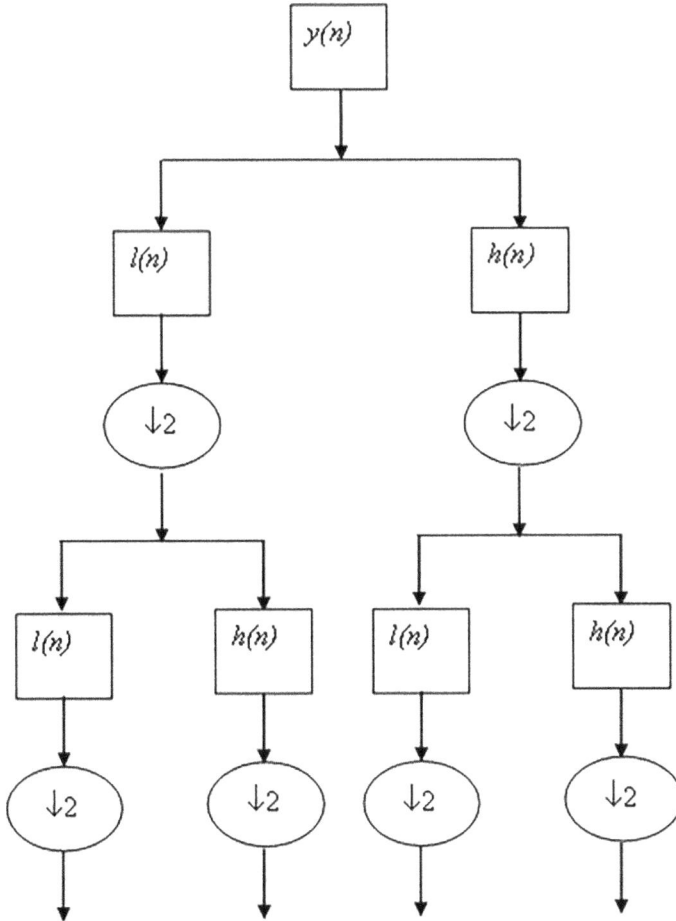

FIGURE 1.4   A wavelet packet transform tree structure.

The coefficients of WT, WPT, STFT and FFT form feature vectors and show the energy distribution of the signal. The dimensionality of extracted feature vectors introduces the computational burden to the classifier. Therefore, feature dimension reduction techniques may be applied to reduce the dimension or statistics over the coefficients used. The following are a few of the statistical features to represent the energy distribution of coefficients: mean of the absolute value, average power, standard deviation, ratio of the absolute mean value, skewness, kurtosis, temporal moments, etc.

1.5   FEATURE DIMENSION REDUCTION

The features extracted from various approaches, specifically FFT, STFT, WT, WPT and from multiple channels may not significantly improve the accuracy of a pattern recognition system. Additionally, having more features increases the computational burden and the cost. Therefore, researchers used techniques to reduce the features in a high-dimensional feature space to lower dimension either using transformation or selection features or selection of channel especially in case of data from numerous channels. However, some of these reduction techniques depend on the method of classification. This section describes the various methods of feature dimensionality reduction.

1.5.1 FEATURE SELECTION

The feature is affected by various factors including acquisition hardware, interferences etc. The selection of one or more features from the predetermined feature set and choice of feature(s) for an accurate discrimination of a task is crucial. Most multichannel recognition systems, especially in BCI, do not use all extracted features from the input data. The selection represents a discrete choice from a set of possibilities to enhance the classification performance in order to discard irrelevant, redundant features. Additionally, reducing the number of features usually lowers the computation cost (processing time, memory requirements) and system complexity in classification stage. Therefore, to reduce the dimensionality of the extracted feature vectors, statistics over the set of the features can be used. The architecture of the pattern recognition system can be compatible with different types of features, but it is essential to select appropriate features to enhance the decision making capability since the fusion of features in some cases of pattern recognition system may influence improving/degrading the accuracy of the system. The choice of appropriate signal processing methods at each level of pattern recognition is not obvious.

There are numerous methods to represent patterns as a grouping of diverse features possessing the capability of maximum separability of classes and robustness. In the feature selection, the process is performed based on either (1) representation of a given class of signals or (2) variation between classes. Therefore, feature selection plays an important role in classifying systems. In order to select the appropriate feature for identification of the user's intent, a feature having high inter-variance (varies significantly between the tasks) carries more user's intent information and low intra-variance (less variation due to repetition of same task in multiple times). Further, stability of feature is vital, i.e., variation of feature should not be large for small change in data, in addition to computational accuracy and speed in task execution with the computed feature. Therefore, the translation process is affected significantly based on selected features and its properties. In general, there are two approaches for selection of features, namely, heuristic approach-rule based approach from experience or knowledge and evaluation approach. There are numerous methods under these approaches such as stepwise/sequential forward selection, stepwise/sequential backward selection, branch and bound method, exhaustive search, simulated annealing, evaluation strategies-Genetic algorithm, particle swarm optimization, ant colony optimization, etc. Here, heuristic search techniques-stepwise forward selection, stepwise backward selection and evolutionary approach-Genetic algorithms and particle swarm optimization are discussed next.

Stepwise/Sequential Forward Selection: In this method, the classification accuracy is measured with the addition of feature in the selected feature set. The feature which contributes to best prediction will be considered in the selected feature set, otherwise the feature will be discarded. The feature also will be discarded based on computational complexity. The process starts with an empty selected feature set and next adding the features which result in highest classification accuracy. The algorithm stops based on certain criteria.

Stepwise/Sequential Backward Selection: In this method, the classification accuracy is measured with the feature set containing all the features. Subsequently, classification accuracy is measured by removing every feature from the current feature set. If the removal of feature reduces the classification efficiency, then features will be added to the current feature set. Similarly, if the removal of feature reduces the computational burden, then also feature will not be added to the feature set.

Evaluation Approach: Sequential forward/backward selection does not guarantee the global optimum feature subset. Therefore, selecting M features in a channel-feature set of F $(H{\times}W)$ requires 2^F combinations to obtain the global optimum. This exhaustive search procedure is not computationally effective. The selection of features in evaluation approach can be done using fitness function. The goodness of feature criterion function may be measured independent of the classification method called filters or depending on the classification performance called wrappers. Due to this evaluation method, filters are faster and more generalized for a family of classifiers than wrappers. But the recognition accuracy is inferior compared to that of the wrappers. Therefore, search strategy based on wrapper algorithm may be good. Genetic algorithm, particle swarm intelligence, ant colony optimization, etc. are a few of such feature selection techniques used by the researchers.

Genetic Algorithm (GA): GA is one of the widely used randomized approaches based on natural selection and reproduction using an evolution of population over time. This technique is based on fixed length binary coding scheme "0" or "1" to represent a feature subset in terms of chromosomes/genomes. The value "0" in the string represents the absence of a particular feature in feature subset and "1" represents the presence of a particular feature in the feature set. The features are selected by evaluation of fitness function using genetic operator selection, cross-over, mutation for the survival of the fittest. GA algorithm is initialized with a random population of solution and evolve to the next generation through evaluation of fitness criterion function using genetic operation. This method does not suffer from local optimum solution.

Particle Swarm Optimization (PSO): PSO is another evolutionary search technique in wrapper method. This algorithm is developed to mimic the social and cognitive interaction behavior of birds. The PSO technique uses the parameters, namely population of particles, the position of the particle, topology and fitness. The particle is referred to swarm when (in general) the population size varies from 20–50, which is less than other evolutionary approaches. PSO operates by modeling and directing particles iteratively in multidimensional problem space represented with position and velocity. The position of the particle is the solution offered by the specific particle in the problem space. If one of the particles, $x_{i,j}$ finds a good path to food, then the rest of the particles are attracted to follow irrespective of the distance of the particle with maximum velocity v_{max}. The position of the particles is shared with each other and they adjust their position and velocity for local best ($p_{i,j}$) to reach best positions called global best (g_i). In general, the local best position of each particle is associated with best fitness value/best performance of the particle and the global best is associated to be seen as the best value found by neighboring particles. The goodness of the particle position is computed using a fitness function as in other evolutionary approaches. The formula for PSO is given next.

$$v_{i,j}(t+1) = wv_{i,j}(t) + c_1 r_1 (lbest_{i,j} - x_{i,j}) + c_2 r_2 (gbest_{i,j} - x_{i,j}) \qquad (1.39)$$

$$s(v_{i,j}) = \frac{1}{1+e^{-v_{i,j}}} \qquad (1.40)$$

$$if \; p_{i,j} < s(v_{i,j}) \, then \, x_{i,j} = 1; else \, x_{i,j} = 0 \qquad (1.41)$$

Where,
r_1 and r_2 are random numbers
c_1 and c_2 are cognitive and social parameters
i represents the particle's index
j is the dimension of current search space
t is the current time step
w is the inertial weight. The inertial weight is decreased gradually over a time to narrow the search space.

1.5.2 CHANNEL SELECTION

In channel selection, subset of channels is selected from the set of channels to optimize the computation cost. The channel selection technique can be performed using the techniques of feature selection applying to a training data. Channel selection is also performed manually based on physiology factors.

1.5.3 FEATURE TRANSFORMATION

The feature translation uses all extracted features or a subset of features from the previous stage. It is not necessary that every element of the feature vector evince good separation capability, robustness, but there must be at least one element in the feature vector that shows significant variation with the user's intention beyond the intra-subject variance, segmentation inaccuracy and noise. The intra- and inter-subject variance of feature are computed from multiple trials of each intended task of the user. In order to obtain a user independent pattern recognition system, the signal processing should include multiple subject signals. However, it is not necessary that removing the irrelevant features/transforming the features improve the pattern recognition performance. In certain cases, the feature transformation may be useful. Numerous features obtained from multichannel data can be reduced using transformation techniques. In transformation method, a complete set of new features is generated. In transformation techniques, best transformed features are selected to enhance the classification performance. Principal component analysis (PCA) and linear discriminant analysis (LDA) are the widely used linear feature projection methods. PCA and LDA are discussed next.

Principal Component Analysis (PCA): One of the widely used linear feature transformation techniques is PCA to decorrelate the multivariate data. During dimension reduction, components with high variance are projected to first coordinate system and the least variance in last coordinate. Therefore, projected components with high variance are considered for further processing and other components are discarded. The PCA feature projection method comprises the following procedure:

(i) Subtract the mean from the feature vector to center the data.
(ii) Compute the covariance matrix from the feature vector of step (i).
(iii) Compute Eigenvalues and Eigenvector from the covariance matrix.
(iv) Arrange the Eigenvalues in descending order.
(v) Select first N Eigenvalues and form the projection/transformation matrix with the associated Eigenvector in the order of Eigenvalues.
(vi) Transform the centered data using projection/transformation matrix for the new set of projected feature set.

Linear Discriminant Analysis (LDA): LDA is one of the linear transformation techniques that identify the coordinate system to enhance the class separability. Similar to PCA, the transformation matrix is computed using family of scatter matrix as given next.

$$S_1 = \sum_{i=1}^{K} \sum_{j=1}^{h} r_j^{(i)} \left(v_j - m^{(i)} \right) \left(v_j - m^{(i)} \right)^T \tag{1.42}$$

$$S_2 = \sum_{i=1}^{l} N_i \left(m^{(i)} - m \right) \left(m^{(i)} - m \right)^T \tag{1.43}$$

$$S_3 = \sum_{j=1}^{h} \left(v_j - m \right) \left(v_j - m \right)^T = S_1 + S_2 \tag{1.44}$$

Where,
S_1 and S_2 are within-class scatter matrix and between-class scatter matrix
S_3 is the total scatter matrix, a measure of covariance for all features
v_j is w-dimensional feature vector
$j = 1,2, \ldots .h$ is the number of features
m is the mean vector for all features
$m^{(i)}$ is the mean vector for class i
$r_j^{(i)} = 1$ if $v_j \in I$ and 0 otherwise

In order to maximize the coupling between-class scatter matrix and minimize the within-class scatter matrix, the projection matrix is calculated using Fischer criterion $\dfrac{\left| P^T S_1 P \right|}{\left| P^T S_2 P \right|}$

The projection matrix (P) is determined by computation of Eigenvalues and Eigenvectors of $S_1^{-1} S_2$. The Eigenvector corresponding to k largest Eigenvalues constitutes the projection matrix similar to PCA. The feature data are transformed with the projection matrix form the new feature vector set.

1.6   FEATURE CLASSIFICATION

Moving along the pattern recognition chain, the feature vectors consisting of similar or diverse or transformed sets of features obtained from previous stage are translated into control commands for assistive devices such as wheelchairs, prosthetic devices, orthotic devices, etc. by developing a model. The output of feature translation model is discrete in the case of classification/pattern recognition methods like a prosthetic hand, a selection of letters, etc. and continuous value in case of regression methods like cursor movement (McFarland and Wolpaw 2005).

Classification: The user's intent from hidden feature vector identification is carried out either using a supervised method or unsupervised method. In supervised classification method, the parameters of the model are computed from the training data/feature vectors. In this learning, the class for the current training feature vector is always attached during training and the task is to find the input-output relationship with iterative procedure. The feature vectors are learned in a way that later allows one to classify a new feature vector, even in the presence of noise and minor variations that inevitably occur in the real world. A large number of supervised pattern recognition algorithms are available to model classifiers such as multilayer perceptron, support vector machine, k-nearest neighbor, etc. The supervised classifiers are discussed in detail in the next section.

In unsupervised classification, the training feature vectors are unlabeled unlike supervised classification. Classification is done using clustering algorithms to cluster the similar feature vectors. The clustering is influenced by explicitly specifying the number of clusters and cluster size. The unsupervised classification returns the cluster index as a result of classification. There are different methods to identify the similarity in clustering algorithms. Two methods of similarity measure are discussed next along with the simple clustering algorithm.

Euclidean distance: The Euclidean distance measured between two features indicates the similarity by small value of distance and dissimilarity of the large value of distance.

Normalized dot product: The cosine angle is computed between two feature vectors, like Euclidean distance, larger value of dot product is interpreted as good similarity and low value as dissimilarity between vectors under consideration.

Other different approaches such as intra-cluster distance, inter-cluster distance, covariance matrix, scatter matrix, etc. are used as a measure of the similarity index in clustering algorithms.

In *simple clustering* algorithm (Tou and Gonzalez 1974), the N feature patterns are clustered as given in this list.

(1)   Consider the center of the first feature as the center of the first cluster. Assume the non-negative threshold as a measure of the similarity index (SI).

(2) Calculate the distance between the first and second feature vectors. If the distance is less than the SI, assign a second feature vector to a cluster of the first feature vector. Otherwise, assume the new cluster with the new center.

(3) Calculate the distance of the third feature vector with the first and second feature vectors, if there are two clusters. Otherwise, measure only the distance of the first feature vector with the third vector. Similar to the earlier step, if distance is less than SI, the new feature vector belongs to the existing cluster. If the distance is less than SI in both the clusters, the new feature vector is assigned to the closer cluster.

(4) Repeat the preceding step for different feature vectors and stop the clustering when all the feature pattern is assigned to a cluster.

There are different algorithms such as maximum-distance clustering algorithm, *K*-means algorithm, etc. to cluster the feature vector existing in the literature. In case data are not available for training, unsupervised learning techniques can be used (Schalk et al. 2008).

Regression: In this method, a model is estimated from the input-output mapping of training vectors. The model is derived such that approximates the relationship between inputs and output as closely as possible to predict function values for new observed data.

For human computer interface application, there are different control schemes that have been applied to control the devices. Till date, numerous supervised learning based pattern recognition algorithms are attempted by different researchers in HCI for control of devices. The reader is encouraged to read references, related to the field of application. Also, the classification models discussed here can also be used as a regression function for continuous outputs.

1.6.1 PATTERN RECOGNITION USING SUPERVISED LEARNING TECHNIQUES

This section focuses on commonly used supervised learning techniques applied to control assistive devices from the bioelectric signals. Algorithms of classification include simple rule-based approach, linear discriminant analysis, neural network, support vector machine, hidden Markov models, etc.

Rule-Based: This is a simple classification method based on *If-then* rules that classify the features based on the threshold values. The number of rules grows as the complexity of the increases. This may be useful for applications with low complexity and simple on-off control.

Linear Discriminant Analysis (LDA): Linear discriminant analysis (LDA) is one of the simplest classification techniques and one of the most popular statistical classifiers commonly used by almost all the researchers in identifying the intended motion hidden in the user's bioelectric data. This LDA function was introduced in Fischer 1936. In this technique, the classifier partitions an input space into various subspaces by creating decision boundaries between the output classes based on training data from each class. The decision boundary is obtained by calculating the average of feature vector for each class and finding the mean of the average. The advantages of LDA classifiers are quick training, not getting trapped in local minima and not requiring an iterative investigation to establish their ideal structure. The disadvantage of LDA is on the assumption that interclass boundaries are linear. In addition, it also suffers from the problem of singularity Khushaba et al. (2009).

In multi-class LDA classification, there will be as many classification functions as there are number of classes. In multiclass, the classification function model is obtained by considering the problem as two class problem. The discriminant function parameters are found on the basis of one against all, that is decision boundary is obtained between one class samples with all other class samples. In this, the classification function is given by the following equation (1.45).

$$c_i = \sum_j \gamma_j x_j \qquad (1.45)$$

Where,
γ_j, is the co-efficient for the jth feature variable
c_i is the ith class function

Each feature vector is assigned to a class for which c_i has the highest linear discriminant function value.

Quadratic Discriminant Analysis (QDA): QDA is different from LDA in the estimation of covariance matrices. In QDA, the covariance matrix is estimated for each class. In a binary class problem, the classification is obtained using the square root of the Mahalanobis distance difference between two classes.

$$D = \left(d_j - d_i \right) \qquad (1.46)$$

$$d_i = \left(\left(x - m^{(i)} \right)^T C_i^{-1} \left(x - m^{(i)} \right) \right)^{1/2} \qquad (1.47)$$

Where,
d_i is the square root of Mahalanobis distance to class i
$m^{(i)}$ is mean of class i
C_i is covariance of class i

The given feature vector is classified as class i, if the D is greater than 0, otherwise as class j. In case of multiclass classifier, the class for the feature vector is similar to the feature vector of linear discriminant problem. QLDA perform better than LDA, in case of large variance differences (Mark and Dunn 1974). This technique is more sensitive to the dimension of the feature vector, i.e., the performance improves with increase in dimension to a certain optimal number, then the performance degrades (Mark and Dunn 1974).

k-Nearest Neighbor: Simple *k*-nearest neighbor (*k*NN) is a distance based classifier and is not based on parameters. Therefore, this classifier is a data driven classifier. The class of test feature vector is determined from the class of *k*-nearest training samples. The *k*-nearest points are estimated by using majority voting or by similarity measure. In majority voting method, the number of points of each class closer to the test feature vector is counted and the class is assigned for which test feature vector has the maximum number of closer points. In similarity measure, the similarity value is estimated based on k-nearest points and the class with the highest similarity value is assigned to feature vector. In this kNN, Euclidean distance is measured between the feature vectors with all stored prototype vectors/training samples to measure the closeness of feature vector with training data. Selection of k value is challenging; a small k results in inaccurate classification. But large value of k results in risk of more bias. The optimal number of k is given based on training data points $N^{3/8}$ for the case, when both the difference of the covariance matrix and difference between the sample proportion are small or large. If, the difference proportion is in the opposite direction, k is $N^{2/8}$(Enas and Choi 1986).

Though this method of classification is easy to perform, it suffers from the disadvantage of larger memory and processing requirements than other methods. The storage of training data and measuring distance between test data and each of the stored data has made this classifier unattractive to the real-time application.

Neural Network (NN): The linear method of classification may not be effective at some times (Muller et al. 2003). Neural network (NN) is widely used in pattern recognition applications due to its capability of solving non-linear problems. In an NN, each neuron consists of summing node to sum the inputs, activation function to transform the input to produce output. There are different activation functions such as threshold function, sigmoidal function, piecewise linear, Gaussian, competitive function, etc. The activation functions are selected appropriate to the type of application. The network is obtained by the interconnection of neurons. The NN varies with the number of neurons interconnected in input, hidden layer and output layer.

A typical NN applied for a classification problem is a two layered NN and is obtained with one hidden layer and one output layer as shown in Figure 1.5. The

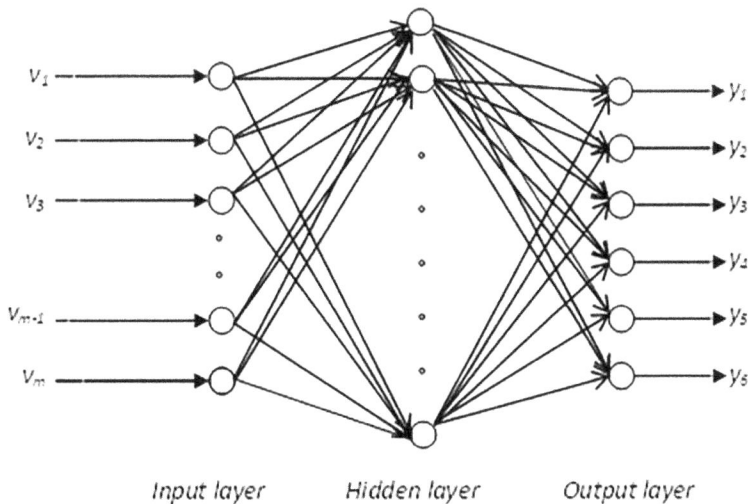

Input layer   Hidden layer   Output layer

FIGURE 1.5   Neural network structure for classification.

number of neurons in the input is determined from the size of the feature vectors. The number of neurons in the output layer equals the number of outputs associated with each input. The number of hidden layers and neurons in the hidden layer varies with the complexity and typically is determined by an iterative procedure. The nodes of neuron are equal to sum of the weighted inputs. Weights are obtained by iterative learning procedure. One of the widely used algorithms for estimating the weights is back-propagation (BP) training (Fielding et al. 2007).

In BP, the network is initialized with random weights. In case of difference between the desired output and actual output, the weights are tuned to correct the error. Training is measured by the number of epochs. Passing a complete set of all the training set once through the network is called one epoch. The training sets are passed, until the termination criterion or minimum error criterion is met. The trained network will be tested with the untrained data to measure the network classification performance. The training of the network is influenced by parameters such as learning rate, error criterion, initial weights, etc.

Support Vector Machines (SVM): The SVM is another linear classification model, introduced by Vapnik (1998). The SVM classifier is based on statistical learning theory and structural risk minimization theory. In case of a two-class linear problem, the feature vector may belong to one of the possible classes. Classification is based on the margins, distance between hyperplane dividing the classes. The linearly separable *d*-dimensional plane for SVM is given by equation (1.48)

$$c(x) = \sum_{j=1}^{n} \left(w_j y_j \right) + a \qquad (1.48)$$

In SVM, feature vector is closer to the hyperplane between two classes for defining margin. The generalization of the classifier is improved to increase in separation. In Figure 1.6, circles and triangles represent two different classes. The circles and

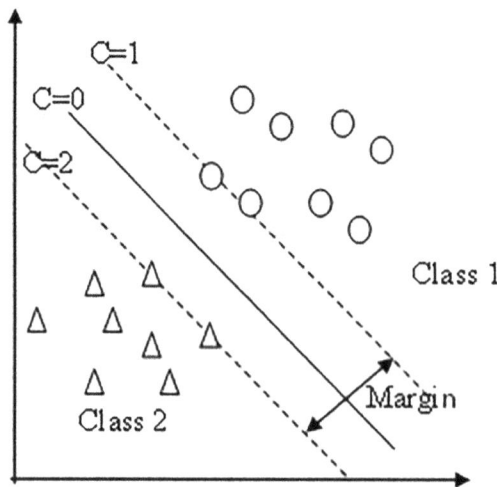

FIGURE 1.6   A two-class SVM classification.

triangles on the dotted line are used to define the upper margin and lower margin. The distance between two margins is 2/‖w‖, so the distance between the margin and the hyperplane is 2/‖w‖. This can be realized by minimizing the cost function shown next.

$$F = \frac{1}{2}\|w\|^2 + b\sum_{j=1}^{n}\xi_j \quad \text{with the constraints} \qquad (1.49)$$

$$y_j c(x) \geq 1 - \xi_j; \ \xi_j > 0$$

Where,
ξ_i, is the slack variable
a,b are constants
w is weight component

The weight component provides Euclidean distances between the training data in each class and the separating hyperplane. Training data that are not in the correct side contribute to the weight component. The second component of the cost function is the slack variable, a measure of the Euclidean distance between margins. Thus, the SVM must select support vectors from the training data set that minimize these two components of the cost function. SVM is popular, because it generalizes well due to the presence of the margins and faster training. However, the generalization is regularized by b. The larger b value offers better classification rate on training data. Small value of b results in higher misclassification rate. Normally, the regularization parameter is varied in linear steps to obtain good performance.

In multiclass SVM classification, similar to LDA, a one against all approach is used. The optimal solution can be obtained by using different kernel function such as polynomial, radial basis, Gaussian, etc. which map the date to higher-dimensional space to obtain hyperplane.

Simple Logistic Regression (SLR): Logistic regression is one of the supervised learning approaches (Landwehr et al. 2005). This supervised classifier is presented with a set of training data/instances $x_1, x_2, \ldots\ldots x_n$ which is defined over the m number of features i.e. $\{v_1, v_2, v_3, \ldots\ldots v_m\}$. The class is identified using a logistic regression model, with the posterior class probabilities $P(C = i|X = x)$ for I classes.

In this classifier, a given feature set is classified into one of the I classes of motion as given in equation (1.50).

$$i * class = \arg\max_i P(C = i \mid X = x) \qquad (1.50)$$

Where,
C is a class variable of motion
x is the feature set which represents the class and
$P(C = i|X = x)$ is the posterior class probability for an instance x

Logistic regression models the posterior probabilities for I classes using linear function in x ensuring that they sum to 1 and remain in [0, 1]. The linear regression model is specified in terms of I-1 log-odds that separate each class from the "base class" I, such that,

$$\log \frac{P(C = i \mid X = x)}{P(C = I \mid X = x)} = \alpha_i^T, i = 1, 2, \dots I - 1 \tag{1.51}$$

Where,

$$P(C = i \mid X = x) = \frac{e^{\alpha_i^T x}}{1 + \sum_{k=1}^{I-1} e^{\alpha_k^T x}}, \text{ for } i = 1, 2 \dots (I\text{-}1) \tag{1.52}$$

$$P(C = i \mid X = x) = \frac{1}{1 + \sum_{k=1}^{I-1} e^{\alpha_k^T x}} \tag{1.53}$$

and, α_i is the parameter vector in logistic regression model

This logistic model produces linear boundaries between the regions in the feature space corresponding to different classes. In logistic regression, model fitting means, estimating the parameter vector α_i. The standard procedure in statistics is to look for the maximum likelihood in logistic regression using numeric optimization algorithms that approach the maximum likelihood solution iteratively and reach it in the limit that is in practice. Friedman et al. (2000) propose the LogitBoost algorithm for fitting additive logistic regression models by maximum likelihood. This logistic regression model generally has the following form.

$$P(C = i \mid X = x) = \frac{e^{G_i(x)}}{\sum_{k=1}^{I} e^{G_k(x)}}, \sum_{k=1}^{I} G_k(x) = 0 \tag{1.54}$$

Where,

$$G_i(x) = \sum_{m=1}^{M} g_{mi} = (x)$$

and g_{mi} are functions of the input feature set

The condition $\sum_{k=1}^{1} Gk(x) = 0$ is for the stability only; adding a constant to all $G_k(x)$ does not change the model. The following steps give the pseudo code for the LogitBoost algorithm Friedman et al. (2000).

Step 1: Start with weights $w_{ji} = 1/n, j = 1, \dots \dots \dots, n, i = 1, \dots \dots \dots, I,$
    $G_i(x) = 0$ and $p_i(x) = 1/I$ $\forall i$.
Step 2: Repeat for $m = 1, \dots \dots \dots M$:
    (a) Repeat for $i = 1, \dots \dots \dots I$:

(i) Compute working responses and weights in the ith class

$$z_{ji} = \frac{y_{ji}^* - p_i(x_j)}{p_i(x_j)(1 - p_i(x_j))} \qquad (1.55)$$

$$w_{ji} = p_i(x_j)(1 - p_i(x_j)) \qquad (1.56)$$

(ii) Fit the function $g_{mi}(x)$ by a weighted least-squares regression of z_{ji} to x_i with weights w_{ji}.

(b) Set $g_{mi}(x) \leftarrow \dfrac{I-1}{I}\left(g_{mi}(x) - \dfrac{1}{I}\sum_{k=1}^{I} g_{mk}(x)\right)$,

$$G_i(x) \leftarrow G_i(x) + g_{mi}(x) \qquad (1.57)$$

(c) Update $p_i(x) = \dfrac{e^{G_i(x)}}{\sum_{k=1}^{I} e^{G_i(x)}} \qquad (1.58)$

Step 3: Output the classifier $\arg\max_i G_j(x)$

Where,
y_{ji}^* are the class probabilities of instance x_j observed in the training data (i.e., one if training data x_j is labelled with class i and zero otherwise) and are given by equation (1.59).

$$y_{ji}^* = 1 \, if \, y_j = i, else \, 0 \qquad (1.59)$$

$p_i(x_j)$ is the probability estimate of the class i for instance x_j given by g
z_{ji} encode the error of the currently fit model on the training data and then try to improve the model by adding a function g_{mi} to G_i, fit to the response by least-squared error.

This LogitBoost algorithm can be used to learn linear, logistic regression model, by fitting a standard least-squares regression function g_{mi} in *step 2(a)ii* of the algorithm.
    Real-time data includes various features, only a few of which are actually relevant to the true target concept. If the non-relevant features are included in, i.e., a logistic regression model, they will usually allow the training data to be fitted with a smaller error, because there is by chance some correlation between the class name and the values of these features for the training data. These non-relevant features, however, will not increase the accuracy and can sometimes even significantly reduce accuracy. Furthermore, including features that are not relevant will make it a lot harder to understand the structure of the domain by looking at the final model. Therefore, it is important to select the most relevant features in the logistic regression models. Further, instead of building g_{mi} by performing multiple regressions based on all features present in the feature space, it is also possible to use a simple regression function that performs a regression on only one feature present in the training data and

fitting simple regression to each feature in the feature space using least-squares as the error criterion, and then selecting that gives the smallest error. Thus, selecting the features that give the least squared error will result in automatic feature selection, because the model will only include the most relevant features present in the training data. The implementation of classifier using simple logistic function is referred to as *simple logistic regression* classification.

In this work, the parameter vector of each target function is estimated (learned) based on simple logistic regression from training data using LogitBoost algorithm. The parameters of the model are estimated using WEKA software. From the measured parameters, the value of $\alpha_i^T x$ can be used to determine the class for the test feature vector. The test feature vector will be assigned to a class for which $\alpha_i^T x$ has the highest value. Thus, the reduction of features in each model reduces the number of computations and hence the time of operation of the controller.

Decision Tree (DT): A decision tree (DT) is a knowledge based supervised statistical classification technique. In this expert system, simple if-then rule is applied for classification. One of the advantages of tree classification is that it can be constructed efficiently and easily. Decision tree begins with a set of cases or training data and creates a data structure that can be used to classify test data. Each training data is described by a set of attributes or features which can have numeric or symbolic values. Associated with each training data is a label representing the name of a class it belongs to. Tree induction splits feature space repeatedly, and stops splitting when the feature subspace contains training data with mostly identical class labels. Splitting consists of selection of feature and decision on threshold value for the feature. A path in a DT basically corresponds to Boolean expression of the form "Feature > threshold" "Feature ≤ threshold", so a tree can be seen as a set of rules to identify feature set. For classification, a new feature set is sorted down to a leaf in the tree and predicts class based on the majority class of the test data in that region.

DT learning algorithm is to develop a tree, with root, which is divided into a number of branches with nodes and further subdivided till it reaches leaves as shown in Figure 1.7, where branches connect the nodes from root to leaf. In Figure 1.6, squares symbolize the leaves as decision points that represent the class to a set of data and ovals represent node of a tree to chance, while outgoing branch corresponds to the possible range of chances. The most important feature in the decision tree algorithm is its ability to automatically select the feature which is appropriate at each node. The major problem is a selection of the best value for the threshold at each node. To develop a tree for real-world problems, researchers used various decision tree algorithms such as divide and conquer, ID3, C4.5 and classification and regression trees (CART). Among decision tree algorithms, Quinlan's ID3 and its successor C4.5 are the most popular learning algorithm for induction of decision trees for classification.

The J48 used by WEKA is a modified version of C4.5 and ID3 for DT classification Witten and Frank (2000). C4.5 and its predecessor ID3 use formulas based on information theory to evaluate the goodness of a test; in particular they choose the test that extracts the maximum amount of information from a set of training data,

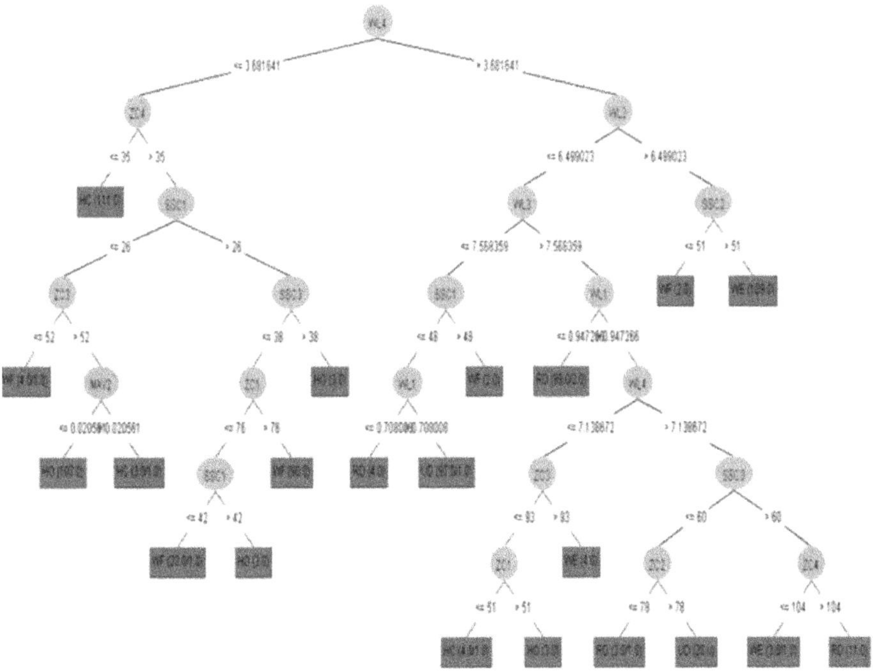

FIGURE 1.7 Decision tree of a subject using feature ensemble-1.

given the constraint that only one attribute will be tested. The basic ideas behind ID3 are that:

- Each node corresponds to a non-categorical attribute and each arc to a possible value of that attribute. A leaf of the tree specifies the expected value of the categorical attribute for the data described by the path from the root to that leaf.
- Each node associated should be a non-categorical attribute with most informative among the attributes not yet considered in the path from the root.
- Entropy is used to measure how informative a node is.

Input to ID3 consists of a collection of instances or training data M, probability distribution $p_1, p_2, \ldots\ldots\ldots p_1$ of the instances are labelled with class $1, 2, \ldots\ldots .I$. Then entropy is defined by equation (1.60).

$$entropy(M) = \sum_{j=1}^{I} -p_i \cdot \log p_i \qquad (1.60)$$

And select the split that gives the highest information gain defined by equation (1.61).

$$IG(S) = entropy(M) - \sum_{j=1}^{k} \frac{|M_i|}{|M|} entropy(M_i) \qquad (1.61)$$

For a split S that splits the set of examples M at the node into the subsets $M_1, \ldots \ldots$ M_k. This equation represents gain information due to attribute. This information gain is used to rank the attributes and build the decision tree where at each node is located the attribute with greatest gain among the attributes not yet considered in the path from the root. Thus, it is necessary to find the best value for the threshold which produces the highest information gain. Thresholds of a set candidate is obtained as midway between corresponding values of the adjacent examples that differ in their target classification by sorting of training instances according to the attribute. These candidate thresholds can then be evaluated by computing information associated with each one. The information gain is computed for each of the candidate features and the one with the highest information gain is selected.

C4.5 is an extension of ID3 that accounts for missing values, continuous attribute value ranges, pruning of decision trees, etc. The J48 algorithm is a WEKA implementation of C4.5 that gives several options to obtain pruned or un-pruned C4.5 decision tree. J48 employs information gain as a splitting criteria and sub-tree raising type of pruning where nodes may be moved downwards towards the root of the tree, replacing other nodes along the way. Reduced error pruning is used to make an actual decision about which parts of the tree to rise. Mostly, sub-tree rising is not advisable due to computational complexity Witten and Frank (2000). A sample of decision tree classifier of a subject with time domain feature ensemble-1 is shown in Figure 4.2. This decision tree is obtained with default settings of WEKA.

Logistic Model Trees (LMT): Logistic model trees (LMT) combines logistic regression models with tree induction and is referred to as logistic model trees. LMT is a regression tree with logistic regression functions at the leaves. Unlike ordinary decision tree (DT) the leaves have logistic regression models for all nodes.

LMT basically consists of a tree structure that is made up of inner or non-terminal nodes N and a set of leaves or terminal nodes T. Let $S = B_1 \times \ldots \ldots \times B_m$ denote the feature space, spanned by m features $V = \{v_1, v_2, v_3, \ldots \ldots \ldots v_m\}$ that are present in the training data. Then the tree structure gives a disjoint subdivision of S into regions S_t and every region is represented by a leaf in the tree as shown in equation (1.61).

$$S = \bigcup_{t \in T} S_t, \quad S_t \cap S_{t'} = \varphi \ for \ t \neq t' \qquad (1.62)$$

Unlike standard decision trees, the leaves $t \in T$ have an associated logistic regression function g_t instead of class name. The regression function g_t takes into account an arbitrary subset of features $V_t \subset V$ of all features present in the data and models the class membership probabilities as given in equation (1.63).

$$P(C = i \mid X = x) = \frac{e^{G_i(x)}}{\sum\limits_{k=1}^{l} e^{G_k(x)}} \qquad (1.63)$$

Where,

$$G_i(x) = \beta_0^i + \sum_{f \in F_i} \beta_v^i \cdot v \qquad (1.64)$$

or, equivalently

$$G_i(x) = \beta_0^i + \sum_{k=1}^{m} \beta_{v_k}^i \cdot v_k \qquad (1.65)$$

If $\beta_v^i = 0$ for $v_k \notin V_t$. The model represented by the whole logistic model tree is then given by equation (1.66).

$$g(x) = \sum_{t \in T} g_t(x) \cdot I(x \in S_t) \qquad (1.66)$$

Where,
$I(x \in S_t)$ is 1 if $x \in S_t$ and 0 otherwise

In LMT, logistic regression and decision trees are special cases. LMT growth would first involve building a standard classification tree and afterwards building a logistic regression model at every node. For building simple logistic regression at the child node, simple logistic regression at the parent node is used as a basis for fitting simple logistic regression at the child node using LogitBoost algorithm. LogitBoost algorithm iteratively changes the simple linear regression class functions $G_i(x)$ to improve the fitting of the data by adding a simple regression function g_{mi} to $G_{i,}$ to fit to the response variable i.e. changing one of the coefficients in the linear function G_i or introducing a new feature variable/coefficient pair. After splitting a node, LogitBoost iterations continue running by fitting g_{mi} to the response variable of the training data at the child node only.

A split for the data at the root is constructed. Growth of the tree continues by sorting the appropriate subsets of data to the child nodes and building the logistic model at the child node. The split will simultaneously optimize the purity in the response variables for all classes. This necessitates a measure of global impurity over all classes and then selects the split that gives the largest decrease in global impurity. Splitting criterion is derived from C4.5 algorithm.

1.7 FACTORS INFLUENCING TRANSLATIONAL PATTERN RECOGNITION

There are various factors that produce varying recognition accuracy in addition to the different methods of signal processing. Some of the factors are discussed in this section.

Post Processing: In addition to the various stages discussed, researchers used post processing to achieve good recognition accuracy. Majority voting is one of the postprocessing stages for making the final decision from a stream of class decisions.

Factors Influencing Translational Model: In supervised learning, the parameters of the model are estimated based on training data, assuming the training and future data will be similar. However, bioelectric signals may change due to factors such as the subject tiring over a period of time, diseases, psychology of the user, etc. The variations of signals influence the feature vector. The feature variability is estimated by variance, the mean of the square of the difference between each feature sample to mean of the feature. The variation of data is mitigated using a large number of training data. The generalization of the classification model improves at a cost of more parameters. The problem of classification using small sets of training data is called overfitting. Hence, there is a tradeoff between number of model parameters and accuracy in decoding with varying data. In general, factors that influence the accuracy in decoding of the user's intent are (1) capability of classifier to select the appropriate feature to identify the user's intent, (2) adaptability of the classifier and (3) effectiveness of the applied method in translating users into device commands (Wolpaw et al. 2002).

Steps to Improve the Classifier Performance: In supervised classification, K-fold cross-validation is performed to identify the parameters that maximize the average predication rate with a smaller set of training data. The K-fold cross-validation is performed for dividing training data into K equal parts, i.e., $n = 1,2,3 \ldots \ldots K$. In this, i^{th} fold cross-validation, is computed using i^{th} data as test data and all other K-1 are used as training data. The predication accuracy is calculated with the test data. Similarly, the remaining K-folds predication accuracy is computed and average over K-folds are calculated.

The adaptability in BCI is addressed by the user being expected to adapt to the features using operant-conditioning system, i.e., adapting the features so that BCI produces correct classification for the intention of the user (Blankertz et al. 2003).

In this supervised classification, ensemble methods are learning algorithms that combine prediction of multiple classifiers in some way (typically with weight or without weight) to classify new data points by taking a vote of their predictions. The idea behind this approach is to train one classifier in every feature subspace and then combine their class predictive distributions. One popular way to select ensemble of feature space in the literature is the random method. This method generates feature subspaces by selecting randomly a number of feature subsets in the original feature space and then constructs a set of classifiers based on these selected subspaces. One of the most active areas of research in ensemble is to study methods for constructing good ensembles for effective classification.

1.8   CONCLUSION

The objective of this chapter is to introduce various signal pattern recognition to control devices in real time for the purpose of HCI from bioelectric signals. The various stages of the pattern recognition module play a vital role in order to improve the accuracy. Till date researchers of pattern recognition are attempting to find a robust classifier to suit the user requirement under varying data.

REFERENCES

Anderson, C.W., Stolz, E.A. and Shamsunder, S. (1998). Multivariate autoregressive models for classification of spontaneous electroencephalographic signals during mental tasks. *IEEE Trans. Biomed. Eng.*, Vol. 45, pp. 277–286.

Biopac Systems, Inc. (2010). *EMG Frequency Signal Analysis*. www.biopac.com/Manuals/app_pdf/app118.pdf.

Blankertz, B., Dornhege, G., Krauledat, M., Müller, K.-R., Kunzmann, V., Losch, F. and Curio, G. (2006). The Berlin brain—computer interface: EEG-based communication without subject training. *IEEE Trans. Neural. Syst. Rehabil. Eng.*, Vol. 14, No. 2, pp. 147–152.

Blankertz, B., Dornhege, G., Scharfer, C., Krepki, R., Kohlmorgen, J., Muller, K.R., Kunzmann, V., Losch, F. and Curio, G. (2003). Boosing bit rates and error detection for the classification of fast-paced motor commands based on single-trial EEG analysis. *IEEE Trans. Rehabil. Eng.*, Vol. 11, pp. 100–104.

Boostani, R. and Moradi, M.H. (2003). Evaluation of the forearm EMG signal features for the control of a prosthetic hand. *J. Physiol. Meas.*, Vol. 24, No. 2, pp. 309–319.

Burke, D.P., Kelly, S.P., de Chazal, P., Reilly, R.B. and Finucane, C. (2005). A parametric feature extraction and classification strategy for brain-computer interfacing. *IEEE Trans. Neural. Syst. Rehabil. Eng.*, Vol. 13, pp. 12–17.

Cheng, M., Gao, X., Gao, S. and Xu, D. (2002). Design and implementation of a Brain—computer interface with high transfer rates. *IEEE Trans. Biomed. Eng.*, Vol. 49, No. 10, pp. 1181–1186.

Du, S. and Vuskovic, M. (2004). Temporal vs. spectral approach to feature extraction from prehensile EMG signals. *Proc. IEEE International Conference on Information Reuse and Integration*, Las Vegas, pp. 344–350, Nevada.

Enas, G.G. and Choi, S.C. (1986). Choice of the smoothing parameter and efficiency of k-nearest neighbor. *Comput. Math. Appl.*, Vol. 12, pp. 235–244.

Fielding, A.H. (2007). *Cluster and Classification Techniques for the Biosciences*. Cambridge: Cambridge University Press, The Edinburgh Building.

Fischer, R.A. (1936). The use of multiple measurements in taxonomic problems. *Ann. Eugen.*, Vol. 7, pp. 179–188.

Friedman, J., Hastie, T. and Tibshirani, R. (2000). Additive logistic regression: A statistical view of boosting. *Ann. Stat.*, Vol. 38, No. 2, pp. 337–374.

Huang, H.-P. and Chen, C.-Y. (1999). Development of a myoelectric discrimination systems for a multi-degree prosthetic hand. *Proc. IEEE International Conference on Robotics and Automation*, Detroit, pp. 2392–2397, Michigan.

Hudgins, B., Parker, P. and Scott, R.N. (1993). A new strategy for multifunction myoelectric control. *IEEE Trans. Biomed. Eng.*, Vol. 40.

Jung, T.-P., Makeig, S., Humphries, C., Lee, T.-W., McKeown, M.J., Iragui, V. and Sejnowski, T. (2000). Removing electroencephalographic artifacts by blind source separation. *Psychophysiology*, Vol. 37, No. 2, pp. 163–178.

Khushaba, R.N., Al-Jumaily, A. and Al-Ani, A. (2009). Evolutionary fuzzy discriminant analysis feature projection technique in myoelectric control. *Pattern Recognit. Lett.*, Vol. 30, No. 7, pp. 699–707.

Krusienski, D.J., Schalk, G., McFarland, D.J. and Wolpaw, J.R. (2007). A mu-rhythm matched filter for continuous control of a brain computer interface. *IEEE Trans. Biomed. Eng.*, Vol. 54, pp. 273–280.

Landwehr, N., Hall, M. and Frank, E. (2005). Logistic model trees. *J. Mach. Learn.*, Vol. 59, No. 1–2, pp. 161–205.

Marks, S. and Dunn, O.J. (1974). Discriminant functions when covariance matrices are unequal. *J. Am. Stat. Assoc.*, Vol. 69, pp. 555–559.

McFarlan, D.J. and Wolpaw, J.R. (2005). Sensorimotor rhythm-based brain computer interface (BCI): Feature selection by regression improves performance. *IEEE Trans. Neural. Syst. Rehabil. Eng.*, Vol. 14, pp. 372–379.

Muller, K.-R., Anderson, C.W. and Birch, G.E. (2003). Linear and nonlinear methods for brain computer interfaces. *IEEE Trans. Neural. Syst. Rehab. Eng.*, Vol. 11, pp. 165–169.

Oskoei, M.A. and Hu, H. (2008). Support vector machine based classification scheme for myoelectric control applied to upper limb. *IEEE Trans. Biomed. Eng.*, Vol. 55, No. 8, pp. 1956–1965.

Park, S.-H. and Lee, S.-P. (1998). EMG pattern recognition based on artificial intelligence techniques. *IEEE Trans. Rehabil. Eng.*, Vol. 6, No. 4, pp. 400–405.

Pfurtscheller, G., Neuper, C., Schologl, A. and Lugger, K. (1998). Separability of EEG signals recorded during right and left motor imagery using adaptive autoregressive parameters. *IEEE Trans. Rehabili. Eng.*, Vol. 6, pp. 316–325.

Proakis, J.G. and Manolakis, D.G. (2007). *Digital Signal Processing-Principles, Algorithms and Applications*. New York: Macmillan.

Schalk, G., Leuthardt, E.C., Brunner, P., Ojemann, J.G., Gerhardt, L.A. and Wolpaw, J.R. (2008). Real-time detection of event-related brain activity. *Neuroimage*, Vol. 43, pp. 245–249.

Tkach, D., Huang, H. and Kuiken, T.A. (2010). Study of stability of time-domain features for electromyographic pattern recognition. *J. Neuroeng. Rehabil.*, Vol. 7, No. 1, pp. 1–13.

Tou, J.T. and Gonzalez, R.C. (1974). *Pattern Recognition Principles*. Reading, MA: Addision-Wesley.

Vapnik, V.N. (1998). *Statistical Learning Theory, Ser. Adaptive and Learning Systems for Signal Processing, Communications, and Control*. New York: Wiley.

Witten, I.H. and Frank, E. (2000). *Data Mining: Practical Machine Learning Tools with Java Implementations*. San Francisco: Morgan Kaufmann.

Wolpaw, J.R., Birbaumer, N., McFarland, D.J., Pfurtscheller, G. and Vaughan, T.M. (2002). Brain-computer interfaces for communication and control. *Clin. Neurophysiol.*, Vol. 113, No. 6, pp. 767–791.

Zardoshti-Kermani, M., Wheeler, B.C., Badie, K. and Hashemi, R.M. (1995). EMG feature evaluation for movement control of upper extremity prosthesis. *IEEE Trans. Rehabil. Eng.*, Vol. 3, No. 4, pp. 324–333.

2 Automated Recognition of Alzheimer's Dementia

A Review of Recent Developments in the Context of Interspeech ADReSS Challenges

*Muhammad Shehram Shah Syed,
Zafi Sherhan Syed, Margaret Lech,
and Elena Pirogova*

2.1 INTRODUCTION

Alzheimer's dementia (AD) is a progressive neurological disease that can severely impair the physical and cognitive well-being of an individual. It can be caused by a variety of factors and manifest itself in different forms. Alzheimer's disease is the most common type of dementia, accounting for more than 60% of reported cases [1, 2]. It affects more than 50 million individuals across the world [1]. Alzheimer's dementia adversely impacts the quality of life of patients and their caregivers, frequently influencing their mental well-being [3]. Thus, an even larger population is being affected indirectly by this condition.

Although the symptoms of dementia differ depending on an individual and the stage of the disease, it is typically characterized by a decline in cognitive functions, such as attention, decision making, language, memory, orientation, and thinking process. Dementia is also associated with causing functional limitations in self-care and daily living tasks. The damage caused by dementia is irreversible and eventually leads to the patient's demise. However, early detection of AD can significantly extend the lifespan and improve the quality of living for patients and their carers.

Due to the lack of accurate indicators, clinical identification of dementia is difficult [4]. In fact, a conclusive diagnosis can only be made post-mortem when the brain tissue is examined during autopsy [5]. Currently used clinical assessment for determining patients' cognitive and functional abilities is typically based on their medical history. This can be supplemented by cognitive evaluation tests, such as the Cambridge Assessment of Memory and Cognition (CAMCOG) [6], Clock Drawing Test [7], Mini-Mental State Examination (MMSE) [8], and Montreal Cognitive

DOI: 10.1201/9781003201137-3

Assessment (MoCA) [9]. Recent advances in artificial intelligence (AI) led to calls for investigating the use of neurological imaging for automated screening of dementia [10–12].

New developments in social signal processing and automated mental health screening techniques have presented clinicians with new aids supporting the diagnosis of disorders manifested as changes in human behaviour [13–15]. Although a plethora of research literature is available on developing automated screening methods for dementia [16–18], their evaluation and comparison are difficult due to a lack of uniform experimental conditions. To address this issue, over the last two years, Luz et al. [19, 20] organized the Alzheimer's Dementia Recognition through Spontaneous Speech challenges at the Interspeech conferences. The aim was to "provide a forum for those different research groups to test their existing methods (or develop novel approaches) on a new shared standardized dataset". This initiative has encouraged researchers to establish new approaches for automated dementia screening under the same experimental conditions facilitating a straightforward comparison. This book chapter reviews the solutions proposed by participants of the 2020 [19] and 2021 [20] editions of the ADReSS challenge. We conclude with a summary of key contributions to the body of knowledge on the speech-based recognition of Alzheimer's dementia and suggest future directions in this area of research.

2.2   THE ADRESS CHALLENGE 2020

The 2020 ADReSS challenge consisted of categorical prediction (classification) and continuous prediction (regression) tasks. The Alzheimer's dementia classification task (Task A) required the challenge participants to train a machine learning classifier to identify whether the subject belongs to the dementia or the healthy (control) category. The MMSE score prediction task (Task B), on the other hand, required participants to train a regression model that can predict a continuous value of the patient's Mini-Mental State Examination (MMSE) score.

2.2.1   DESCRIPTION OF THE DATASET

The dataset used in the ADReSS challenge was sourced from the Dementia-Bank repository [21]. It consists of speech recordings and transcripts of subjects describing the "Cookie Theft" picture from the Boston Diagnostic Aphasia Exam [22, 23]. In addition to speech recordings, the organizers of the challenge provided textual transcripts of speech based on the annotation Codes for the Human Analysis of Transcripts (CHAT) coding system [24]. The data also included timestamps identifying whether the speech belongs to the subject or the interlocutor. Most importantly, the dataset was balanced in terms of diagnostic categories, i.e., age and gender. It eliminated the potential for a prediction bias due to the relative under- or over-representation of genders, age groups, or classes. The dataset was partitioned into training (development) and test subsets.

The development partition consisted of 108 recordings for which ground truth labels were also provided. The labels included dementia status, i.e., healthy or AD, for the binary classification task and the MMSE value for the regression task. The test partition consisted of 48 recordings, but in this case, the labels were not provided to comply with the ADReSS challenge regulations. The test subset was used to predict the unknown labels. Participants of the ADReSS challenge could make five attempts to predict the binary labels in the classification task and five attempts to predict the MMSE values in the regression task.

2.2.2 DESCRIPTION OF THE CHALLENGE BASELINE

Although the ADReSS challenge organizers provided separate baselines for the audio and textual modality, the challenge was to outperform the textual baseline as it was higher than the audio baseline. The baseline prediction from the audio modality used five standard feature sets, including emobase, ComParE, eGeMAPS, MRCG [25], and a minimal set of low-level acoustic parameters, such as vocalization duration, vocalization count, pauses, and speech rate and their statistical functionals [26]. Meanwhile, for the prediction from the textual modality, the CLAN [27] program was used to compute a set of 34 standard linguistic parameters, such as parts-of-speech, type-token-ratio, duration, total utterances, etc.

To set the binary classification (Task A) baseline, five different classification algorithms were used, including linear discriminant analysis (LDA), decision trees (DT), k-nearest neighbour with $k = 1$ (1-NN), support vector machine classifier (SVC), and random forests (RF). The binary classification baseline was then set to the best performing method, which was the LDA classifier yielding the average accuracy of 75%. Similarly, for the regression task (Task B), five different algorithms were explored. These included decision trees (DT), linear regression (LinReg), Gaussian process regression (GPR), least-squares boosting (LSBoost), and support vector machine regression (SVR). The regression baseline was then set to the best performing algorithm, which in this case was the decision tree regressor achieving an average Root Mean Squared Error (RMSE) of 5.20.

2.2.3 SUMMARY OF RESULTS

In Table 2.1, we provide a summary of results for the two tasks of the ADReSS challenge based on the proceedings of Interspeech 2020 where solutions from 12 teams (excluding the challenge baseline) were published. Out of these, ten teams achieved a better performance than the challenge baseline of 75.00% for the classification task, whereas nine teams were able to improve the baseline of 5.20 for the regression task. The best performing model for Task A was proposed by Yuan et al. [28], which achieved an accuracy of 86.60%. Meanwhile, Koo et al. [29] achieved the best RMSE score of 3.74 for the regression task. Interestingly, Yuan et al. did not report results for Task B, whereas Koo et al. stood only joint 6th in terms of ranking for Task A. On the other hand, our solutions [30] for the two tasks ranked top-2 for both tasks of the ADReSS challenge.

2.2.4 REVIEW OF SOLUTIONS PROPOSED FOR THE ADReSS CHALLENGE

In [40], Cummins et al. used three types of acoustic feature representations. These included bag-of-audio words feature aggregation method for acoustic low-level

TABLE 2.1

Summary of Results for the Performance of Systems Published as part of Interspeech 2020 Proceedings. Task A: Classification of Alzheimer's Dementia and Task B: Prediction of MMSE Scores.

Publication	Task A		Task B	
	Score	Rank	Score	Rank
Farzana et al. [31]	–	–	4.34	3
Pappagari et al. [32]	75.00	12	5.37	11
Luz et. Al. [19]	75.00	11	5.20	10
Martinc et al. [33]	77.08	10	4.44	4
Edwards et al. [34]	79.17	9	–	–
Rohanian et al. [35]	79.20	8	4.54	5
Searle et al. [36]	81.00	7	4.58	7
Koo et al. [29]	81.25	6	3.74	1
Pompili et al. [37]	81.25	6	–	–
Sarawgi et al. [38]	83.00	5	4.60	8
Balagopalan et al. [39]	83.33	4	4.56	6
Cummins ct al. [40]	85.20	3	4.65	9
Syed et al. [30]	85.42	2	4.30	2
Yuan et al. [28]	89.60	1	–	–

descriptors, an end-to-end convolutional neural network that learns to classify using raw audio waveforms, and a Siamese network that learns to classify using Mel spectrogram representation of the subjects' speech signal. For the linguistic modality, their proposed system utilizes Global vectors for word representation (GloVe) [41], word embeddings followed by temporal modelling based on Bidirectional Long-Short Term Memory (BLSTM) with attention mechanism and Hierarchical attention networks. The main findings from their work are that linguistic modality outperforms audio with a classification accuracy of 81.30% vs 70.80% and an RMSE score of 4.66 vs 6.45 for the regression task for the test partition.

Searle et al. [36] performed several experiments to assess the performance of transformer-based pretrained text embeddings such as Bidirectional Encoder Representations from Transformers (BERT) [42] and its variants for the ADReSS challenge classification and regression tasks. The Term-Frequency-Inverse Document Frequency (TF-IDF) features offered the best classification performance for the test partition, with an accuracy of 81.00%, whereas DistilBERT [43], a distilled version of BERT, provided the highest performance for the regression task, with an RMSE of 4.58—both of which are better than the challenge baseline. The classification task result was particularly interesting since it demonstrated that the basic TF-IDF approach can capture relevant information to provide the best classification result, thus avoiding the requirement for more complex transformer-based text embeddings.

Yuan et al. [28] used transformer-based text models to encode words, word repetitions, and pause durations from the transcripts for the classification task. This elegant approach forced their machine learning model to differentiate between healthy subjects and those with dementia based on disfluencies within features the model learns from the textual content of speech transcripts. The BERT and Enhanced language representation models with informative entities (ErniE) [44] were finetuned. A significant variance in the classification performance for the training partition between 75.00% and 86.00% in a LOO-CV setting was observed. It was attributed to the limited number of training examples available in the dataset. As a solution, 50 different models were trained, and the majority voting was used to yield a final set of predictions for the test partitions. Their best-performing model provided a classification accuracy of 89.60% for the test partition.

Koo et al. [29] proposed a multimodal system based on audio and textual information. To construct feature vectors from the subjects' speech recordings, ComParE and eGeMAPS functionals using the openSmile toolkit were computed. In addition to this, deep acoustic embeddings from the VGGish network [45] were calculated. A range of features for text modality, including transformers-based deep contextual embeddings, GloVe-based word embeddings, and several handmade features for the task at hand (classification or regression), were tested. These handcrafted features were aimed to quantify the psycholinguistic traits, lexical complexity, and repetitiveness from manually transcribed voice transcripts. Koo et al.'s machine learning models were based on Convolutional Recurrent Neural Networks, and they tested the performance of these models in unimodal and multimodal settings. The highest performing model for the classification task obtained an accuracy of 81.25%, whereas their best result for the regression task achieved an RMSE of 3.74.

Farzana et al. [31] seek to model verbal disfluency markers using textual features, including fine-tuned contextual embeddings based on the DistilBERT model and a wide range of handcrafted features of their solution for the regression task of the ADReSS challenge. The handcrafted features represented lexico-syntactic, psycholinguistic, and discourse characteristics of the subjects' speech. Given the large dimensionality of handcrafted features, the authors trained a random forest regressor and selected a smaller set of features based on the mean decrease impurity metric. In addition to text-based features, Farzana et al. also used MFCCs to generate feature vectors for speech recording of each subject. Their best RMSE score was 4.34, which was achieved by selected handcrafted features.

Rohanian et al. [35] proposed a multimodal fusion system with a gating mechanism as part of their solution to the ADReSS challenge. They used a large set of acoustic features from the Cooperative Voice Analysis Repository for Speech Technologies (COVAREP) toolkit to quantify disfluency from the audio modality. Meanwhile, they applied GloVe embeddings and a deep neural network model to model disfluencies from the text modality. Their best result for the test partition achieved the classification accuracy of 79.20% and the RMSE score of 4.54—both of which were better than the challenge baseline.

Sarawgi et al. [38] used ComParE functionals to generate feature vectors for speech recordings from subjects in the audio modality and various handcrafted features in the text modality. They also computed intervention features using the meta-data available in transcripts. These features informed how frequently the interlocutor intervened to assist subjects during the recording session. The most interesting aspect of their work was that they (i) first trained a multilayer perception network to distinguish between healthy and dementia patients and (ii) then finetuned this model for the regression task to predict MMSE scores. Sarawgi et al.'s best results show that their proposed system achieves a classification accuracy of 83.00% and an RMSE score of 4.60.

Edwards et al. [34] conducted several experiments with acoustic and textual features to build their proposed system. They demonstrated that feature selection methods can be applied to feature sets used in the baseline paper to produce improved classification results for the training partition for the audio modality. To generate a numerical representation of the text contained in speech transcripts, they used various word embeddings such as Word2Vec [46], GloVe, Sent2Vec [47], fasttext [48], Electra [49], and RoBERTa [50]. Edwards et al. also computed various types of handcrafted features to measure linguistic characteristics of speech in addition to these features. The main contribution of their work is that the features computed on the word-level representation of text in speech transcripts resulted in better classification performance than features computed on the phoneme-level representation of texts. Their best result based on an ensemble of fast text embeddings trained with the phoneme-level representation of transcripts and the posterior of the prediction from the LDA classifier for the selected ComParE features achieved a classification accuracy of 79.17% for the test partition.

Martinc et al. [33] carried out a large number of experiments as part of their proposed multimodal approach to addressing the ADReSS challenge. They used the average value of MFCC features for the audio modality, as well as active data representation on eGeMAPS and MFCC features. They used TF-IDF vectoriser and Doc2Vec embeddings [51], as well as various types of handcrafted features, for the text modality. Martinc et al. also experimented with machine learning algorithms such as support vector machines, XGBoost [52], random forests, and logistic regression. They claim to have conducted 65,535 experiments to identify the best-performing models. The best model achieved an accuracy of 77.80% on the test partition using Gunning fog index, the number of unique words, duration, 4-grams with four characters, suffix tokens, POS tags, and universal dependency features trained using a logistic regression classifier. On the other hand, the best RMSE score of 4.44 was achieved whilst using an SVM model trained with a collection of features which included number of unique words, bigrams, 4-grams for characters, suffix tokens, POS tags, and grammatical dependency.

Pompili et al. [37] used pre-trained embeddings for both audio and textual modalities as part of their proposed multimodal system for the ADReSS challenge. They used x-vector- and ivector-based acoustic embeddings previously trained on SRE and VoxCeleb datasets for the audio modality, whereas pre-trained BERT embeddings along with parts of speech (POS) tags were used as the primary feature for textual modality. Their best-performing model resulted from a fusion of three neural models for text modality. It achieved a classification accuracy of 81.25%. It is worth noting that pre-trained acoustic embeddings only achieved a test partition accuracy of 54.17%, which is an interesting result since it suggests that x-vector- and ivector-based embeddings may not be useful for the task at hand.

As part of their solution to the ADReSS challenge, Balagopalan et al. [39] proposed a unimodal system based on text modality. They used BERT embeddings along with a collection of handcrafted features to assess the lexico-syntactic and semantic properties of spoken words as recorded in speech transcripts. Given the high dimensionality of these features, Balagopalan et al. chose a smaller set of 'k' features using the F-measure from the ANOVA test (the parameter k was optimized using 10-fold cross-validation). While the selected handcrafted features achieved a classification accuracy of 81.30%, BERT embeddings performed slightly better with an accuracy of 83.33%. These findings are interesting because they demonstrate that contextual embeddings provided by BERT can capture linguistic characteristics of texts that can be used to distinguish between speech uttered by healthy individuals and those with dementia. Finally, they reported the RMSE of 4.56 for the test partition using selected features and ridge regression as their best results for the regression task.

Pappagari et al. [32] proposed a multimodal system based on x-vector embeddings and silence duration features from the audio modality and finetuned BERT embeddings from the text modality as part of their solution for the ADReSS challenge. They reported that x-vectors achieved an accuracy of 62.50% and the RMSE of 6.14, BERT embeddings achieved the accuracy of 72.92% and the RMSE of 5.86, but a fusion of acoustic and textual features led to the best results, improving classification accuracy to 75.00% and reducing the RMSE score to 5.37. It is worth noting that silence duration was not deemed useful because it provided accuracy near the chance level.

2.3 THE ADRESSO CHALLENGE 2021

The ADReSSo challenge consists of three tasks, namely: Alzheimer's dementia classification task (Task A), the MMSE score regression task (Task B), and the cognitive decline inference task (Task C). In Task A, the participants of the challenge were required to produce a model that can differentiate between individuals from the dementia group and the control group. In Task B, participants had to infer the patient's MMSE score, whereas, in Task C, participants were expected to predict cognitive status changes over time.

2.3.1 DESCRIPTION OF THE DATASET

Organizers of the ADReSSo challenge provided two datasets for the three tasks. Here, Tasks A and B shared a dataset that consisted of speech recordings of subjects performing the picture description task, i.e., the Cookie Theft picture from the Boston Diagnostic Aphasia Examination [22, 23]. In total, this dataset consisted of 237 files, with 166 files provided with their labels as part of the development partition and 71 provided without their labels for the test partition. The dataset for Task C consisted of 105 speech recordings from subjects with dementia performing a semantic fluency task [53]. Here, 71 files were provided for the development partition and 32 for the test partition. Similar to the dataset used for the ADReSS challenge, the ADReSSo challenge dataset was carefully matched in terms of gender and age distribution to reduce performance bias due to these confounding factors. It should be mentioned that although speech diarization timestamps for the subject and the interlocutor were provided for the speech recordings, we did not find the timestamps to be accurate. Similar to the ADReSS challenge, a participating team had the opportunity to submit five prediction attempts for each task. Unlike the ADReSS challenge, transcripts were not provided to participants. They were encouraged to use automated speech recognition tools to generate speech transcripts themselves if they wanted to explore features based on textual modality.

2.3.2 Description of the Dataset

The baseline paper for the ADReSSo challenge by Luz et al. provided separate benchmarks for audio and text modalities. For audio modality, they computed eGeMAPS features and applied the active data representation method [17] to generate the acoustic representation for each audio recording. For textual modality, they used Google's speech recognition service to generate speech transcripts. These transcripts were then converted into CHAT format and the CLAN tool [27] was used to compute a wide range of features that quantify linguistic and discourse characteristics along with lexical and morphological descriptions of words in the transcripts.

For classification, they explored the efficacy of five different methods i.e., decision trees, nearest neighbours, linear discriminant analysis, tree bagger, and support vector machine. The hyperparameters for these algorithms were optimized using Leave-One-Subject-Out cross-validation (LOSO-CV). In this approach, a model is trained on all data except from one "hold-out" subject. The data from the hold-out subject is used to test the trained model. Meanwhile, for the task of regression, they used linear regression, decision trees, support vector regressor, random forest regression, and Gaussian process regression with the hyperparameters being optimized using LOSO-CV. The performance metric for Task A was classification accuracy, for Task B it was RMSE, and for Task C it was the average value of the f1 score (avg-f1 score). Their best result for Task A (accuracy of 77.46%) was achieved using textual features with the support vector machine classifier. For Task B, their best system achieved a score of 5.28 which uses textual features with SVR for regression. Finally, the best system for Task C used textual features with a decision tree classifier and achieved the avg-f1 score of 66.67%.

2.3.3 Summary of Results

As summarized in Table 2.2, a total of 11 papers (excluding the baseline paper) describing system solutions were published as part of the Interspeech 2021 ADReSSo challenge. Out of the 11 teams which reported results for Task A, all except Balagopalan et al. [54] achieved better performance than the baseline of 78.87%. Interestingly, three teams [55–57] were tied at first place in terms of the achieved classification accuracy of 84.51%—the official metric for this task. Meanwhile, five teams reported the results for Task B, with all six achieving a smaller RMSE score than the challenge baseline of 3.85. Here, Pappagari achieved the best score of 3.85.

TABLE 2.2
Summary of the Performance Results for the Developed Systems Published in the Interspeech 2021 Proceedings. Task A: Classification of Alzheimer's Dementia, Task B: Prediction of MMSE Scores, and Task C: Classification of Cognitive Decline.

Publication	Task A		Task B		Task C	
	Score	Rank	Score	Rank	Score	Rank
Balagopalan et al. [54]	67.61	9	–	–	–	–
Luz et al. (baseline) [20]	78.87	8	5.28	6	66.67	3
Gauder et al. [58]	78.90	7	–	–	–	–
Rohanian et al. [59]	84.00	6	4.26	2	62.00	4
Qiao et al. [60]	83.00	5	–	–	–	–
Zhu et al. [61]	83.10	4	4.44	4	70.91	2
Chen et al. [62]	81.69	3	–	–	–	–
Perez-Toro et al. [63]	80.28	2	4.56	5	–	–
Wang et al. [64]	80.28	2	–	–	–	–
Pan et al. [55]	84.51	1	–	–	–	–
Pappagari et al. [56]	84.51	1	3.85	1	–	–
Syed et al. [57]	84.51	1	4.35	3	73.80	1

Finally, only three teams reported the results for Task C, and our work [57] achieved the best avg-f1 score of 73.80% compared to the baseline score of 66.67%.

2.3.4 REVIEW OF SOLUTIONS PROPOSED FOR THE ADRESSO CHALLENGE

In [54], Balagopalan et al. described the findings of experimental research into the efficacy of handcrafted acoustic features and deep acoustic embeddings. They computed handcrafted features as given in the prior work of Fraser et al. [16] and used the wav2vec2 model for embeddings. It was observed that, while handcrafted features performed better in terms of precision, embedding-based features performed better for other classification criteria, such as accuracy, f1 score, and recall. Their best-performing model was based on feature fusion of handcrafted embeddings and employed the SVM classifier with ANOVA F-value-based feature selection as part of the machine learning workflow. This model attained an accuracy of 67.61% for the test partition which was lower than the challenge baseline of 78.87%.

Gauder et al. [58] presented results from their experimentation with different types of acoustic features, such as eGeMAPS, TRILL [65], Allosaurus [66], and wav-2vec2 features. Since these features were computed at the frame level, they used a deep neural network with 1-D convolutional layers to yield 128-dimensional embedding. This method, in a sense, implements transfer learning of the knowledge contained in three embeddings to solve the Alzheimer's dementia prediction problem. The best individual features were wav2vec2, achieving an accuracy of 75.30% on the development partition (6-fold cross-validation) and 78.90% for the test partition. Meanwhile, TRILL features achieved an accuracy of 72.90% for the development partition and 69.00% for the test partition. Gauder et al. reported that feature fusion did not improve the performance of their system and surmised that this was because wav2vec2 contain all information conveyed by eGeMAPS, TRILL, and Allosaurus features. This observation is interesting since many researchers (including us) find that feature fusion consistently improves classification performance.

Rohanian et al. [59] presented two deep learning models that use Automatic Speech Recognition (ASR) transcribed speech and acoustic data to identify subjects with dementia and the degree of severity of this disease. For acoustic features, the COVAREP toolkit [67] was applied to compute LLDs and seven functionals to generate a feature vector for the audio recording. For ASR, they used IBM Watson speech-to-text services. The automatically generated transcripts were provided as inputs to compute GloVe and BERT embeddings, thus converting sequences of words into vectors of parameters. In addition, they used features to measure disfluency, pause information, and language model probability features from their previous work [68]. The best result for Task A achieved the accuracy of 84.00% using BLSTM with Gating that utilized GloVe, COVAREP, disfluency, language model probability features, and pause features. This system achieved the RMSE of 4.26 for Task B and the average f1 score of 62.00% for task C. It should be noted that this system did not outperform the challenge baseline of 66.67% for Task C.

Qiao et al. [60] proposed a solution based on deriving features from the process of ASR, textual disfluency, linguistic complexity, and sophistication, along with word embeddings from contextual models. For ASR, they used AppTek's Automatic Speech Recognition technology via a cloud API service. To model disfluency, pause duration per sentence, speed of articulation, the frequency count of filled pause types, and the average word-level confidence scores as a proxy of pronunciation quality were estimated using the CoCoGen toolkit [69]. It provided information about the computation of syntactic complexity, lexical richness, n-grams, and various information-theoretic parameters. For the word embeddings, BERT and ERNIE models were used. The baseline models were applied given that their initial experimentation showed no reliable differences in terms of classification accuracies between baseline and large models. Qiao et al. investigated feature-level fusion as well as decision-level fusion but found that a stacking ensemble is most robust against overfitting across both development and test partitions. Some interesting observations from their work are that amongst disfluency features, the mean syllable duration is larger for the AD group than for the control group and that subjects from the AD group utter a smaller number of syllables per minute. The authors also reported that the AD group has a larger pause time per sentence, longer pauses (2 seconds), and shorter pauses (2 seconds). From the ASR, it was also found that the AD group had on average smaller pronunciation confidence.

Zhu et al. [61] proposed the WavBERT model for dementia detection. This model seeks to exploit semantic and non-semantic information through fine-tuned wav2vec and BERT models. They first tested the vanilla WavBERT model (WavBERT1.0) that provides ASR transcripts generated through wav2vec to BERT, which in turn is fine-tuned for the dementia detection task (Task A). An expanded version of the WavBERT model (WavBERT2.0) was then proposed that uses intermediate information from wav2vec to determine the locations and the length of pauses and passes this non-semantic information to BERT along with semantic information. The WavBERT2.0 achieved the classification accuracy of 83.10% and RMSE of 4.44, while the basic WavBERT1.0 yielded only 73.24% classification accuracy and RMSE of 4.6. Zhu et al. also investigated expanding the WavBERT further (WavBERT3.0) by converting wav2vec embeddings to BERT such that the input embeddings preserve non-semantic information at the embedding level. This was achieved by using a network consisting of two 1-D convolutional layers with layer-normalization operation in between. The best f1 score of 70.91% for Task C was obtained when adding pause information from WavBERT2.0 and embedding conversion from WavBERT3.0.

In [62], Chen et al. conducted a systematic comparison of approaches to detect cognitive impairment based on spontaneous speech. For the audio modality, they used MFCCs, GeMAPS, eGeMAPS, ComParE, and IS10 features. To generate transcripts from speech recordings, they used two models. The first was the English ASR model from DeepSpeech toolkit, whereas the second was Google speech-to-text. Using the transcripts generated by these two methods, they computed three types of textual features, i.e., LIWC, BERT word, and BERT sentence embeddings. To generate transcript-level features for BERT word embeddings, they computed the average, maximum, minimum, and standard deviation functionals and concatenated them into feature vectors. They experimented with LR, SVM, DR, and MLP using LOSO-CV but found LR to be the best classification method for the task at hand. In terms of fusion methodology, they experimented with feature level fusion and decision level fusion (both label and confidence-based) but found confidence-based fusion to provide the best results. Their best result was the accuracy of 81.69% for the test partition that was achieved through an ensemble of top-10 models.

As part of their solution to the ADReSSo challenge, Perez-Toro et al. [63] undertook an analysis of speech and language. For audio modality, they used prosody features and x-vector embeddings [70] and embeddings computed from a DNN trained to recognize emotions on the 3-D Pleasure-Arousal-Valence scale [71]. In addition to these, they also computed statistical features based on Voice Activity Detection. For the linguistic modality, Amazon Web Services Transcribe was used to generate speech transcripts and then computed BERT and Electra-based word embeddings, as well as Perplexity features. For Task A, they used SVM with an RBF kernel whereas, for Task B, they used the Logistic Regression classifier. The fusion of Prosody, Arousal, Dominance, and BERT features provided the classification accuracy of 80.28% for the test partition on Task A. Meanwhile, for Task B, their best performance model for the test partition was a fusion of Perplexity and BERT features which yielded an RMSE of 4.56. Other interesting observations were:

i. Better classification performance was observed with data from both the subjects and the interviewer possibly because the interviewer adapts differently to subjects with AD, and the interviewer's data does not improve the performance for the regression task. The authors surmise that this could be since interviewers cannot gauge the severity of the disease.

ii.   A caveat of their experimental analysis is that they used timestamps as
      provided in the metadata. We and other participants of the challenge have
      found these timestamps to be inaccurate. Perez-Toro et al. also pointed out
      that different acoustic conditions could add bias to the classification based
      on audio modality only.

Wang et al. [64] explored the use of unimodal and multimodal attention networks
with audio and textual features as part of their solution to the ADReSSo challenge.
They used Emobase, IS10, VGGish, and x-vector features to represent acoustic features. For the text modality, they used universal sentence encoder embedding [72]
and a suite of features generated from the CLAN toolkit by running EVAL and
FREQ commands. Experimental results show that a multiheaded attention network
trained with CLAN features from text modality and x-vector embeddings from the
acoustic modality provides the best results overall, i.e., accuracy of 80.28% for the
test partition. They also report that while CLAN features performed better than USE
embeddings, x-vector features performed better than handcrafted acoustic features.

As part of their solution to the ADReSSo challenge, Pan et al. [55] explored two
ASR paradigms, i.e. wav2vec2 and time-delay neural networks (TDNN) trained
on their in-house dataset. Model 1 uses wav2vec features with tree bagger used as
the classifier. Model 2 uses ASR transcript generated by wav2vec2, which are then
passed as input to BERT for linguistic information modeling and subsequent classification. Model 3 is a fusion of features from wav2vec2 and BERT, with BERT
used for the classification as well. Models 4 and 5 were based on integrating confidence scores with the hidden states of BERT with two "maximal word lengths".
They report that Model 5 achieved the best result for the test partition with 84.51%
accuracy.

Addressing the ADReSSo challenge, Pappagari et al. [56] used prosodic features
computed using DigiPsych toolkit, ASR embeddings from SpeechBrain [73], and
finetuned an x-vector model [70]. To explore the text modality, they used three types
of ASR models. The first model used the automatic speech recognition in rever-
berant environments (ASpIRE) recipe in Kaldi [74]. The second ASR model was
an interpolated version of ASpIRE with a language model trained using the development partition of ADReSSo challenge dataset. The interpolated model hypoth-
esized that it has a better chance of accurate speech transcription. The third ASR
model (ASR3) used Amazon Web Service transcription service, whereas the fourth
ASR model was based on the model from otter.ai. Transcripts generated from these
models were used to finetune a pre-trained BERT model. To cater BERT for binary
classification and regression tasks, they added a linear layer with two or one output
respectively for the two tasks. Although BERT and x-vector models were finetuned,
they used logistic regression and XGBoost classifier in an 8-fold cross-validation
set-up for the two machine learning tasks. They reported that SpeechBrain's embeddings performed better than x-vectors whereas the BERT model finetuned with AWS
transcripts achieved the best performance for the text modality on the development
partition. However, the textual modality with classification/regression performance
of 81.30%/5.23 was better suited for the task at hand than the audio modality with
the classification/regression scores of 71.30%/6.44. The best classification result was

the accuracy of 84.51% achieved by global fusion of all features (with transcripts generated from AWS). For the regression task, the best result was an RMSE of 3.85 achieved by finetuning BERT with transcripts generated through otter.ai.

In [57], we described our proposed systems for the three tasks of the ADReSSo challenge. These systems were based on our previous research experience [30] with the task at hand where we had identified that: (i) the textual modality outperforms the acoustic modality; (ii) deep textual embeddings are effective feature representations yielding top performances for the classification and regression tasks when applied in combination with a logistic regression classifier or partial least squares regressor. We began by exploring three ASR methods that included wav2vec2, Silero STT, and Microsoft Streams for automated transcription. We determined that Microsoft Streams provided the lowest world error rates; therefore, we used it for subsequent experiments. We used nine pre-trained transformer-based models to compute deep textual embeddings and used those as feature vectors after using functionals of descriptive statistics for feature aggregation. Inspired by the previous work of Yuan et al. [28], we also integrated pause information into these embeddings. Our best result for Task A was an accuracy of 84.51% which is the joint best performance for this task. Our proposed system for Task B achieved the RMSE of 4.35 and ranked as the 3rd best solution. Finally, our proposed system offers the best result for Task C with an average f1 score of 73.80%.

2.4   CONCLUSION AND FUTURE OUTLOOK

This section highlights the key deliverables from the two editions of the ADReSS challenges which contribute to the body of knowledge for automated recognition of Alzheimer's dementia.

2.4.1   TEXTUAL MODALITY VS AUDIO MODALITY

We note that a number of participants of these challenges concluded that the textual modality offers better performance than the audio modality [30, 40, 19, 20]. In fact, Yuan et al. [28], who achieved the best classification performance between the two challenges, proposed a system based on the use of textual modality alone. The observations suggest that textual modality is likely to offer better performance than acoustic.

2.4.2   ASR TRANSCRIPTS ARE SUITABLE FOR MANUAL TRANSCRIPTION

A key factor of the ADReSSo challenge 2021 was that speech transcripts were not provided as part of the challenge dataset. Instead, participants of the challenge were expected to use automated speech recognition methods to generate transcripts themselves if they were interested in exploring features from textual modality for the task at hand. This specialty of the ADReSSo challenge enabled us to understand the feasibility of ASR transcripts as an alternative to manually annotated transcripts.

We note that all but two teams explored textual modality out of the 11 manuscripts published in the Interspeech 2021 proceedings. Amongst these, three teams

[61, 55, 57] used Google's speech recognition services, three teams used the wav2vec ASR models [63, 56] and two teams used Amazon web services [63, 56]. Some of the other ASR tools used for the challenge include AppTek's ASR, English DeepSpeech model from Mozilla [75], Otter.ai ASR, and Silero speech-to-text toolkit [76], and the ASpIRE time-delay neural network model in Kaldi [74]. Given that different types of textual features were computed for transcripts generated by these ASR methods, it is impossible to compare their performance across the board directly. However, one can draw conclusions based on performance reported by participating teams. Thus, it can be observed that commercial ASR services were a popular choice, with eight out of the nine teams opting to use a commercial service for speech transcription. We reported [57] that initial experiments showed Microsoft's Stream ASR service produced better transcription results than the Silero toolkit and the wav2vec2 model available with the Hugging-Face toolkit. Also, the results reported by Pappagari et al. [56] showed that ASR services from Amazon web services and Otter.ai produced better results than using the ASpIRE model. Another interesting point to note is that IBM Watson STT, as used by Rohanian et al., does not filter out hesitation or disfluency markers which could be useful for screening Alzheimer's dementia since disfluency is a known trait for this condition.

2.4.3 FEATURE- AND DECISION-LEVEL FUSION PROVIDE PERFORMANCE IMPROVEMENT

Based on this review it can be suggested that feature- and decision-level fusion methods generally improved the performance of machine learning models for tasks from both editions of the ADReSS challenge. The fundamental explanation for the success of such methods is that they help combine information from individual models to build a consensus about prediction labels. Fusion methods are well known for their high prediction efficiency from Kaggle challenges, and they constitute a common practice in that community.

However, it also should be mentioned here that fusion does not always improve performance. We found that one needs to select models based on their sensitivity and specificity scores [57]; otherwise, the performance may even degrade after fusion. We also reported that fusion might not improve performance if there is a large class imbalance in the dataset since it can penalize the number of predictions from minority classes. Also, Gauder et al. [58] reported that fusion did not yield an improvement in classification performance based on their experiments.

2.4.4 THE EFFICACY OF DEEP TEXTUAL EMBEDDINGS FROM CONTEXTUAL MODELS

Our review shows that deep textual embeddings based on contextual models have become a popular choice for participants of the ADReSS challenges. In fact, solutions based on models such as BERT, Electra, RoBERTa, and Distil-BERT have performed better than traditional word vector representations such as Word2Vec and GloVe.

In essence, findings presented at both editions of the ADReSS challenge show that a combination of deep textual embeddings and functionals-based feature aggregation can yield meaningful representations of speech transcripts as was demonstrated by the top-3 performances for classification and regression tasks. Noteworthy is the work by Searle et al. [36] that revealed that TF-IDF-based features could achieve better performance than BERT representation. Although this observation may be due to specific experimental conditions, it necessitates an empirical analysis of methods for generating numerical representations of speech text documents.

REFERENCES

1. World Health Organization, "The epidemiology and impact of dementia—current state and future trends," *Technical Report*, 2018 [Online]. Available: www.who.int/health-topics/dementia

2. Alzheimer's Association, "2021 Alzheimer's disease facts and figures," *Technical Report*, 2021 [Online]. Available: www.alz.org/media/documents/alzheimers-facts-and-figures.pdf

3. S. Alfakhri, A. W. Alshudukhi, A. A. Alqahtani, A. M. Alhumaid, O. A. Al-hathlol, A. I. Almojali, M. A. Alotaibi, and M. K. Alaqeel, "Depression among caregivers of patients with dementia," *Inquiry*, vol. 55, no. 1, pp. 1–6, 2018.

4. J. Larner, "Getting it wrong: The clinical misdiagnosis of Alzheimer's disease," *International Journal of Clinical Practice*, vol. 58, no. 11, pp. 1092–1094, 2004.

5. S. Love, "Neuropathological investigation of dementia: A guide for neurologists," *Neurology in Practice*, vol. 76, no. Suppl. 5, pp. v8–v14, 2005.

6. B. Schmand, G. Walstra, J. Lindeboom, S. Teunisse, and C. Jonker, "Early detection of Alzheimer's disease using the Cambridge Cognitive Examination (CAMCOG)," *Psychological Medicine*, vol. 30, no. 3, pp. 619–627, 2000.

7. E. Pinto and R. Peters, "Literature review of the Clock Drawing Test as a tool for cognitive screening," *Dementia and Geriatric Cognitive Disorders*, vol. 27, no. 3, pp. 201–213, 2009.

8. M. F. Folstein, S. E. Folstein, and P. R. McHugh, "'Mini-mental state'. A practical method for grading the cognitive state of patients for the clinician," *Journal of Psychiatric Research*, vol. 12, no. 3, pp. 189–198, 1975.

9. Z. S. Nasreddine, N. A. Phillips, V. Bedirian, S. Charbonneau, V. Whitehead, I. Collin, J. L. Cummings, and H. Chertkow, "The montreal cognitive assessment, MoCA: A brief screening tool for mild cognitive impairment," *Journal of the American Geriatrics Society*, vol. 53, no. 4, pp. 695–699, 2005.

10. R. Mistur, L. Mosconi, S. de Santi, M. Guzman, Y. Li, W. Tsui, and M. J. de Leon, "Current challenges for the early detection of alzheimer's disease: Brain imaging and CSF studies," *Journal of Clinical Neurology*, vol. 5, no. 4, pp. 153–166, 2009.

11. C. Laske, H. R. Sohrabi, S. M. Frost, K. Lopez-De-Ipina, P. Garrard, M. Buscema, J. Dauwels, S. R. Soekadar, S. Mueller, C. Linnemann, S. A. Bridenbaugh, Y. Kanagasingam, R. N. Martins, and S. E. O'Bryant, "Innovative diagnostic tools for early detection of Alzheimer's disease," *Alzheimer's and Dementia*, vol. 11, no. 5, pp. 561–578, 2015.

12. M. Mortamais, J. A. Ash, J. Harrison, J. Kaye, J. Kramer, C. Randolph, C. Pose, B. Albala, M. Ropacki, C. W. Ritchie, and K. Ritchie, "Detecting cognitive changes in preclinical Alzheimer's disease: A review of its feasibility," *Alzheimer's and Dementia*, vol. 13, no. 4, pp. 468–492, 2017.

13. Z. S. Shah, K. Sidorov, and D. Marshall, "Psychomotor cues for depression screening," in *IEEE International Conference on Digital Signal Processing*, IEEE, 2017, pp. 1–5.

14. Z. S. Syed, K. Sidorov, and D. Marshall, "Automated screening for bipolar disorder from audio/visual modalities," in *ACM International Workshop on Audio/Visual Emotion Challenge*, ACM, 2018, pp. 39–45.

15. Z. S. Syed, S. A. Memon, and A. L. Memon, "Deep acoustic embeddings for identifying parkinsonian speech," *International Journal of Advanced Computer Science and Applications*, vol. 11, no. 10, pp. 726–734, 2020.

16. K. C. Fraser, J. A. Meltzer, and F. Rudzicz, "Linguistic features identify Alzheimer's disease in narrative speech," *Journal of Alzheimer's Disease*, vol. 49, no. 2, pp. 407–422, 2015.

17. F. Haider, S. de la Fuente, and S. Luz, "An assessment of paralinguistic acoustic features for detection of Alzheimer's dementia in spontaneous speech," *IEEE Journal of Selected Topics in Signal Processing*, vol. 14, no. 2, pp. 272–281, 2020.

18. Z. S. Syed, M. S. S. Syed, M. Lech, and E. Pirogova, "Automated recognition of Alzheimer's dementia using bag-of-deep-features and model ensembling," *IEEE Access*, vol. 9, pp. 88377–88390, 2021.

19. S. Luz, F. Haider, S. de la Fuente, D. Fromm, and B. MacWhinney, "Alzheimer's dementia recognition through spontaneous speech: The ADReSS challenge," in *INTERSPEECH*, IEEE, 2020, pp. 2172–2176.

20. S. Luz, F. Haider, S. de la Fuente, D. Fromm, and B. MacWhinney, "Detecting cognitive decline using speech only: The ADReSSo challenge," in *INTERSPEECH*, IEEE, pp. 3780–3784, 2021.

21. DementiaBank Consortium Group, "DementiaBank database" [Online]. Available: https://dementia.talkbank.org/

22. J. T. Becker, F. Boiler, O. L. Lopez, J. Saxton, and K. L. Mcgonigle, "The natural history of Alzheimer's disease: Description of study cohort and accuracy of diagnosis," *Archives of Neurology*, vol. 51, no. 6, pp. 585–594, 1994.

23. H. Goodglass, E. Kaplan, and B. Barresi, *BDAE-3: Boston Diagnostic Aphasia Examination*, 3rd ed. Lippincott Williams & Wilkins, Philadelphia, PA, 2001.

24. B. Macwhinney, *The CHILDES Project: Tools for Analyzing Talk*, 3rd ed. American Psychological Association, 1992.

25. J. Chen, Y. Wang, and D. Wang, "A feature study for classification-based speech separation at low signal-to-noise ratios," *IEEE/ACM Transactions on Audio Speech and Language Processing*, vol. 22, no. 12, pp. 1993–2002, 2014.

26. S. Luz, "Longitudinal monitoring and detection of Alzheimer's type dementia from spontaneous speech data," in *IEEE Symposium on Computer-Based Medical Systems*, IEEE, 2017, pp. 45–46.

27. B. MacWhinney, "Tools for analyzing talk part 2: The CLAN program," 2017. Available: https://dali.talkbank.org/clan/

28. J. Yuan, Y. Bian, X. Cai, J. Huang, Z. Ye, and K. Church, "Disfluencies and fine-tuning pre-trained language models for detection of Alzheimer's disease," in *INTERSPEECH*, IEEE, 2020, pp. 2162–2166.

29. J. Koo, J. H. Lee, J. Pyo, Y. Jo, and K. Lee, "Exploiting multi-modal features from pre-trained networks for Alzheimer's dementia recognition," in *INTERSPEECH*, IEEE, 2020, pp. 2217–2221.

30. M. S. S. Syed, Z. S. Syed, M. Lech, and E. Pirogova, "Automated screening for Alzheimer's dementia through spontaneous speech," in *INTERSPEECH*, IEEE, 2020, pp. 2222–2226.

31. S. Farzana and N. Parde, "Exploring MMSE score prediction using verbal and non-verbal cues," in *INTERSPEECH*, IEEE, 2020, pp. 2207–2211.

32. R. Pappagari, J. Cho, L. Moro-Velazquez, and N. Dehak, "Using state of the art speaker recognition and natural language processing technologies to detect Alzheimer's disease and assess its severity," in *INTERSPEECH*, IEEE, 2020, pp. 2177–2181.

33. M. Martinc and S. Pollak, "Tackling the ADReSS challenge: a multimodal approach to the automated recognition of Alzheimer's dementia," in *INTERSPEECH*, IEEE, 2020, pp. 2157–2161.

34. Edwards, C. Dognin, B. Bollepalli, and M. Singh, "Multiscale system for Alzheimer's dementia recognition through spontaneous speech," in *INTERSPEECH*, IEEE, 2020, pp. 2197–2201.

35. M. Rohanian, J. Hough, and M. Purver, "Multi-modal fusion with gating using audio, lexical and disfluency features for Alzheimer's Dementia recognition from spontaneous speech," in *INTERSPEECH*, IEEE, 2020, pp. 2187–2191.

36. T. Searle, Z. Ibrahim, and R. Dobson, "Comparing natural language processing techniques for Alzheimer's dementia prediction in spontaneous speech," in *INTERSPEECH*, IEEE, 2020, pp. 2192–2196.

37. A. Pompili, T. Rolland, and A. Abad, "The INESC-ID multi-modal system for the ADReSS 2020 challenge," in *INTERSPEECH*, IEEE, 2020, pp. 2202–2206.

38. U. Sarawgi, W. Zulfikar, N. Soliman, and P. Maes, "Multimodal inductive transfer learning for detection of Alzheimer's dementia and its severity," in *INTERSPEECH*, IEEE, 2020, pp. 2212–2216.

39. A. Balagopalan, B. Eyre, F. Rudzicz, and J. Novikova, "To BERT or not to BERT: Comparing speech and language-based approaches for Alzheimer's disease detection," in *INTERSPEECH*, IEEE, 2020, pp. 2167–2171.

40. N. Cummins, Y. Pan, Z. Ren, J. Fritsch, V. S. Nallanthighal, H. Christensen, D. Blackburn, B. W. Schuller, M. Magimai-Doss, H. Strik, and A. Harma, "A comparison of acoustic and linguistics methodologies for Alzheimer's dementia recognition," in *INTERSPEECH*, IEEE, 2020, pp. 2182–2186.

41. J. Pennington, R. Socher, and C. D. Manning, "GloVe: Global vectors for word representation," in *Conference on Empirical Methods in Natural Language Processing*, 2014, pp. 1532–1543. Available: https://aclanthology.org/D14-1162/

42. J. Devlin, M.-W. Chang, K. Lee, and K. Toutanova, "BERT: Pre-training of deep bidirectional transformers for language understanding," *arXiv*, 2018.

43. V. Sanh, L. Debut, J. Chaumond, and T. Wolf, "DistilBERT, a distilled version of BERT: Smaller, faster, cheaper and lighter," *arXiv*, 2019.

44. Z. Zhang, X. Han, Z. Liu, X. Jiang, M. Sun, and Q. Liu, "ErniE: Enhanced language representation with informative entities," in *Annual Meeting of the Association for Computational Linguistics*, 2020, pp. 1441–1451. Available: https://aclanthology.org/P19-1139/

45. S. Hershey, S. Chaudhuri, D. P. Ellis, J. F. Gemmeke, A. Jansen, R. C. Moore, M. Plakal, D. Platt, R. A. Saurous, B. Seybold, M. Slaney, R. J. Weiss, and K. Wilson, "CNN architectures for large-scale audio classification," in *IEEE International Conference on Acoustics, Speech and Signal Processing*, IEEE, 2017, pp. 131–135.

46. T. Mikolov, K. Chen, G. Corrado, and J. Dean, "Efficient estimation of word representations in vector space," in *International Conference on Learning Representations*, 2013, pp. 1–12. Available: https://research.google/pubs/pub41224/

47. M. N. Moghadasi and Y. Zhuang, "Sent2Vec: A new sentence embedding representation with sentimental semantic," in *IEEE International Conference on Big Data*, IEEE, 2020, pp. 4672–4680.

48. Meta.com, "FastText library" [Online]. Available: https://github.com/facebookresearch/fastText

49. K. Clark, M. T. Luong, Q. V. Le, and C. D. Manning, "Electra: Pre-training text encoders as discriminators rather than generators," *arXiv*, 2020.

50. Y. Liu, M. Ott, N. Goyal, J. Du, M. Joshi, D. Chen, O. Levy, M. Lewis, L. Zettlemoyer, and V. Stoyanov, "RoBERTa: A robustly optimized BERT pretraining approach,", *arXiv*, 2019.

51. Q. Le and T. Mikolov, "Distributed representations of sentences and documents," in *International Conference on International Conference on Machine Learning*, 2014, pp. 1188–1196. Available: https://proceedings.mlr.press/v32/le14.html

52. T. Chen and C. Guestrin, "XGBoost: A scalable tree boosting system," in *ACM SIGKDD International Conference on Knowledge Discovery and Data Mining*, ACM, 2016, pp. 785–794.

53. L. Benton, "Differential behavioral effects in frontal lobe disease," *Neuropsychologia*, vol. 6, no. 1, pp. 53–60, 1968.

54. A. Balagopalan and J. Novikova, "Comparing acoustic-based approaches for Alzheimer's disease detection," in *INTERSPEECH*, IEEE, 2021, pp. 3800–3804.

55. Y. Pan, B. Mirheidari, J. M. Harris, J. C. Thompson, J. Jones, J. S. Snowden, D. Blackburn, and H. Christensen, "Using the outputs of different automatic speech recognition paradigms for acoustic- and BERT-based Alzheimer's dementia detection through spontaneous speech," in *INTERSPEECH*, IEEE, 2021, pp. 3810–3814.

56. R. Pappagari, J. Cho, S. Joshi, L. Moro-Velazquez, P. Zelasko, J. Villalba, and N. Dehak, "Automatic detection and assessment of Alzheimer disease using speech and language technologies in low-resource scenarios," in *INTERSPEECH*, IEEE, 2021, pp. 3825–3829.

57. Z. S. Syed, M. S. S. Syed, M. Lech, and E. Pirogova, "Tackling the ADRESSO challenge 2021: The MUET-RMIT system for Alzheimer's dementia recognition from spontaneous speech," in *INTERSPEECH*, IEEE, 2021, pp. 3815–3819.

58. L. Gauder, L. Pepino, L. Ferrer, and P. Riera, "Alzheimer disease recognition using speech-based embeddings from pre-trained models," in *INTERSPEECH*, IEEE, 2021, pp. 3795–3799.

59. M. Rohanian, J. Hough, and M. Purver, "Alzheimer's dementia recognition using acoustic, lexical, disfluency and speech pause features robust to noisy inputs," in *INTERSPEECH*, IEEE, 2021, pp. 3820–3824.

60. Y. Qiao, X. Yin, D. Wiechmann, and E. Kerz, "Alzheimer's disease detection from spontaneous speech through combining linguistic complexity and (dis)fluency features with pretrained language models," in *INTERSPEECH*, IEEE, 2021, pp. 3805–3809.

61. Y. Zhu, A. Obyat, X. Liang, J. A. Batsis, and R. M. Roth, "WavBERT: Exploiting semantic and non-semantic speech using Wav2vec and BERT for dementia detection," in *INTERSPEECH*, IEEE, 2021, pp. 3790–3794.

62. J. Chen, J. Ye, F. Tang, and J. Zhou, "Automatic detection of Alzheimer's disease using spontaneous speech only," in *INTERSPEECH*, IEEE, 2021, pp. 3830–3834.

63. P. A. Perez-Toro, S. P. Bayerl, T. Arias-Vergara, J. C. Vasquez-Correa, P. Klumpp, M. Schuster, E. Noth, J. R. Orozco-Arroyave, and K. Riedhammer, "Influence of the interviewer on the automatic assessment of Alzheimer's disease in the context of the ADReSSo challenge," in *INTERSPEECH*, IEEE, 2021, pp. 3785–3789.

64. N. Wang, Y. Cao, S. Hao, Z. Shao, and K. P. Subbalakshmi, "Modular MultiModal attention network for Alzheimer's disease detection using patient audio and language data," in *INTERSPEECH*, IEEE, 2021, pp. 3835–3839.

65. J. Shor, A. Jansen, R. Maor, O. Lang, O. Tuval, F. D. C. Quitry, M. Tagliasacchi, I. Shavitt, D. Emanuel, and Y. Haviv, "Towards learning a universal non-semantic representation of speech," in *INTERSPEECH*, IEEE, 2020, pp. 140–144.

66. Q. Kong, Y. Cao, T. Iqbal, Y. Wang, W. Wang, and M. D. Plumbley, "PANNs: Large-scale pretrained audio neural networks for audio pattern recognition," *IEEE/ACM Transactions on Audio, Speech, and Language Processing*, vol. 28, pp. 2880–2894, 2020.

67. Degottex, J. Kane, T. Drugman, T. Raitio, and S. Scherer, "COVAREP—A collaborative voice analysis repository for speech technologies," in *IEEE International Conference on Acoustics, Speech and Signal Processing (ICASSP)*, IEEE, 2014, pp. 960–964.

68. M. Rohanian and J. Hough, "Re-framing incremental deep language models for dialogue processing with multi-task learning," in *International Conference on Computational Linguistics*, 2021, pp. 497–507. Available: https://aclanthology.org/2020.coling-main.43/

69. E. Kerz, Y. Qiao, D. Wiechmann, and M. Strobel, "Becoming linguistically mature: Modeling English and german children's writing development across school grades," in *Workshop on Innovative Use of NLP for Building Educational Applications*, 2020, pp. 65–74. Available: https://aclanthology.org/2020.bea-1.6/

70. D. Snyder, D. Garcia-Romero, G. Sell, D. Povey, and S. Khudanpur, "X-vectors: Robust DNN embeddings for speaker recognition," in *IEEE International Conference on Acoustics, Speech and Signal Processing (ICASSP)*, IEEE, 2018, pp. 5329–5333.

71. A. Mehrabian, *Basic Dimensions for a General Psychological Theory Implications for Personality, Social, Environmental, and Developmental Studies*. Oelgeschlager, Gunn & Hain, 1980.

72. D. Cer, Y. Yang, S.-Y. Kong, N. Hua, N. Limtiaco, R. S. John, N. Constant, M. Guajardo-Cespedes, S. Yuan, C. Tar, B. Strope, and R. Kurzweil, "Universal sentence encoder for English," in *Conference on Empirical Methods in Natural Language Processing: System Demonstrations*, 2018, pp. 169–174. Available: https://aclanthology.org/D18-2029/

73. M. Ravanelli, T. Parcollet, P. Plantinga, A. Rouhe, S. Cornell, L. Lugosch, C. Subakan, N. Dawalatabad, A. Heba, J. Zhong, J.-C. Chou, S.-L. Yeh, S.-W. Fu, C.-F. Liao, E. Rastorgueva, F. Grondin, W. Aris, H. Na, Y. Gao, R. D. Mori, and Y. Bengio, "SpeechBrain: A general-purpose speech toolkit," 2021 [Online]. Available: https://github.com/speechbrain/speechbrain

74. D. Povey, A. Ghoshal, G. Boulianne, L. Burget, O. Glembek, N. Goel, M. Hannemann, P. Motlicek, Y. Qian, P. Schwarz, J. Silovsky, G. Stemmer, and K. Vesely, "The Kaldi speech recognition toolkit," in *IEEE Workshop on Automatic Speech Recognition and Understanding*, IEEE, 2011, pp. 1–4.

75. Hannun, C. Case, J. Casper, B. Catanzaro, G. Diamos, E. Elsen, R. Prenger, S. Satheesh, S. Sengupta, A. Coates, and A. Y. Ng, "Deep speech: Scaling up end-to-end speech recognition," *arXiv*, 2014.

76. Silero Team and S. Team, "Silero Models: Pre-trained enterprise-grade STT/TTS models and benchmarks," 2021 [Online]. Available: https://github.com/snakers4/silero-models

3 Electrogastrogram Signal Processing

Techniques and Challenges with Application for Simulator Sickness Assessment

Nadica Miljković, Nenad B. Popović, and Jaka Sodnik

3.1 INTRODUCTION TO ELECTROGASTROGRAPHY (EGG)

Electrogastrography (EGG)[1] was initially introduced in 1922 by Walter Alvarez [1] and rediscovered in 1957 by Ronald Clark Davis [2], [3]. Alvarez reported his work after a few years spent on resolving technical challenges to capture electrical activity of stomach smooth muscles. In years to come, equipment for non-invasive electrophysiological recordings became lighter and more robust to noise by introducing transistor technology and amplifiers with relatively large Common Mode Rejection Ratio (CMRR). This technical change led to a wider application and adoption of EGG technique and its popularization in the 1990s. Interest in EGG rose again in recent years with its promising application for assessment of sickness in virtual environments [2], [4]–[9].

Stomach electrical activity consists of Electrical Control Activity (ECA) or slow waves generated by the pacemaker cells and Electrical Response Activity (ERA) or spike potentials related to the peristaltic stomach motility [4], [6]. Surface EGG contains only ECA and normally does not include ERA as a consequence of the abdominal wall acting as a low pass filter by eliminating spike potentials [10]. Hence, we further discuss only ECA or slow waves recorded by the non-invasive EGG technique.

Briefly, EGG is a method for recording electrical potentials originating from the smooth stomach muscle by application of silver/silver-chloride (Ag/AgCl) surface electrodes on the skin over the stomach. Commonly, after passing through Ag/AgCl transducer, the signal is further conditioned by analogue circuits that amplify and filter the EGG signal before digitalization. This analogue processing step is important as EGG frequency content ranges from 1 to 10 cycles per minute (cpm) with three

DOI: 10.1201/9781003201137-4

different bounds: 1–2 cpm bradygastria, 2–4 cpm normal rhythm, and 4–10 cpm tachygastria, while EGG amplitudes range from 0.1 mV to 0.5 mV [13]. High pass filter is used to reduce the Direct Current (DC) component and very low frequency content (~10 mHz) whereas low pass filter is used as an anti-aliasing filter before the Analog to Digital Conversion (ADC) required for further digital processing [4]. Sample electrode positioning with the corresponding EGG signal in the time domain are presented in Figure 3.1.

By far, EGG has proven applicability for studying (1) effects of the nutrients on EGG signal that is required for the diagnosis of long lasting eating disorders; (2) sham feeding where it was concluded that cephalic vagal reflex depends on subjective savory; (3) stress and anxiety effects on EGG after physiological or psychological stressors i.e., studying brain-gut interactions; (4) the outcomes of therapeutic electrical stimulation in persons with gastroparesis; (5) gastric disorders, especially dysrhythmias, as a result of diabetes, collagen disease, and functional dyspepsia; (6) dysrhythmias related to the early phase of Parkinson's disease; and (7) motion sickness and nausea, as well as other applications where the changes of EGG power and frequency content are of interest [4]–[6], [14]–[17].

Despite relatively wide adoption in research practice, EGG is not yet used in standard clinical practice and despite initial progress with measurement equipment, noise sensitivity of EGG signal deteriorates wider acceptance of this promising technique. This chapter focuses on methods that can address this challenge by appropriate employment of EGG signal analysis workflows for EGG-based simulator sickness assessment.

As a rule EGG signal processing methods are performed offline. The reason probably lays in the fact that EGG signal is too slow (one period of normal rhythm lasts for about 20 s). Consequently, online analysis is almost impossible, especially in the frequency domain. In this chapter, we present offline methods for

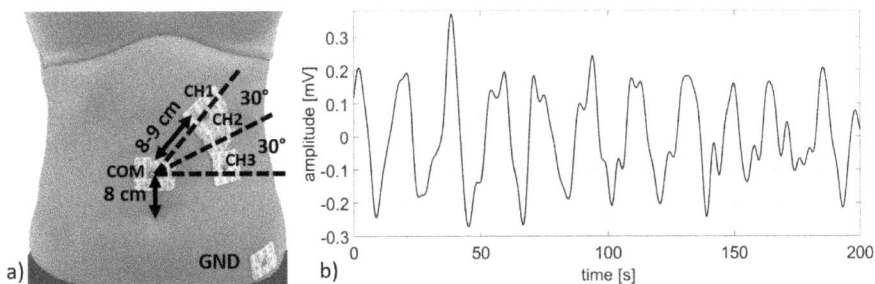

FIGURE 3.1 a) Photograph of the subject's torso with surface Ag/AgCl electrodes placed over the stomach for the recording of 3-channel EGG (CH1—channel 1, CH2—channel 2, CH3—channel 3, COM—common electrode, GND—ground electrode). Electrodes are placed according to the recommendations from [11]. b) Sample EGG signal with visible normal gastric rhythm after meal intake and during relaxation in supine position. Presented signal was recorded at the University of Belgrade—School of Electrical Engineering and is available in an open database [12] (filename: "ID_postprandial.txt", channel 1).

EGG analysis for sickness assessment in driving simulators with some recommendations for quasi online analysis by introducing short-term parameters [8]. It should be noted that motion sickness itself can be assessed in real time by for example measuring ElectroDermal Activity (EDA) and skin temperature [19], but here we aimed at a different approach by analyzing electrical gut potentials especially as EGG is the most commonly used physiological measure of simulator sickness [18].

3.1.1 Motivation for Introducing EGG for Sickness Assessment

The first study reporting the EGG-based assessment of motion sickness induced by the optokinetic rotating drum was published in 1985 [20]. The relation between tachygastria (higher gastric rhythm) and motion sickness was consistently found and confirmed in a series of succeeding studies [2], [5]. It is important to note that an increase of gastric rhythm is also an indicator of other states such as depression and schizophrenia [16], so it does not present a unique change in EGG signal.

Motion sickness can be defined as a natural response of a body experiencing an unnatural environment and although there is no consensus on the motion sickness origins, nausea presents a common symptom and motion sickness can ultimately lead to vomiting [9]. The first and genuine objective method, among available procedures that utilize physiological response for sickness assessment (e.g., heart rate variability, skin conductance, skin temperature, respiration rate [21], [22]), is measurement of electrical potentials of stomach smooth muscles i.e., electrogastrography as it has proven efficacy for assessing nausea and vomiting [14], [23]. The supremacy of EGG for sickness assessment comes from the drawback of the commonly used questionnaires such as Pensacola Motion Sickness Questionnaire (MSQ), Simulator Sickness Questionnaire (SSQ), Motion Sickness Assessment Questionnaire (MSAQ), Virtual Reality Symptom Questionnaire (VRSQ), the Misery scale, the Well-being scale, and Motion Sickness Susceptibility Questionnaire (MSSQ), all being post-experience-based scales except for the Fast Sickness Scale (FSS) which is on the other hand disruptive as it contains 20 verbal ratings delivered during the sickness exposure [9], [24].

We elaborate on adoption of EGG for simulator sickness assessment and propose dedicated signal processing procedures for adequate noise elimination and feature extraction procedures in driving simulators.

3.1.2 EGG and Driving Simulators

Driving simulators are used worldwide for training new drivers, evaluating driving skills, rehabilitation, observing drivers' behavior, and race-car driving training or simply for entertainment [25]. In comparison to the actual vehicles, simulators provide a safe and controlled experimental environment for fast and reliable data acquisition with high reproducibility. In the simulated environment it is very easy to manipulate traffic, weather, and road conditions where everything surrounding the ego vehicle responds to the driver's actions and behaviors. Positions, velocities, accelerations, and mutual interaction of all objects are predefined, controlled,

FIGURE 3.2   Driving simulator interior (Nervtech Ltd., Ljubljana, Slovenia).

and measurable. Through this control it is possible to create very efficient training procedures with highly risky and safety-critical driving situations which could not be replicated safely in a real environment [26]. Simulators also enable studying of hazard anticipation and reactions, and may even expose drivers to accidents and crashes [27].

Simulator studies often include application of external sensors for measuring psychophysiological signals to detect different drivers' states such as stress level, attention, cognitive workload, or fatigue [28]. It is possible to acquire all signals accurately, efficiently, and synchronously. All influencing factors can be clearly identified and all drivers' action can be properly explained. In the automotive industry, driving simulators offer a perfect tool for evaluating new in-vehicle information systems and user interfaces, as well as their potential impact on driving performance [29], [30]. Sample driving simulator environment is presented in Figure 3.2.

In comparison to the real vehicles, physical, perceptual, and behavioral fidelity of driving simulators are limited and may in some cases result in an unwanted risky behavior and unnaturalistic drivers' reactions. The false sense of safety and excessive self-confidence could limit the overall positive effect of simulator-based driving trainings and behavioral studies [31]. Even more important limitation is simulator discomfort caused by the simulator sickness sometimes referred to as cybersickness [32].

As stated previously, there is no unified theory explaining the simulator sickness phenomenon, but the three commonly used theories are [33]:

- Sensory conflict theory [34]: visual, vestibular, and non-vestibular signals do not match with individual's expectations. The sickness symptoms are caused due to the conflict between present sensory input and its immediate past input.
- Neural mismatch model [35]: the symptoms originate from the discrepancy between the input stored in the human neural network (past experiences) and the current sensory input.
- Postural instability theory [36]: the symptoms are caused by long exposure to the postural instability without a proper adjustment for maintaining balance.

All three theories refer to both simulator and motion sickness in general. Both types of sicknesses elicit symptoms such as nausea, dizziness, fatigue, fullness of head, and in some cases even vomiting. Simulator sickness is highly affecting mainly older people or drivers facing very challenging and dynamic driving with complex junctions and roundabouts.

Simulator sickness symptoms can be assessed subjectively or measured objectively through different physiological signals. The most commonly used tool for subjective assessment is SSQ [37] which consists of 16 items and collects responses before and after the experiment. The main outcome of the SSQ is typically grouped into three categories: nausea, oculomotor disturbances, and disorientation. It is routinely applied as a paper-and-pencil questionnaire and in some cases responses can be collected orally or by software form. Among the physiological measurements commonly used signals for assessment of simulator sickness are heart rate, skin conductance, respiration rate, and also EGG. Here, we provide a detailed overview of EGG pathways and pitfalls for simulator sickness assessment by discussing the most important aspects of EGG signal analysis.

3.2   EGG ARTIFACT CANCELLATION

Despite all the progress made with the technological aspects of measurement and analysis of electrophysiological signals in general, obstacles in wider EGG adoption still prevail. Due to the high vulnerability to artifacts, EGG-based methodology has not yet reached its full potential. The most critical part related to the elimination of EGG artifacts is preservation of the relevant frequency content (bradygastria, normogastria, and tachygastria frequency ranges). Hence, artifact cancellation presents the first and the most important step in the EGG signal processing workflow. The selection of filtering methods is of great importance as inaccurate preprocessing can distort EGG signal and even lead to misleading conclusions [2], [10], [38], [39].

In this chapter we use interference, artifact, and noise as synonyms and adopt the statement: "Any signal other than that of interest could be termed as an interference, artifact, or simply noise" [40]. Here, we acknowledge some common terms to be associated with the particular synonym, such as "movement artifact", "power inter-ference", or "random noise".

For digital filtering, linear Butterworth band pass filter is commonly applied as it has "maximally flat" magnitude [41], [42]. Butterworth filter does not have pass band nor stop band ripple and has a clear roll-off. When compared to the Chebyshev Type I filter, Butterworth filter has a slower roll-off and requires higher orders [41]. This gain flatness in the stop and pass bands together with satisfactory phase response present attractive traits of the Butterworth filter for biosignal filtering [7], [42]. Routinely, the filtering process is conducted in the forward and backward direc-tions to avoid phase distortion and to ensure zero phase [10], [11]. Commonly, the third and the fourth order of the Butterworth filter are employed [10], [11]. Although Butterworth filter is regularly used as a band pass filter, one can apply two cas-cade separate low and high pass filters to independently select filter type and order.[2] Besides linear digital filters, other methods are used to attenuate EGG artifacts. For example, for signal recorded with sampling frequency fs, frequency content can be reduced/filtered to up to $fsn/2$ ($fsn < fs$) by down-sampling the signal to the new sam-pling frequency fsn [10]. Index blocked discrete cosine transform filter, wavelet filter, Savitzky-Golay filter, and moving median filters were used for EGG artifact elimi-nation too [10], [45]. Luckily, the majority of EGG artifacts do not overlap with the EGG frequency content. However, motion artifacts[3] are rarely stationary and cover a wide range of frequency content that overlaps with EGG in the spectral domain causing its elimination to be more complex or even impossible [2].

The best method to eliminate movement artifacts is to design protocols where body movements, talking, and coughing are avoided as much as possible [2], [16]. Unfortunately, this is not always possible, especially not in a dynamic setting as subjects are expected to drive a simulated vehicle. Furthermore, even in controlled scenarios the movements may not be avoided as the human body tends to move all the time [46], [47]. Adaptive filtering procedures for movement artifact detection and deletion with application of additional sensors such as accelerometers have been developed [7], [16], [39], [48], as well as artifact elimination procedures based on the Independent Component Analysis (ICA) and Empirical Mode Decomposition (EDA) applied on multi-channel EGG recordings [49]. The ultimate goal of all these procedures is wider EGG application and clinical standardization.

3.2.1 THE QUESTION OF EGG STANDARDIZATION

There is no standardized procedure for EGG measurement and analysis (electrode placement, recording duration, test meals—protocol, number of recording locations, recommended noise elimination procedure, etc.) [14], [16]. However, the current body of knowledge and evidence-based research have laid the foundations for future standard formation by identifying basic problems and proposing improvements [4]. For an electrophysiological test, the standard should contain a definition of all steps related to the recording protocol such as instrumentation specification, measurement duration, and electrode setting procedure details, as well as recommendations for analysis and interpretation of the obtained results. An example of the existing stan-dard is the SENIAM (Surface EMG for a Non-Invasive Assessment of Muscles) protocol (www.seniam.org/pdf/contents8.PDF) [50] for non-invasive recording of Electromyogram (EMG) or electrical activity of skeletal muscles. Other than EMG, standards have been defined for the application of electrocardiogram (ECG) and electroencephalogram (EEG) too [51], [52].

To reject or further process the complete measurement, it is necessary to perform an evaluation of the signal quality. Consequently, the standard may include visual observation of the signal for evaluation purposes, as is the case for the standard for measuring ECG signal [53]. Depending on the method being standardized, the standard may contain also an instrumentation specification as it is the case for ECG and EMG standards [50], [53]. Furthermore, protocol, methods for signal interpretation, the calculation of parameters, experts' certification, etc. should be standardized too [50], [53]. Another important aspect for future EGG standard definition may be the precision or the level of description of details for instrumentation and processing methods. The precision should enable the reproducibility of EGG measurements and analysis (for more details, please, see the self-published Letter to Editor [54]). There are ongoing discussions and initiatives for standardization of other electrophysiological measurements focused towards more transparent research and greater accuracy in the published results. For example Agreed Reporting Template for Electrophysiology Methods—International Standard (ARTEM-IS) yields standardization in the process of accurate reporting of EEG methodology in neuroscience [55]. ARTEM-IS draft template should be both machine- and human-readable and currently contains the following nine sections: title, hardware, data acquisition, preprocessing, experimental design, measurement, channel selection, visualization, and other. Another good example is the open-access Platform for the Exchange of Experimental Research Standards (PEERS) database for research in biomedicine with Wiki-like functionality with the aim to assist researchers with complex workflows and to share scientific results in a standardized manner [56]. By following an existing need for standardization in EGG measurements and analysis, as well as initiatives in similar fields [4], [6], [14], [54], we argue that future EGG standardization will be welcomed by the research and clinical community.

Standards are not constants, and they present a living body of knowledge and should be constantly reviewed and polished up by the relevant international bodies. For example, in EMG-based research the Consensus for Experimental Design in Electromyography (CEDE) project was preceded by the SENIAM project after more than two decades of SENIAM introduction [50], [57]. Similar models of EGG standard developments can be applied and expected in the future.

3.2.2   EGG ARTIFACTS

Common interference in biosignals, in general, include: (1) the narrow band noises caused by the baseline wander and slight movements (such as power interference at 50/60 Hz); (2) the broad band signals originating from the skeletal muscles, bulk movements, and medical equipment (occupying higher frequency ranges than EGG i.e., > 20 Hz); and (3) the impulse noise caused for example by technical artifacts [58]. All the known EGG artifacts are either narrow band or impulse noises: pulse rate, EMG, myoelectrical signals from the distal parts of the gastrointestinal tract, respiratory artifacts, movement artifacts, and ElectroMagnetic Interference (EMI) originating from the power grid (50/60 Hz) [4]. As EGG is yet to be entirely explored [59], there is a chance that new influences may emerge. Currently, we know that EGG signal can be additionally influenced by various diseases and states of the organism such as overweight (affects the weakening of signal strength) [60], increased size of the heart chambers or myocardial wall (increases the amplitude of ECG signals and the strength of ECG noise in the EGG signal) [61], [62], and the presence of cardiac arrhythmias (affects ECG characteristics and the shape of ECG noise in the EGG signal) [63]. In some cases the noise in EGG signal can prevent signal measurement and/or analysis.

3.2.3 OVERVIEW OF (PARTIALLY) AUTOMATED PROCEDURES FOR ARTIFACT REJECTION

The overview of developed methods for automated artifact rejection is summarized in Table 3.1. In [64] we presented the method for noise extraction from EGG signal by EMD and autocorrelation. EMD was used to decompose signal into Intrinsic

---

TABLE 3.1

Outlook of the Existing Automated Procedures for Noise Rejection in the Time and Frequency Domains. Except for the Results Presented by Wolpert et al. 2020 [16], All Methods Are Applicable for Simulator Sickness Assessment as They Do Not Alter Further Feature Extraction. Abbreviations Are EMD (Empirical Mode Decomposition) and SNR (Signal-to-Noise Ratio).

Method	Reference	Noise type	Main finding
EMD and autocorrelation	[64]	Stationary and non-stationary simulated white noise	Artifact elimination depends on the SNR.
Video analysis	[65]	Non-stationary movement artifact	Artifacts influence EGG-based features.
Electrode selection and activation time reviewing	[45]	Stationary and non-stationary artifacts, as well as EGG activation	Although procedures can be automated, manual correction for complex cases remains.
Total power comparison for channel selection	[7]	Non-stationary movement artifact	Artifacts with large power can be eliminated.
Comparison of EGG amplitude with amplitude threshold	[10]	Non-stationary artifacts	Artifacts with large power can be eliminated.
Dominant frequency power comparison for channel selection	[16]	Stationary and non-stationary artifacts	Favors dominant frequency—cannot be used for studying gastric dysrhythmias and arrhythmias.[4]
Artifact detection—comparison with the threshold and nonmonotonic test	[16]	Stationary and non-stationary artifacts	Guided visual inspection is always advised as there is no guarantee that artifacts can be eliminated.

Mode Functions (IMFs), while autocorrelation was applied to determine relevant IMFs to reconstruct the signal. This method was tested on semisynthetic signals with finite-amplitude additive white Gaussian noise to create a set of signals with different Signal-to-Noise Ratios (SNR). Obtained results were promising in terms of successful and automatic DF detection in timeseries compromised with random noise. Future research should confirm these findings and include *in vivo* recorded signals, as well as examination of the algorithm performance.

Video-based approach for movement artifact cancellation consisted of simultaneous EGG measurements and video recordings of subject's abdomen. The main idea was to capture subject's movements to determine artifact onset and offset. Working principle of the algorithm consisted of comparison of the current video frame with the referent i.e., averaged video frame. The substantial difference between the frames determined the movement artifact occurrence in the current frame interval. Subsequently, identified time intervals compromised by the movement artifacts were excluded from further analysis. By deleting only 0.02% of the samples, normogastric power share increased by 7%. Further investigation and confirmation of the obtained results should be performed in a larger sample [65].

Another interesting approach was proposed by Yassi et al. [45] and consisted of the Gastrointestinal Electrical Mapping Suite (GEMS) interface that combines various preprocessing algorithms and automatic methods for EGG analysis enabling the user to study high-resolution and multichannel EGG recordings. After visual inspection of the signal, users can choose channels with the high-quality signals (i.e., less pronounced noise) and apply desired analysis. The package includes the following methods for artifact cancelation: wavelet transformation, moving median filter, Butterworth filter, and Savitzky-Golay filter. The system design was tailor-made for non-experts in signal processing.

Another automated procedure for artifact rejection is based on the empirical reasoning [7]. The idea is to eliminate the signals that were at large compromised by noise and select the most optimal channel from multi-channel EGG recording. The term "optimal" referred to the channel with the least noise. We assumed that a large increase in signal strength originates exclusively from the noise, because the electrode setting itself is realized to cover the so-called pacemaker region of the stomach [11]. Consequently, large differences between the channels confirmed by the cross-correlation coefficients cannot persist if the signal is clean [11]. Further improvement of this method should include algorithm evaluation in different EGG measurement conditions and for different artifacts. This kind of multichannel measurement differs from methods in which matrix electrodes (two-dimensional array of electrodes) are used to estimate the spatial distribution of the biosignals with the assumption that individual measuring points carry different information [66], [67].

Extensive work was performed by Komorowski who developed two cascade automated methods for the artifact detection [10]. The first method is applied on separate channels and amplitude threshold is used to identify segments with artifacts to exclude them from the analysis. Firstly, the EGG signal is divided into segments and the average of standard deviations of all segments is calculated. Resulting value is multiplied by 4.2 (default scale) in order to obtain artifact threshold that is further compared to the absolute EGG amplitude. The second method is carried independently and the artifact threshold is obtained from the average standard deviation multiplied by 1.6 for each channel. For segments exceeding the artifact threshold, the exclusion is applied on all channels. The first approach can be used for short-term recordings when excessive deletion would influence results and quality of the analysis, while the second approach can be preferred in situations where noise-free signal is an imperative.

Wolpert et al. [16] performed channel selection with the opposite approach. Instead of determining erroneous and excessive parts, they selected the channel with the strongest peak in normogastric frequency range (0.033–0.067 Hz) and used it for further analysis. The disadvantage of such an approach is that it works well in cases when normal rhythm is expected, but in cases of rhythmical irregularities it would perform poorly. This procedure is partly automated as a complete workflow of EGG processing also includes manual inspection. After channel selection, the artifact identification based on the amplitude threshold (±3 standard deviations) was performed. For each step a guided observation i.e., visual inspection of the EGG power spectrum was advised to verify applied procedures and processing methods [16].

Although all described procedures and attempts (Table 3.1) to automate the process proved its usability, visual inspection of the signal is still used to qualitatively inspect the EGG signal and to identify the artifacts, as well as to select the portions of EGG signal for further analysis or deletion [39].

3.2.4   Why Manual Inspection and Sample Deletion Still Prevail

Visual inspection can be performed in time and frequency domains, and commonly involves identification of normal gastric rhythm as a reference for further analysis. For simulator sickness assessment, the resting period with normal gastric rhythm is used as a reference recording [7], [8] and as a rule it should contain gastric rhythm with the dominant frequency in normal gastric range. This baseline recording is valuable for examining effects of sickness to EGG traces in the time domain as changes in the rhythm and amplitude can be visually discerned in relation to the reference signal [39].

There are no clear recommendations for duration of portions of EGG signal for visualization in time domain. However, it would be reasonable to guess that two cycles (~40 s) present short segment while 20 cycles (~400 s) can be too long to look at. In our work, visual inspections of five to ten cycles which correspond to 100–200 s were performed. This time frame is in agreement with recommended visualization interval of EGG in frequency domain by application of Fast Fourier Transform (FFT) of 120 s [39].

Changes in EGG power and rhythm can be visualized in the frequency domain also by application of the Running Spectral Analysis (RSA) that generates pseudo 3D plot. Sample EGG RSA graphs in two subjects who underwent circular vection drum for about 15 min are presented in Figure 3.3 [68]. Individual differences between two subjects that correspond to qualitatively reported sickness symptoms (on a scale from 0 to 7) are visible as increase in tachygastria rhythm that was more pronounced in a volunteer with more severe sickness symptoms. In both cases, persistence of dysrhythmia after the kinetic drum rotation (marked with

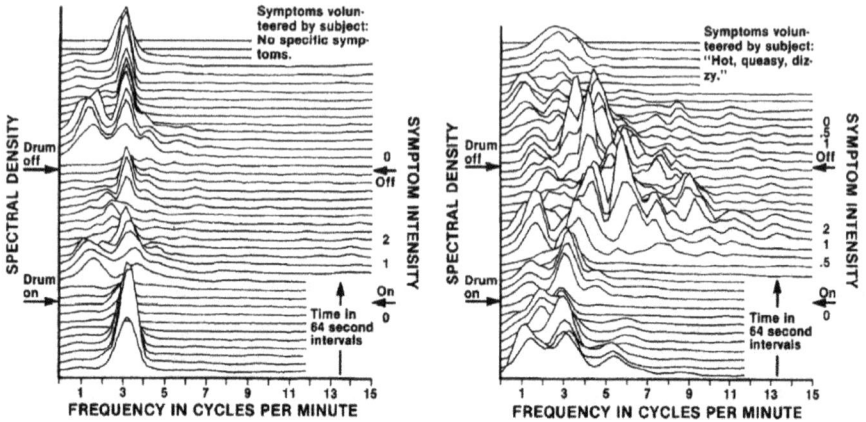

FIGURE 3.3 Running Spectral Analysis (RSA) in subject with no specific symptoms and in subject who reported motion sickness symptoms caused by the rotation in vection drum on the left-hand and the right-hand panels, respectively. Sickness symptoms intensity was self-reported in a scale from 0 (no symptoms) to 7 (near vomiting). The images are taken from Stern, Robert M., et al. "Spectral analysis of tachygastria recorded during motion sickness". *Gastroenterology* 92.1 (1987): 92–97. Copyright © 1987 Published by Elsevier Inc. With permission. License number: 5136990349712.

"off" in Figure 3.3) was observed. This sample analysis of visual representation in Figure 3.3 can be performed if RSA is applied on a previously filtered EGG signal (with no present artifacts). For proper application of RSA, standardized procedures for noise elimination are required. The RSA procedure remained the same after more than three decades since Stern et al. proposed RSA for motion sickness assessment (Figure 3.3) [68].

Ejection of the noisy parts of the signal is a commonly applied step for artifact elimination [14], [16], [48], [69]. This is performed with the assumption that the artifact duration is significantly shorter in comparison to the EGG signal duration. As EGG signal is a slow signal (one cycle of normal EGG rhythm lasts for about 20 s) in relation to the noise commonly originating from the subject's movement (of up to few seconds long intervals), it is not expected that the deletion of the noisy samples will significantly affect the signal if the movement occurs by accident in a static recording protocol. For instance, in the case study with video-based artifact detection in a static setting when the subject was placed in supine position the noise duration was 0.02% of the total signal duration [65]. Although the deleted segment presented a very small portion of the recorded signal it resulted in the normogastria increase from 33% to 40% making the sample deletion procedure an efficient approach in EGG analysis. On the other hand, if dynamic recording protocol takes place (e.g., recording EGG during active driving) then the large portion of the recorded EGG signal would be covered with the motion artifact and noise recognition and deletion would affect the entire EGG. Thus, the whole session could be discarded. To the best of our knowledge, there are no conducted studies showing to what extent the length, intensity, and frequency of the deleted noise in the EGG signal affect the quality of the signal. Acceptable duration of the deleted noise segments depends on the EGG application. For example, for studying spectral components an appropriate number of samples is needed to reach the desired resolution. On the other hand, methods are being developed so that a signal does not necessarily need to satisfy Nyquist-Shannon theorem if the nature of the EGG signal or the frequency content of the EGG signal is known i.e., stationary or deterministic for the selected intervals or during the whole recording session. Furthermore, gradient-based iterative algorithms have been used for ECG reconstruction after compression and can be applied for EGG restoration. These algorithms are used for signal recovery of missing samples caused by the sensor damage when samples are lost in signal transmission or simply deleted due to the excessive noise [70], [71]. Methods for signal reconstruction present a promising approach as total examination time cannot be always prolonged—it is reported that subjects can feel uncomfortable and become uncooperative after 60 min of EGG examination [6]. In the worst case, the whole recording session can be discarded. In [7] we excluded three out of 13 subjects (23%) and in [8] three out of 20 (15%) due to the excessive artifacts in EGG signal while using driving simulation. It should be noted that participants were actively engaged in driving activities in [7] in comparison to the simulated driving of an autonomous vehicle in [8] which subjects were not actively engaged in the simulated automated driving corresponding to the Society of Automotive Engineers (SAE) International level 5.

For the most part, fully or partly automated procedures for artifact deletion have been proposed in Table 3.1 and manual approach still remains (in case of partially automated procedure, the method for artifact detection is commonly completely automated) [10].

3.3 EGG-BASED FEATURE EXTRACTION FOR SICKNESS ASSESSMENT

Computer-aided analysis of EGG signals should involve the following steps: (1) artifact cancellation with application of digital filters, (2) analysis of selected noise-free segments, and (3) calculation of EGG features. Step (2) commonly incorporates application of FFT or other techniques (e.g., autoregressive model for estimation of Power Spectral Density, PSD) in order to determine the frequencies in the EGG signal or simply to perform an adequate transformation [10], [39]. This step can also include calculation of EGG envelope in time domain [8]. Instantaneous phase and amplitude envelopes of the EGG can be also obtained by applying Hilbert transform [16].

For features calculated in the spectral domain, the EGG window duration should incorporate at least a few EGG cycles. Different durations of EGG have been used for calculating RSA: most commonly 1–2 min and in some cases 4 min long EGG windows were applied [2], [10], [14], [39], [66]. For calculation of percentage of time with dominant frequency, EGG signal was segmented by the application Hamming, Hanning, or Tukey-Blackman windows for preventing spectral leakage with the overlap of 75% [14], [39].

Although there is no standardized processing workflow, the majority of authors agree that windows from 1–4 min of duration should be applied for obtaining features in the spectral domain. For example, EGG DWPack software enables the selection for calculation of spectral parameters of either 1 min or 4 min long windows with the overlap of segments of 10 s and 120 s, respectively [10]. We emphasize that windows of 4 min duration cannot detect changes lasting shorter than 60 s—which is the relevant duration for establishing diagnosis in a clinical setting [39].

Analysis of at least 30 min of the overall EGG duration was recommended for an adequate assessment of EGG for baseline, during, and after intervention [2], [10], [14]. In our analysis of EGG-based simulator sickness assessment, we used 5 min [7] and 15 min [8] long segments with satisfactory results.

After application of selected processing methods on determined signal duration or selected windows, other techniques can be used to extract EGG-based features. Overview of EGG-based features with expected changes indicating motion sickness are presented in Table 3.2.

TABLE 3.2

Overview of EGG-Based Features. Abbreviations Are DF (Dominant Frequency) and FFT (Fast Fourier Transform).

Feature	Description	References	Expected change during motion sickness
Power percentage	Percentage distribution of EGG power in frequency range of interest	[2], [10], [14], [39]	Increase in tachygastria and decrease in normogastria
Percentage of time with DF	Percentage of time with the DF in the frequency range of interest. Focuses only on the DF in the range of interest	[2], [6], [10], [16], [39]	Increase in tachygastria and decrease in normogastria
Power ratio	Can be calculated for the entire frequency range or for the range of interest	[39]	Increase during motion sickness in comparison to the baseline
DF	Frequency with the greatest power in FFT during a specific time	[2], [6], [10], [14], [16], [39]	Increase
Dominant power	Power of DF	[2], [6], [10], [14]	Increase
Power instability coefficient	Standard deviation divided by the mean value of dominant power	[10], [14]	Increase
Stability of dominant frequency	Defines DF variation in normogastric range and presents the mean frequency divided by the standard deviation	[2], [10], [14], [39]	Decrease
Averaged amplitude	Calculated in time domain	[16]	Increase
Standard deviation of cycle duration	Calculated in time domain	[16]	Increase
Falling-edge variable threshold	Calculates EGG activation time	[45], [72]	Increase

TABLE 3.3

EGG-Based Features for Studying Simulator Sickness in [7], [8]. Abbreviations Are PSD (Power Spectrum Density), T (Time Domain), and F (Frequency Domain).

Feature	Description	Reference	Change with nausea	Dynamic	Domain	Online
DF	Dominant frequency	[7], [8]	Increase	Yes	F	No
Power percentage	The percentage of PSD in the normogastric range (2–4 cpm)	[7]	Increase	Yes	F	No
MF	Median frequency of the EGG signal	[7], [8]	Increase	Yes	F	No
CF	Crest Factor of the PSD	[7], [8]	Decrease	Yes	F	No
MFM	Maximum magnitude of PSD	[8]	Increase	No	F	No
FSD	Percentage of the high PSD than MFM/4	[8]	Increase	No	F	No
Amplitude increase duration	Time during which EGG envelope exceeds the threshold	[8]	Increase	Yes	T	Yes
RMS (including relative RMS increase)	Root Mean Square of the EGG amplitude	[7], [8]	Increase	Yes	T	Yes

In addition to features presented in Table 3.2 and depending on the signal length for parameter calculation, additional estimates such as mean, median, and standard deviation of dominant frequency or dominant power can be used. Also, one can calculate the overall parameters as the average PSD of all segments for the examination period, as well as relative and absolute ratios [10].

We added expected changes as a consequence of motion sickness in Table 3.2 that correspond to gastric rhythm irregularities and power increase while considering general increase of variability (amplitude and frequency content) caused by the sickness occurrence. Features related to the multi-channel recording such as the percentage of EGG coupling between two channels and the slow wave propagation [2], [6], [10] were not considered here.

To complement existing commonly derived parameters (e.g., Dominant Frequency—DF), we introduced three novel EGG parameters presented in Table 3.3 (Root Mean Square—RMS, Median Frequency—MF, and Crest Factor—CF) after a thorough analysis and consideration of existing parameters and their shortcomings [7], [8]. The detailed presentation of introduced parameters for sickness assessment is provided in the following sections. For recordings in a dynamic environment, reference measurements were performed in static conditions (baseline) before the start of the driving simulation. Hence, the novel parameters were evaluated in both static and dynamic measurement conditions.

3.3.1 Time Domain Features for Simulator Sickness Assessment

Root mean square is commonly used in EMG analysis to estimate signal strength and skeletal muscle contraction levels [73], [74] and we used this reasoning for introducing RMS for power estimate of the electrical potentials originating from the smooth muscle in time domain. RMS presents a good choice as certain stimuli can result in EGG amplitude changes [75]. In studies where measurement of EGG amplitude increase is of interest, as in studies aiming to assess simulator sickness [7], [8], the RMS parameter can be and was applied with success.

Another time domain parameter is amplitude increase duration [8] that was calculated on EGG envelope obtained by a median filter with a 150 s long window with two thresholds. The first threshold was determined as the mean value of the EGG baseline amplitude and the second threshold at 0.6 fraction of the envelope during driving simulation. To calculate the time duration with power increase both thresholds needed to be crossed. This method was designed for offline signal analysis and for online application adaptive thresholds can be designed in a manner similar to the Pan-Tompkins algorithm for online QRS detection [76].

For automated computation of EGG activation onset and offset, a Falling-Edge Variable Threshold (FEVT) method was successfully applied for serosal recording [45], [72]. We note that the FEVT applied time-varying threshold on smoothed and preprocessed signal that bears resemblance to the Pan-Tompkins algorithm [76] whereas adaptive FEVT technique could be developed for quasi online EGG activation detection.

3.3.2 Frequency Domain Features for Simulator Sickness Assessment

Median frequency is commonly used for EMG analysis to assess skeletal muscle fatigue based on the evaluation of shifts in the signal PSD towards higher or lower frequencies [67], [77], [78]. We introduced MF for characterization of the EGG signal frequency content. MF was applied due to its higher sensitivity to frequency shifts in PSD in comparison to the dominant frequency. Namely, the dominant frequency depends only on the position of the peak with the highest power in the EGG spectrum and the appearance of spectral components of lower power that correspond to dysrhythmias may not affect its value. On the other hand, MF, as a parameter that divides the spectrum into two parts with the same area under the curve, depends on the entire signal spectrum and can indicate the appearance of frequency components throughout the complete frequency range. Such nature of the MF parameter is evidently suitable for simulator sickness assessment in a dynamic environment where alterations of the EGG signal spectrum can be expected [79], [80].

The crest factor describes the prominence of the dominant peak in relation to the rest of the signal [81]. Owing to its ability to evaluate peak expression, it was chosen for describing the spectrum shape of EGG signals. Variations of CS can indicate differences among EGG spectra shapes, which cannot be detected by dominant frequency due to the dependence on the single peak position and not on the whole spectrum. Additionally, the crest factor may indicate changes in the signal (peak sharpening) that MF may fail to detect, because peak sharpening can occur without changing the overall spectrum. Therefore, CF can be used to investigate the

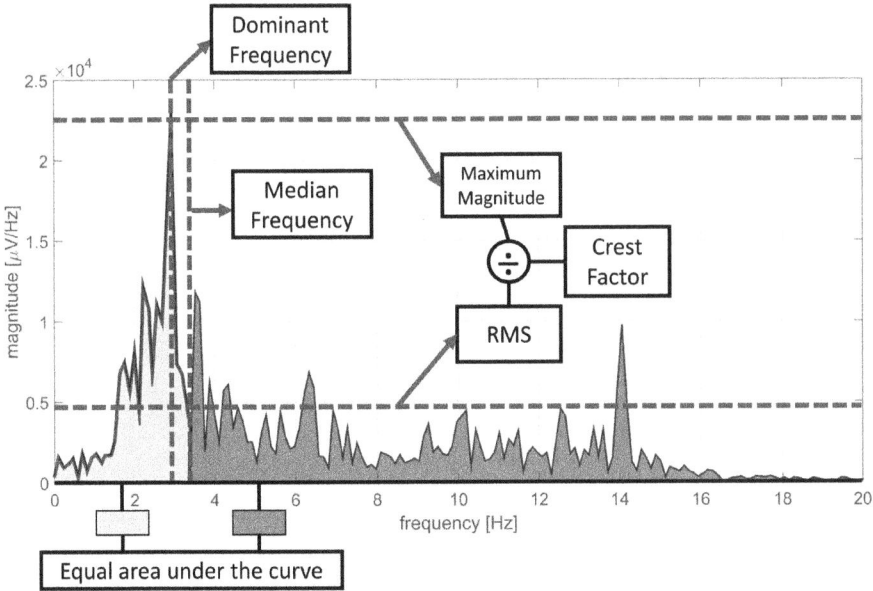

FIGURE 3.4   EGG features in frequency domain. RMS stands for Root Mean Square.

influence of virtual reality and driving simulation on alterations in the EGG spectrum where shifts to the tachygastria frequency range are expected. The graphical representation of the proposed parameters is presented in Figure 3.4.

3.3.3   Usefulness of the Proposed Features

To the best of our knowledge, the usefulness factor of the EGG features has not been defined so far. In electrophysiology, usefulness has already been introduced as a term that determines the qualitative assessment of the practical usability of signal parameters [77] and it may be a good approach to rate all features by defining a utility factor. Additionally, feature importance can be estimated with for example Random Forest Machine Learning (RF ML) algorithm that is broadly used for a large number of features with complex interactions [82], [83]. The approach that combines Support Vector Machine (SVM) and Genetic Algorithm (GA) has been applied in EGG with some success [84]. Other feature selection methods used for biosignal classification such as filters, wrappers, embedded methods, and hybrid approaches also can be exploited for application in EGG [85].

3.4   PRACTICAL CONSIDERATIONS FOR EGG-BASED ASSESSMENT OF SIMULATOR SICKNESS

Although current body of knowledge is scarce, published results offer some guidelines for EGG application for simulator sickness assessment. Based on experiences in static and dynamic recordings in driving simulators [7], [8], we propose separate workflows for signal analysis in static and dynamic conditions.

General approach to EGG analysis would benefit from good practices recognized and related to the data science life cycle such as availability of data, software, and other elements of research to enable reproducibility and replicability [86]. Open signals measured in healthy subjects can serve as benchmark data (control group) for comparison with signals measured in subjects experiencing simulator sickness or in patients with gastric disorders. In general, examples of benchmark data and their importance in the field of electrophysiology can be found in many areas, such as EEG [87] and EMG [73], [88]. Therefore, practices where authors share software code for the EGG analysis and EGG data obtained in driving simulator studies in accordance with existing practices [10]–[12], [16], [45] should be recognized and encouraged.

3.4.1 PROPOSED WORKFLOWS FOR EGG-BASED ASSESSMENT OF SIMULATOR SICKNESS IN STATIC AND DYNAMIC CONDITIONS

The workflows for EGG analysis in static and dynamic recording setups are presented in Figure 3.5. EGG signal processing workflow for static measurements consists of signal visualization, noise elimination and preprocessing, signal check

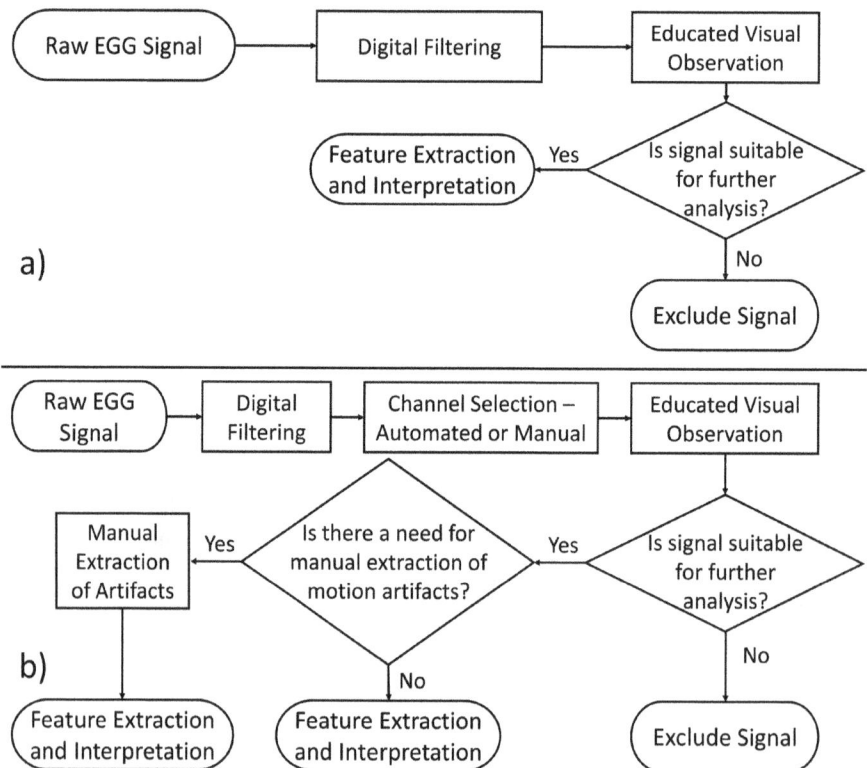

FIGURE 3.5 Proposed EGG data processing workflows for EGG recorded in (a) static and (b) dynamic conditions.

procedure, and feature extraction for further application of statistical analysis, feature selection, classification, and model validation. As dynamic environments may add unwanted motion components, we introduced a modified workflow consisting of channel selection and other procedures for motion artifact rejection.

For both static and dynamic protocols, processing workflow starts with the digital filtering. As suggested in [11] for static recording single channel setup is sufficient, while for dynamic it is beneficial to include more than one channel [7], which is why there is an additional step in the workflow—channel selection. Educated visual observation is required in both cases and it is used to determine whether the signal is suitable for further analysis. While this is expected to be the step prior to the feature extraction and interpretation in the static environment, in the dynamic environment, due to the increased occurrence of motion artifacts, one additional step is suggested—if needed manual extraction of samples compromised by noise should be performed [7], [69]. It should be stated that for the long-term parameters (calculated offline on substantial portion of the signal) it is crucial to follow the proposed workflows, while for the short-term features that could be determined and assessed quasi online, the workflows are not applicable.

3.4.2 INDIVIDUAL DIFFERENCES AND OTHER FACTORS AFFECTING EGG PROCESSING WORKFLOWS: CHALLENGES AND OPPORTUNITIES

In order to introduce a complete set of possible features one should have in mind pathology-specific parameters as for example frequent vomiting syndrome, central nervous system disorders in children, chronic idiopathic intestinal pseudo-obstruction, nervous bowel syndrome, gastroparesis, nausea in pregnancy, and Helicobacter pylori that can induce changes in EGG features and alter the processing workflows [59], [89]–[93]. Expected deviations are mainly reflected in: tachygastria, bradygastria, inability to determine gastric rhythm, lack of dominant frequency, and lack of EGG power increase after meals. Certain diseases that are not related to the gastrointestinal system, but present state of the organism can affect changes in the EGG signal too.

The fact that stressors such as anxiety and emotions can influence gastric emptying has been known for centuries. The first insight into the brain-gut interactions was revealed in the 19th century [5]. Consequently, EGG can be used to detect changes produced by the stressors or its complex relations to other symptoms [94]. For example, in [95] a subject with acrophobia reported higher nausea sensation while experiencing the fear of height in a Virtual Reality (VR) roller coaster.

To conclude, one must have in mind pathologies and organism states that can influence EGG signal in order to keep all these factors under control. The usage of driving simulators and automated vehicles by patients is not a common procedure, but recent results suggested that it can be beneficial. Namely, fully automated vehicles (SAE International levels 4–5) can be valuable for active aging as drivers can maintain independence in mobility. Classen et al. found that elderly had higher motion sickness symptoms in simulators in comparison with the automated shuttle drive after approximately 10 min of exposure [96]. Despite demonstrated discomfort, studies with patients with neurological disorders revealed the simulator potential application for assessment of fitness to drive in neurology by providing a quantitative assessment tool in high risk driving scenarios [97], [98]. Currently, studies with patients in driving simulators do not include EGG measurement and analysis. We argue that EGG measurements in patients can be beneficial to develop dedicated simulations with enhanced patients' comfort. The detailed discussion on individual differences and other environmental factors influencing EGG signal and their interaction with simulator sickness assessment are out of the scope of this chapter.

3.5 CONCLUSIONS ON EGG SIGNAL PROCESSING METHODS FOR SIMULATOR SICKNESS ASSESSMENT

EGG is a promising technique for simulator sickness assessment. However, an informed precaution is needed for proper scientific conclusions. We propose dedicated workflows that meet environmental demands—whether driving simulators are used in static (automated driving without human engagement and interventions—SAE International levels 4–5) or dynamic recordings (classical manual driving procedure—SAE International levels 0–3).

3.5.1 CURRENT PATHWAYS

Despite the fact that there is no consensus on the most appropriate method to record and analyze EGG signals, significant steps were undertaken in the scientific community towards automated procedures. Once applicable, these procedures can be straightforwardly applied for calculation of short-term and long-term parameters to objectively assess sickness in different environments.

Driving simulation studies are commonly affected by the simulation sickness which decreases the performance of affected test subjects or in the worst case even prevents them from using the simulator. While active monitoring of EGG signals enables prompt detection of the nausea and sickness during driving simulation, the latter could be actively controlled for participants' discomfort by changing the driving environment, turning on or off a motion system when available, using external screens instead of VR headset, etc. In the driving simulator EGG signals could further be synchronized with the driving data (e.g., vehicle acceleration and velocity, road configuration, weather conditions) and also other driver's physiological measurements to observe their interdependencies and causal relationships. EGG adoption has already been performed in combination with other quantitative physiological procedures such as blinking and breathing [99]. In accordance with the multi-signal studies, procedures to obtain more signals from one recording channel are being developed [100], [101]. This research direction can utilize one channel noisy recording to separate EGG, breathing, electrical muscle activity, and heart rate.

3.5.2 CURRENT PITFALLS

Current pitfalls are similar as the ones determined more than 20 years ago in [102] and include: (1) visual inspection of signal is still required; (2) noise presence affects EGG signal, especially movement artifacts making EGG preprocessing and analysis challenging; (3) some level of signal processing knowledge is still essential for EGG interpretation, which limits clinical application; and (4) there is no consensus on the feature selection procedures for EGG signal characterization.

Educated visual observation is suggested prior to further analysis of EGG signal [16], [69] and this is mainly the consequence of movement artifacts which cannot be completely extracted via digital filtering due to the fact that frequency content overlaps with EGG. Its occurrence cannot be prevented, due to the vulnerability of the skin-electrode impedance causing even slight body movements to induce substantial changes that subsequently lead to artifacts. Experienced investigators can identify this kind of artifact based on its amplitude and sudden onset, as well as by checking its presence on all recording channels. The downside of this approach is the inability to automate analysis procedures, as well as dependence on the subjective assessment. It should be mentioned that besides required experience in the EGG field, considerable signal processing knowledge is required.

Moreover, the most frequently used parameter—DF—is not ideal for describing altered EGG spectrums, due to the fact that its value depends only on one peak. In our research related to the EGG-based simulator sickness assessment, we proposed two different parameters that could be applied: median frequency and crest factor [7], [8]. Although we acquired some promising results, there is a need for studies that will evaluate proposed parameters in a larger sample to reach potential consensus regarding the selection of features for quantification of altered EGG spectrums.

Concerns were raised also in regards to the EGG spectral analysis, its inherent variability, and intersubject variability [94]. There are studies reporting that physiological features are not necessarily correlated with the motion sickness and this can be seen in EGG amplitudes as it varies among individuals and depends on adipose tissue and other anatomical differences [103]. Future steps towards standardization, automation of procedures, and recommendations for amplitude normalization are required for wider adoption of EGG as an assessment tool.

3.5.3 Future R&D Directions

Current processing methods provide a promising framework for future application of automated procedures for simulator sickness evaluation. The need for novel and efficient processing methods embedded in an easy-to-use interface has been previously recognized by Komorowski [10]. The design of software exceeds the scope of this chapter, although it presents an important aspect, especially in multi- and interdisciplinary research settings where the end user does not necessarily have extensive engineering and data science knowledge. Here, we elaborate only on methods for noise elimination and feature extraction. We showed that short-term parameters can provide immediate assessment required for the real-time simulation changes in the feedback loop. On the other hand, long-term parameters may contribute to the better understanding of consequences, nature, and origins of simulator sickness in relation to driving scenarios, haptic interface, user experience, and comfort, as well as on the other aspects of driving simulators.

The main idea behind EGG feature extraction is to mitigate sickness, to boost user experience, and to propose design strategies for ship, flying, and driving simulators, and for automated vehicles and aircrafts. Further exploration of the proposed processing workflows can be used to objectively assess cognitive load related to online meetings and the so-called Zoom fatigue. Despite telco effectiveness in terms of cost and incomparable supremacy in the times of crisis, the negative influences on health are very well recognized and incorporate sickness occurrence as well [104]–[106].

Although EGG was recorded by the end of the 1990s on a BION-11 spaceflight mission to study space sickness, data from the space stations on EGG related changes are still scarce [107], [108]. Rather interesting results were revealed in [108] where the adaptation to space station was noticed—sickness disappeared after three days spent on a space station. To the best of our knowledge adaptation to driving simulators has not been studied previously and possibly presents a very promising research direction.

An attractive method in biosignal research is in application of machine learning algorithms that can be applied either for feature selection such as RF ML or for classification of nausea. Initial results of multi-class classification of three ranges of gastric rhythms appear promising and can be explored for more complex problems in the future [109].

3.6 ACKNOWLEDGMENTS

We owe exceptional gratitude to Assoc. Prof. Zaharije Radivojević, PhD for his thorough and interesting questions on EGG for simulator sickness assessment research related to the Nenad B. Popović's Doctoral Dissertation titled "Methods for assessment of electrical activity of smooth muscles", University of Belgrade—School of Electrical Engineering, 2021 (Mentor: Assoc. Prof. Nadica Miljković) that inspired us to write this chapter.

The work on this chapter was partly supported by Grant No. TR33020 funded by the Ministry of Education, Science and Technological Development, Republic of Serbia and by the Slovenian Research Agency within the research program ICT4QoL—Information and Communications Technologies for Quality of Life, grant number P2–0246.

NOTES

1 Electrogastrography (EGG) is a method used to record an EGG signal or electrogastrogram. For both terms (electrogastrography and electrogastrogram), we used EGG abbreviation throughout the chapter.
2 Analogue realization of the band pass filters includes cascading of the low pass and high pass filters. Here, we refer to the application of filtering functions used for digital filters such as *butter()* function in GNU Octave [43] and *scipy.signal.butter()* function in Python (Python Software Foundation, Delaware, USA) [44] where only one order can be set for design of symmetrical bandpass frequency response.
3 Motion artifact and movement artifact are considered and used as synonyms.
4 We adopted the definition that dysrhythmias relate to rhythms outside of the normal range and arrhythmias to the absence of rhythm i.e., slow waves [2]. EGG arrhythmia should not be confused with the ECG arrhythmia that is defined as an anomaly compared to the normal ECG rhythm [58].

REFERENCES

[1] W. C. Alvarez, "The electrogastrogram and what it shows," *J. Am. Med. Assoc.*, vol. 78, pp. 1116–1119, 1922.

[2] J. Yin and J. D. Z. Chen, "Electrogastrography: Methodology, validation and applications," *J. Neurogastroenterol. Motil.*, vol. 19, no. 1, pp. 5–17, 2013, doi:10.5056/jnm.2013.19.1.5.

[3] R. C. Davis, L. Garafolo, and F. P. Gault, "An exploration of abdominal potentials," *J. Comp. Physiol. Psychol.*, vol. 50, no. 5, pp. 519–523, 1957, doi:10.1037/h0048466.

[4] N. B. Popović, "Methods for assessment of electrical activity of smooth muscles," University of Belgrade—School of Electrical Engineering, 2021.

[5] J. T. Cacioppo and L. G. Tassinary (eds.), *Principles of psychophysiology: Physical, social, and inferential elements*. New York: Cambridge University Press, 1990, pp. xiii, 914.

[6] H. Murakami, et al., "Current status of multichannel electrogastrography and examples of its use," *J. Smooth Muscle Res.*, vol. 49, pp. 78–88, 2013, doi:10.1540/jsmr.49.78.

[7] N. B. Popovic, N. Miljkovic, K. Stojmenova, G. Jakus, M. Prodanov, and J. Sodnik, "Lessons learned: Gastric motility assessment during driving simulation," *Sensors*, vol. 19, no. 14, p. 3175, 2019, doi:10.3390/s19143175.

[8] T. Gruden, et al., "Electrogastrography in autonomous vehicles—An objective method for assessment of motion sickness in simulated driving environments," *Sensors*, vol. 21, no. 2, Art. No. 2, 2021, doi:10.3390/s21020550.

[9] J. Iskander, et al., "From car sickness to autonomous car sickness: A review," *Transp. Res. Part F Traffic Psychol. Behav.*, vol. 62, pp. 716–726, 2019, doi:10.1016/j.trf.2019.02.020.

[10] D. Komorowski, "EGG DWPack: System for multi-channel electrogastrographic signals recording and analysis," *J. Med. Syst.*, vol. 42, no. 11, p. 201, 2018, doi:10.1007/s10916-018-1035-1.

[11] N. B. Popovic, N. Miljkovic, and M. B. Popović, "Simple gastric motility assessment method with a single-channel electrogastrogram," *Biomed. Eng. Biomed. Tech.*, vol. 64, no. 2, pp. 177–185. 2019, doi:10.1515/bmt-2017-0218.

[12] N. Popović B., N. Miljković, and M. Popović B., "Three-channel surface electrogastrogram (EGG) dataset recorded during fasting and post-prandial states in 20 healthy individuals," *Zenodo*, 2020. Doi:10.5281/zenodo.3878435.

[13] G. Riezzo, F. Russo, and F. Indrio, "Electrogastrography in adults and children: The strength, pitfalls, and clinical significance of the cutaneous recording of the gastric electrical activity," *BioMed Res. Int.*, vol. 2013, pp. 1–14, 2013, doi:10.1155/2013/282757.

[14] F.-Y. Chang, "Electrogastrography: Basic knowledge, recording, processing and its clinical applications," *J. Gastroenterol. Hepatol.*, vol. 20, no. 4, pp. 502–516, 2005, doi:10.1111/j.1440-1746.2004.03751.x.

[15] A. Diamanti, et al., "Gastric electric activity assessed by electrogastrography and gastric emptying scintigraphy in adolescents with eating disorders," *J. Pediatr. Gastroenterol. Nutr.*, vol. 37, no. 1, pp. 35–41, 2003.

[16] N. Wolpert, I. Rebollo, and C. Tallon-Baudry, "Electrogastrography for psychophysiological research: Practical considerations, analysis pipeline, and normative data in a large sample," *Psychophysiology*, vol. 57, no. 9, p. e13599, 2020, doi:10.1111/psyp.13599.

[17] N. Araki, et al., "Electrogastrography for diagnosis of early-stage Parkinson's disease," *Parkinsonism Relat. Disord.*, vol. 86, pp. 61–66, 2021, doi:10.1016/j.parkreldis.2021.03.016.

[18] A. Dam and M. Jeon, "A review of motion sickness in automated vehicles," in *13th International Conference on Automotive User Interfaces and Interactive Vehicular Applications*, New York, 2021, pp. 39–48. Doi:10.1145/3409118.3475146.

[19] J. Smyth, S. Birrell, R. Woodman, and P. Jennings, "Exploring the utility of EDA and skin temperature as individual physiological correlates of motion sickness," *Appl. Ergon.*, vol. 92, p. 103315, 2021, doi:10.1016/j.apergo.2020.103315.

[20] R. M. Stern, K. L. Koch, H. W. Leibowitz, I. M. Lindblad, C. L. Shupert, and W. R. Stewart, "Tachygastria and motion sickness," *Aviat. Space Environ. Med.*, vol. 56, no. 11, pp. 1074–1077, 1985.

[21] G. Geršak, H. Lu, and J. Guna, "Effect of VR technology matureness on VR sickness," *Multimed. Tools Appl.*, vol. 79, no. 21, pp. 14491–14507, 2020, doi:10.1007/s11042-018-6969-2.

[22] J. Guna, G. Geršak, I. Humar, J. Song, J. Drnovšek, and M. Pogačnik, "Influence of video content type on users' virtual reality sickness perception and physiological response," *Future Gener. Comput. Syst.*, vol. 91, pp. 263–276, 2019, doi:10.1016/j.future.2018.08.049.

[23] K. Stanney, C. Fidopiastis, and L. Foster, "Virtual reality is sexist: But it does not have to be," *Front. Robot. AI*, vol. 7, p. 4, 2020, doi:10.3389/frobt.2020.00004.

[24] P. Green, "Motion sickness and concerns for self-driving vehicles: A literature review," *The University of Michigan Transportation Research Institute (UMTRI)*, UMTRI-2016, 2016.

[25] O. Carsten and A. H. Jamson, "Driving simulators as research tools in traffic psychology," *Handbook of Traffic Psychology*, pp. 87–96, 2011. Doi:10.1016/B978-0-12-381984-0.10007-4.

[26] A. Hoeschen, et al., *TRAINER: Inventory of driver training needs and major gaps in the relevant training procedures*, Brussels, Belgium: European Commission, 2001.

[27] R. W. Allen, G. D. Park, M. L. Cook, and D. Fiorentino, "The effect of driving simulator fidelity on training effectiveness," *N. Am.*, p. 15, 2007.

[28] T. Gruden, K. Stojmenova, J. Sodnik, and G. Jakus, "Assessing drivers' physiological responses using consumer grade devices," *Appl. Sci.*, vol. 9, no. 24, art. No. 24, 2019, doi:10.3390/app9245353.

[29] G. Jakus, C. Dicke, and J. Sodnik, "A user study of auditory, head-up and multimodal displays in vehicles," *Appl. Ergon.*, vol. 46, pp. 184–192, 2015, doi:10.1016/j.apergo.2014.08.008.

[30] J. Sodnik, C. Dicke, S. Tomažič, and M. Billinghurst, "A user study of auditory versus visual interfaces for use while driving," *Int. J. Hum.-Comput. Stud.*, vol. 66, no. 5, pp. 318–332, 2008, doi:10.1016/j.ijhcs.2007.11.001.

[31] W. D. Kappler, "Views on the role of simulation in driver training," in *Proceedings of the 12th European Annual Conference on Human Decision Making and Manual Control*, Kassel, Germany, 1993, pp. 5.12–5.17.

[32] J. D. Winter and R. Happee, "Advantages and disadvantages of driving simulators: A discussion," *Undefined*, 2012, Accessed: Oct. 09, 2021 [Online]. Available: www.semanticscholar.org/paper/Advantages-and-Disadvantages-of-Driving-Simulators%3A-Winter-Happee/b1a70385f980efd993b57142cca100d1e88672b0

[33] N. Dużmańska, P. Strojny, and A. Strojny, "Can simulator sickness be avoided? A review on temporal aspects of simulator sickness," *Front. Psychol.*, vol. 9, p. 2132, 2018, doi:10.3389/fpsyg.2018.02132.

[34] J. T. Reason and J. J. Brand, *Motion Sickness*. Oxford: Academic Press, 1975.

[35] J. T. Reason, "Motion sickness adaptation: A neural mismatch model," *J. R. Soc. Med.*, vol. 71, no. 11, pp. 819–829, 1978.

[36] G. E. Riccio and T. A. Stoffregen, "An ecological theory of motion sickness and postural instability," *Ecol. Psychol.*, vol. 3, no. 3, pp. 195–240, 1991, doi:10.1207/s15326969eco0303_2.

[37] R. S. Kennedy, N. E. Lane, K. S. Berbaum, and M. G. Lilienthal, "Simulator sickness questionnaire: An enhanced method for quantifying simulator sickness," *Int. J. Aviat. Psychol.*, vol. 3, no. 3, pp. 203–220, 1993, doi:10.1207/s15327108ijap0303_3.

[38] N. Jovanović, N. B. Popović, and N. Miljković, "Combined approach for automatic and robust calculation of dominant frequency of electrogastrogram," *ArXiv200909023 Eess*, 2020, Accessed: Dec. 07, 2020 [Online]. Available: http://arxiv.org/abs/2009.09023

[39] K. L. Koch and R. M. Stern, *Handbook of Electrogastrography*. Oxford and New York: Oxford University Press, 2003.

[40] R. M. Rangayyan, "Filtering for removal of artifacts," in *Biomedical Signal Analysis*. New Jersey, USA: John Wiley & Sons, Ltd, 2015, pp. 91–231. Doi:10.1002/9781119068129.ch3.

[41] M. Sandhu, S. Kaur, and J. Kaur, "A study on design and implementation of butterworth, chebyshev and elliptic filter with MatLab," *Int. J. Emerg. Technol. Eng. Res.*, vol. 4, no. 6, p. 4, 2016.

[42] R. E. Challis and R. I. Kitney, "The design of digital filters for biomedical signal processing Part 3: The design of Butterworth and Chebychev filters," *J. Biomed. Eng.*, vol. 5, no. 2, pp. 91–102, 1983, doi:10.1016/0141-5425(83)90026-2.

[43] J. Eaton, D. Bateman, S. Hauberg, and R. Wehbring, "GNU Octave version 5.2.0 manual: A high-level interactive language for numerical computations," 2019. Accessed: Aug. 04, 2021 [Online]. Available: https://octave.org/doc/v5.2.0/

[44] E. Jones, T. Oliphant, and P. Peterson, "SciPy: Open source scientific tools for python," *Comput. Sci.*, 2001.

[45] R. Yassi, et al., "The gastrointestinal electrical mapping suite (GEMS): Software for analyzing and visualizing high-resolution (multi-electrode) recordings in spatiotemporal detail," *BMC Gastroenterol.*, vol. 12, no. 1, p. 60, 2012, doi:10.1186/1471-230X-12-60.

[46] I. D. Loram, C. N. Maganaris, and M. Lakie, "Paradoxical muscle movement in human standing," *J. Physiol.*, vol. 556, no. 3, pp. 683–689, 2004, doi:10.1113/jphysiol.2004.062398.

[47] D. Winter, "Human balance and posture control during standing and walking," *Gait Posture*, vol. 3, no. 4, pp. 193–214, 1995, doi:10.1016/0966-6362(96)82849-9.

[48] A. A. Gharibans, B. L. Smarr, D. C. Kunkel, L. J. Kriegsfeld, H. M. Mousa, and T. P. Coleman, "Artifact rejection methodology enables continuous, noninvasive measurement of gastric myoelectric activity in ambulatory subjects," *Sci. Rep.*, vol. 8, no. 1, p. 5019, 2018, doi:10.1038/s41598-018-23302-9.

[49] S. Sengottuvel, P. F. Khan, N. Mariyappa, R. Patel, S. Saipriya, and K. Gireesan, "A combined methodology to eliminate artifacts in multichannel electrogastrogram based on independent component analysis and ensemble empirical mode decomposition," *SLAS Technol. Transl. Life Sci. Innov.*, vol. 23, no. 3, pp. 269–280, 2018, doi:10.1177/2472630318756903.

[50] H. J. Hermens, B. Freriks, C. Disselhorst-Klug, and G. Rau, "Development of recommendations for SEMG sensors and sensor placement procedures," *J. Electromyogr. Kinesiol.*, vol. 10, no. 5, pp. 361–374, 2000, doi:10.1016/S1050-6411(00)00027-4.

[51] P. Kligfield, et al., "Recommendations for the standardization and interpretation of the electrocardiogram: Part I: The electrocardiogram and its technology a scientific statement from the american heart association electrocardiography and arrhythmias committee, council on clinical cardiology; the american college of cardiology foundation; and the heart rhythm society endorsed by the international society for

computerized electrocardiology," *J. Am. Coll. Cardiol.*, vol. 49, no. 10, pp. 1109–1127, 2007, doi:10.1016/j.jacc.2007.01.024.

[52] D. Dash, et al., "Update on minimal standards for electroencephalography in Canada: A review by the Canadian society of clinical neurophysiologists," *Can. J. Neurol. Sci.*, vol. 44, no. 6, pp. 631–642, 2017, doi:10.1017/cjn.2017.217.

[53] J. Eldridge, et al., "Recording a standard 12—lead electrocardiogram. An approved methodology by the Society of Cardiological Science and Technology (SCST). Clinical guidelines by consensus," 2014, Accessed: Nov. 30, 2020 [Online]. Available: https://pure.ulster.ac.uk/en/publications/recording-a-standard-12-lead-electrocardiogram-an-approved-method-3

[54] N. Popović B. and N. Miljković, "Towards standardization of electrogastrography," 2020. Available: https://zenodo.org/record/3980694

[55] S. J. Styles, V. Kovic, H. Ke, and A. Šoškić, "Towards ARTEM-IS: An evidence-based agreed reporting template for electrophysiology methods—International standard." *PsyArXiv*, 2021. Doi:10.31234/osf.io/myn7t.

[56] A. Sil, et al., "PEERS—an open science 'platform for the exchange of experimental research standards' in biomedicine," *Front. Behav. Neurosci.*, vol. 15, p. 755812, 2021, doi:10.1101/2021.07.31.454443.

[57] M. Besomi, et al., "Consensus for experimental design in electromyography (CEDE) project: Electrode selection matrix," *J. Electromyogr. Kinesiol.*, vol. 48, pp. 128–144, 2019, doi:10.1016/j.jelekin.2019.07.008.

[58] R. Martinek, et al., "Advanced bioelectrical signal processing methods: Past, present and future approach—Part I: Cardiac signals," *Sensors*, vol. 21, no. 15, art. No. 15, 2021, doi:10.3390/s21155186.

[59] A. Al Taee and A. Al-Jumaily, "Electrogastrogram based medical applications an overview and processing frame work," *Hybrid Intell. Syst.*, vol. 923, A. M. Madureira, A. Abraham, N. Gandhi, and M. L. Varela (eds.). Cham: Springer International Publishing, 2020, pp. 511–520. Doi:10.1007/978-3-030-14347-3_50.

[60] G. Riezzo, F. Pezzolla, and I. Giorgio, "Effects of age and obesity on fasting gastric electrical activity in man: A cutaneous electrogastrographic study," *Digestion*, vol. 50, no. 3–4, pp. 176–181, 1991, doi:10.1159/000200759.

[61] T. Feldman, K. M. Borow, A. Neumann, R. M. Lang, and R. W. Childers, "Relation of electrocardiographic R-wave amplitude to changes in left ventricular chamber size and position in normal subjects," *Am. J. Cardiol.*, vol. 55, no. 9, pp. 1168–1174, 1985, doi:10.1016/0002-9149(85)90657-5.

[62] J. E. Madias, "Electrocardiogram in apical hypertrophic cardiomyopathy with a speculation as to the mechanism of its features," *Neth. Heart J.*, vol. 21, no. 6, pp. 268–271, 2013, doi:10.1007/s12471-013-0400-4.

[63] C. Antzelevitch and A. Burashnikov, "Overview of basic mechanisms of cardiac arrhythmia," *Card. Electrophysiol. Clin.*, vol. 3, no. 1, pp. 23–45, 2011, doi:10.1016/j.ccep.2010.10.012.

[64] N. Jovanović, N. B. Popović, and N. Miljković, "Empirical mode decomposition for automatic artifact elimination in electrogastrogram," *2021 20th International Symposium Infoteh-Jahorina (Infoteh)*, IEEE, 2021, pp. 1–6. Doi:10.1109/INFOTEH51037.2021.9400683.

[65] N. B. Popović, N. Miljković, and V. Papić, "Video-based extraction of movement artifacts in electrogastrography signal," in *Book of Abstracts, Belgrade Bioinformatics Conference (BelBi)*, Belgrade, Serbia, 2018, vol. 40, no. 1, p. 26. Doi:10.13140/RG.2.2.19753.29280.

[66] R. Merletti, A. Botter, A. Troiano, E. Merlo, and M. A. Minetto, "Technology and instrumentation for detection and conditioning of the surface electromyographic signal: State of the art," *Clin. Biomech.*, vol. 24, no. 2, pp. 122–134, 2009, doi:10.1016/j.clinbiomech.2008.08.006.

[67] R. Merletti and D. Farina, *Surface Electromyography: Physiology, Engineering, and Applications*. New Jersey, USA: John Wiley & Sons, 2016.

[68] R. M. Stern, K. L. Koch, W. R. Stewart, and I. M. Lindblad, "Spectral analysis of tachygastria recorded during motion sickness," *Gastroenterology*, vol. 92, no. 1, pp. 92–97, 1987, doi:10.1016/0016-5085(87)90843-2.

[69] P. Du, et al., "High-resolution mapping of in Vivo gastrointestinal slow wave activity using flexible printed circuit board electrodes: methodology and validation," *Ann. Biomed. Eng.*, vol. 37, no. 4, pp. 839–846, 2009, doi:10.1007/s10439-009-9654-9.

[70] L. Stanković, M. Daković, and S. Vujović, "Reconstruction of sparse signals in impulsive disturbance environments," *Circuits Syst. Signal Process.*, vol. 36, no. 2, pp. 767–794, 2017, doi:10.1007/s00034-016-0334-3.

[71] S. Vujović, et al., "Sparse analyzer tool for biomedical signals," *Sensors*, vol. 20, no. 9, art. No. 9, 2020, doi:10.3390/s20092602.

[72] J. C. Erickson *et al.*, "Falling-edge, variable threshold (FEVT) method for the automated detection of gastric slow wave events in high-resolution serosal electrode recordings," *Ann. Biomed. Eng.*, vol. 38, no. 4, pp. 1511–1529, 2010, doi:10.1007/s10439-009-9870-3.

[73] S. G. Boe, C. L. Rice, and T. J. Doherty, "Estimating contraction level using root mean square amplitude in control subjects and patients with neuromuscular disorders," *Arch. Phys. Med. Rehabil.*, vol. 89, no. 4, pp. 711–718, 2008, doi:10.1016/j.apmr.2007.09.047.

[74] P. Konrad, *A Practical Introduction to Kinesiological Electromyography*. Scottsdale, AZ: Noraxon INC, 2005.

[75] M. J. Cevette, et al., "Electrogastrographic and autonomic responses during oculovestibular recoupling in flight simulation," *Aviat. Space Environ. Med.*, vol. 85, no. 1, pp. 15–24, 2014, doi:10.3357/ASEM.3673.2014.

[76] J. Pan and W. J. Tompkins, "A real-time QRS detection algorithm," *IEEE Trans. Biomed. Eng.*, vol. BME-32, no. 3, pp. 230–236, 1985, doi:10.1109/TBME.1985.325532.

[77] A. Phinyomark, S. Thongpanja, H. Hu, P. Phukpattaranont, and C. Limsakul, "The usefulness of mean and median frequencies in electromyography analysis," *Comput. Intell. Electromyogr. Anal.—Perspect. Curr. Appl. Future Chall.*, 2012, doi:10.5772/50639.

[78] G. T. Allison and T. Fujiwara, "The relationship between EMG median frequency and low frequency band amplitude changes at different levels of muscle capacity," *Clin. Biomech.*, vol. 17, no. 6, pp. 464–469, 2002, doi:10.1016/S0268-0033(02)00033-5.

[79] K. Imai, H. Kitakoji, and M. Sakita, "Gastric arrhythmia and nausea of motion sickness induced in healthy Japanese subjects viewing an optokinetic rotating drum," *J. Physiol. Sci. JPS*, vol. 56, no. 5, pp. 341–345, 2006, doi:10.2170/physiolsci.RP005306.

[80] H. Zhang, F. Yang, Z. Q. Liu, F. S. Zhang, Y. K. Peng, and T. D. Yang, "Relationship between EGG and the dynamic process of motion sickness induced by optokinetic vection," *Hang Tian Yi Xue Yu Yi Xue Gong Cheng Space Med. Med. Eng.*, vol. 14, no. 1, pp. 45–49, 2001.

[81] N. Miljković, N. Popovic, O. Djordjevic, L. Konstantinovic, and T. Šekara, "ECG artifact cancellation in surface EMG signals by fractional order calculus application," *Comput. Methods Programs Biomed.*, vol. 140, pp. 259–264, 2017, doi:10.1016/j.cmpb.2016.12.017.

[82] L. Breiman, "Random forests," *Mach. Learn.*, vol. 45, no. 1, pp. 5–32, 2001, doi:10.1023/A:1010933404324.

[83] C. Strobl, J. Malley, and G. Tutz, "An introduction to recursive partitioning: Rationale, application, and characteristics of classification and regression trees, bagging, and random forests," *Psychol. Methods*, vol. 14, no. 4, pp. 323–348, 2009, doi:10.1037/a0016973.

[84] M. Curilem, et al., "Comparison of artificial neural networks an support vector machines for feature selection in electrogastrography signal processing," in *2010 Annual International Conference of the IEEE Engineering in Medicine and Biology*, IEEE, 2010, pp. 2774–2777. Doi:10.1109/IEMBS.2010.5626362.

[85] A. Jovic, "1. Feature selection in biomedical signal classification process and current software implementations," in *Intelligent Decision Support Systems: Applications in Signal Processing*. Berlin, Germany: De Gruyter, 2019, pp. 1–30. Doi:10.1515/9783110621105-001.

[86] V. Stodden, "The data science life cycle: A disciplined approach to advancing data science as a science," *Commun. ACM*, vol. 63, no. 7, pp. 58–66, 2020, doi:10.1145/3360646.

[87] Y. Wang, X. Chen, X. Gao, and S. Gao, "A benchmark dataset for SSVEP-based brain—computer interfaces," *IEEE Trans. Neural Syst. Rehabil. Eng.*, vol. 25, no. 10, pp. 1746–1752, 2017, doi:10.1109/TNSRE.2016.2627556.

[88] H. Rehbaum, N. Jiang, L. Paredes, S. Amsuess, B. Graimann, and D. Farina, "Real time simultaneous and proportional control of multiple degrees of freedom from surface EMG: Preliminary results on subjects with limb deficiency," *IEEE Eng. Med. Biol. Soc. Annu. Int. Conf.*, vol. 2012, pp. 1346–1349, 2012, doi:10.1109/EMBC.2012.6346187.

[89] S. Cucchiara, G. Riezzo, R. Minella, F. Pezzolla, I. Giorgio, and S. Auricchio, "Electrogastrography in non-ulcer dyspepsia," *Arch. Dis. Child.*, vol. 67, no. 5, pp. 613–617, 1992, doi:10.1136/adc.67.5.613.

[90] J. W. Walsh, W. L. Hasler, C. E. Nugent, and C. Owyang, "Progesterone and estrogen are potential mediators of gastric slow-wave dysrhythmias in nausea of pregnancy," *Am. J. Physiol.-Gastrointest. Liver Physiol.*, vol. 270, no. 3, pp. G506–G514, 1996, doi:10.1152/ajpgi.1996.270.3.G506.

[91] K. L. Koch, "Gastric dysrhythmias: A potential objective measure of nausea," *Exp. Brain Res.*, vol. 232, no. 8, pp. 2553–2561, 2014, doi:10.1007/s00221-014-4007-9.

[92] M. A. Jednak, et al., "Protein meals reduce nausea and gastric slow wave dysrhythmic activity in first trimester pregnancy," *Am. J. Physiol.-Gastrointest. Liver Physiol.*, vol. 277, no. 4, pp. G855–G861, 1999, doi:10.1152/ajpgi.1999.277.4.G855.

[93] A. Leahy, K. Besherdas, C. Clayman, I. Mason, and O. Epstein, "Abnormalities of the electrogastrogram in functional gastrointestinal disorders," *Am. J. Gastroenterol.*, vol. 94, no. 4, pp. 1023–1028, 1999, doi:10.1111/j.1572-0241.1999.01007.x.

[94] B. Cheung and P. Vaitkus, "Perspectives of electrogastrography and motion sickness," *Brain Res. Bull.*, vol. 47, no. 5, pp. 421–431, 1998, doi:10.1016/S0361-9230(98)00095-1.

[95] N. Miljković, N. B. Popović, M. Prodanov, and J. Sodnik, "Assessment of sickness in virtual environments," in *Proceedings of the 9th International Conference on Information Society and Technology ICIST*, Kopaonik, Serbia, 2019, vol. 1, pp. 76–81.

[96] S. Classen, S. W. Hwangbo, J. Mason, J. Wersal, J. Rogers, and V. P. Sisiopiku, "Older Drivers' motion and simulator sickness before and after automated vehicle exposure," *Safety*, vol. 7, no. 2, art. No. 2, 2021, doi:10.3390/safety7020026.

[97] L. Motnikar, K. Stojmenova, U. Č. Štaba, T. Klun, K. R. Robida, and J. Sodnik, "Exploring driving characteristics of fit- and unfit-to-drive neurological patients: A driving simulator study," *Traffic Inj. Prev.*, vol. 21, no. 6, pp. 359–364, 2020, doi:10.1080/15389588.2020.1764547.

[98] U. Cizman Staba, T. Klun, K. Stojmenova, G. Jakus, and J. Sodnik, "Consistency of neuropsychological and driving simulator assessment after neurological impairment," *Appl. Neuropsychol. Adult*, pp. 1–10, 2020, doi:10.1080/23279095.2020.1815747.

[99] M. S. Dennison, A. Z. Wisti, and M. D'Zmura, "Use of physiological signals to predict cybersickness," *Displays*, vol. 44, pp. 42–52, 2016, doi:10.1016/j.displa.2016.07.002.

[100] S. Pietraszek and D. Komorowski, "The simultaneous recording and analysis both EGG and HRV signals," in *2009 Annual International Conference of the IEEE Engineering in Medicine and Biology Society*, pp. 396–399, 2009, doi:10.1109/IEMBS.2009.5333455.

[101] N. B. Popovic, N. Miljkovic, and T. B. Sekara, "Electrogastrogram and electrocardiogram interference: Application of fractional order calculus and Savitzky-Golay filter for biosignals segregation," in *2020 19th International Symposium INFOTEH-JAHORINA (INFOTEH)*, IEEE, 2020. doi:10.1109/infoteh48170.2020.9066278.

[102] M. A. M. T. Verhagen, L. J. Van Schelven, M. Samsom, and A. J. P. M. Smout, "Pitfalls in the analysis of electrogastrographic recordings," *Gastroenterology*, vol. 117, no. 2, pp. 453–460, 1999, doi:10.1053/gast.1999.0029900453.

[103] A. Koohestani, et al., "A knowledge discovery in motion sickness: A comprehensive literature review," *IEEE Acc.*, vol. 7, pp. 85755–85770, 2019, doi:10.1109/ACCESS.2019.2922993.

[104] B. K. Wiederhold, "Connecting through technology during the coronavirus disease 2019 pandemic: Avoiding 'zoom fatigue'," *Cyberpsychol. Behav. Soc. Netw.*, vol. 23, no. 7, pp. 437–438, 2020, doi:10.1089/cyber.2020.29188.bkw.

[105] P. Mouzourakis, "Videoconferencing: Techniques and challenges," *Interpreting*, vol. 1, no. 1, pp. 21–38, 1996, doi:10.1075/intp.1.1.03mou.

[106] J. N. Bailenson, "Nonverbal overload: A theoretical argument for the causes of zoom fatigue," *Technol. Mind Behav.*, vol. 2, no. 1, 2021, doi:10.1037/tmb0000030.

[107] *Proceedings of the First Biennial Space Biomedical Investigators' Workshop*, 1999. Accessed: Oct. 09, 2021 [Online]. Available: https://ntrs.nasa.gov/citations/20000020485

[108] D. L. Harm, G. R. Sandoz, and R. M. Stern, "Changes in gastric myoelectric activity during space flight," *Dig. Dis. Sci.*, vol. 47, no. 8, pp. 1737–1745, 2002, doi:10.1023/a:1016480109272.

[109] Md. M. S. Raihan, A. B. Shams, and R. B. Preo, "Multi-class electrogastrogram (EGG) signal classification using machine learning algorithms," in *2020 23rd International Conference on Computer and Information Technology (ICCIT)*, IEEE, 2020, pp. 1–6. doi:10.1109/ICCIT51783.2020.9392695.

4 Impact of Cognitive Demand on the Voice Responses of Parkinson's Disease and Healthy Cohorts

Rekha Viswanathan and Sridhar P. Arjunan

4.1 INTRODUCTION

Parkinson's disease (PD) is a neurodegenerative condition which revels through speech impairments even before the clinical diagnosis of the disease [1]. Perceptual evaluation of parkinsonian speech reveals a variety of disturbances—hypophonia (reduced loudness), dysphonia (poor, often breathy voice quality), hypokinetic articulation (imprecise consonant production from reduced articulatory movements), hypoprosodia (impaired pitch variability and range), rush (the speech equivalent of gait festination) and dysfluency. This 'hypokinetic dysarthria' is in some respects a counterpart of limb or gait hypokinesia. There is evidence to suggest that abnormal sensory processing in PD affects loudness and pitch variability [2].

Perceptual evaluation is the standard clinical method for classifying parkinsonian speech, although its subjectivity is a weakness [3]. Automated analysis of speech can be used as an alternative or adjunct to perceptual analysis for more accurate, objective and quantifiable speech evaluation. These derived measurements have complex relationships with subjective perceptual assessments of voice. Analysis of speech signals from connected (e.g., conversational) speech and sustained vowel samples yields acoustic information that correlates with the perceptual features—voice quality (e.g., rough, breathy, strained), loudness, pitch and resonance. On isolated vowels, these voice characteristics are assessed acoustically by the perturbation measures—jitter (frequency variability) and shimmer (amplitude variability)—harmonics to noise ratio (HNR), fundamental frequency (f_0), vocal intensity and formant frequency profiles. Speech production features extracted from the glottal waveform remove the effect of articulation on the acoustic signal. They approximate the volume velocity of the air flowing through the vocal folds and may have an advantage for the analysis of the pathological voice [4],[5]. The cognitive load produced due to task complexity and distractions may affect a person's speech [6–8]. The complexity of the task or the

DOI: 10.1201/9781003201137-5

distractions determines the level of cognitive-linguistic loading that may influence voice and speech [9].

When it comes to the individuals with PD and provided the evidence of early speech-motor symptoms, it is important to understand on how cognitive loading affects the speech-motor symptoms and in most cases it is ambiguous. We summarise the findings from similar studies carried out on PD individuals to strengthen our hypothesis for the current work. The study carried out by Ho et al. showed that the visuo-manual tracking task affected volumetric and temporal measures of speech motor control [10]. Galaz et al. examined differences in prosodic and articulatory measures in emotionally neutral, stress-modified and rhymed reading tasks in PD [11].

In this chapter we discuss the methods which determine the effect of two levels of cognitive loading imposed through the Stroop colour word test on the motor-speech aspects of PD in a practically defined off-state and on-state of Levodopa medication. We also report how different is the motor-speech aspects of controls compared to PD in the respective medication states.

4.2 METHODS

In order to understand the effect of two levels of cognitive loading imposed through the Stroop colour word test on the motor-speech aspects of PD, this work has conducted experimental study of the PD and control participants.

Twenty-six patients (12 females and 14 males) diagnosed with PD were recruited from the Movement Disorders Clinic at Monash Medical Centre. All complied with the Queen Square Brain Bank criteria for idiopathic PD [12]. Twenty healthy similarly aged controls (seven females and 13 males) were also recruited. Presence of any advanced clinical milestones—visual hallucinations, colour blindness, frequent falling, voice therapy, cognitive disability, need for institutional care—was an exclusion criterion for PD recruitment [13].

Presence of any neurological disorders, surgeries in the past three months, voice training, stammering, stuttering, colour blindness, cognitive disability were part of the exclusion criterion for the control cohort. The study was conducted in accordance with the Helsinki Declaration on human experiments (revised 2004) and approved by the Monash Health, Victoria, Australia (LNR/16/MonH/319) and ratified by RMIT University, Victoria, Australia (BSEHAPP22–15KUMAR). All participants in this study gave their written informed consent prior to data recording. During the experiments, the patients were under the observation of nurse and supervised by the consultant neurologist. The dataset used in this study is available at RMIT website [14]. Table 4.1 shows the participants' information.

4.2.1 STUDY PROTOCOL

The Stroop colour word test, developed in 1935, assesses cognitive flexibility and processing anterior to verbal output [15]. It calls for responses at two levels of cognitive loading. In the congruent colour word test (minimal loading), participants

TABLE 4.1
Participant Demographics.

	PD mean (SD)		Control mean (SD)
Age	72 ± 7.47		66.3 ± 6.19
(Years)	(range:57–84)		(range:56–84)
MoCA	27.27 ± 2.60		28.3 ± 1.34
	(range:20–30)		(range:26–30)
PD Duration	4.92 ± 3.14		–
(Years)	(range:1–10)		
Levodopa Dosage/day	490.38 ± 313.21		–
	(range:100–1200)		
Motor UPDRS	**off-state**		2.8 ± 3.78
	27.31 ± 2.59		(range:0–12)
	(range:20–30)		
	on-state		
	20.42 ± 10.08		
	(range:4–43)		

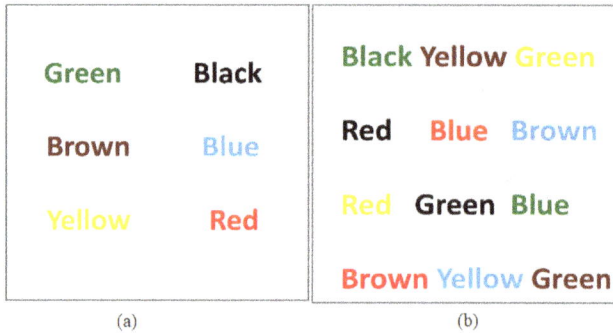

FIGURE 4.1 (a): Stroop word colour congruent test (six colours); (b): Stroop word colour incongruent test (12 colours).

read aloud a set of colour words each printed in the ink colour corresponding to the word (e.g., the word 'red' is printed in red). They then performed the more demanding incongruent (enhanced loading) colour word test. The enhanced cognitive load was the requirement to say the ink colour of words printed in a different colour to that represented by each word (e.g., the word 'red' is printed in yellow and the participant is required to say yellow). The printed colour words are shown in Figure 4.1.

In our study, we used six random colours: green, brown, yellow, black, blue and red for implementing the two levels of cognitive loading. For the congruent test (minimal loading), the participants were asked to read the words as shown in

Figure 4.1.a. continuously. For the incongruent task (enhanced loading), the participants were asked to read the ink colour of the words continuously as: green, brown, yellow, black, red, blue, yellow, black, green, red, blue and brown as shown in Figure 4.1.b. The colour words for both the congruent and incongruent tests were printed in a high quality A4 size sheet in the landscape orientation and were placed on a standard writing/reading table in front of the study participants to read out while recording the tasks.

Participants were instructed to perform the congruent and incongruent Stroop colour word test at their own pace using their most comfortable voice (pitch). The congruent test was carried out first, with a 2-minute rest period before the incongruent task. All participants completed both the tasks in a single trial. For PD subjects, the congruent and incongruent Stroop colour word test was administered in a practically defined off-state (fasting, with anti-parkinsonian medication withheld for at least 12 hours). Each subject's usual morning Levodopa was then given, the on-state taken to be the maximum improvement over the subsequent 30–90 minutes. At this point, congruent and incongruent Stroop testing was repeated. Motor function in off- and on-states was scored by a neurologist on the Unified Parkinson's Disease Rating Scale Part III motor examination (UPDRS-III). Cognitive assessment was performed for all participants using the Montreal Cognitive Assessment (MoCA) before the morning Levodopa.

4.2.2 EXPERIMENTAL SETUP AND FEATURE EXTRACTION

The Stroop colour word test was recorded using an omnidirectional head-worn microphone placed approximately 4 cm from the participant's mouth, which was plugged to an Apple smartphone (iPhone 6s plus) with a sampling rate of 48 kHz and 16-bit resolution in WAV format. The recordings were carried out in a realistically noise restricted room without adjustment for ambient noise.

The instructions provided by the researcher at the beginning of the recordings were trimmed off before pre-processing and the duration of the recordings for the congruent and incongruent Stroop word test was calculated and used as one of the study features. All the recordings were then pre-processed to remove high-frequency noise and low-frequency artefacts using a 2nd order bandpass filter of 80 Hz—24kHz. Four sets of features based on i) fundamental frequency (f_0) contour, ii) low level features and iii) glottal and spectro-temporal features and iv) duration of the recording were evaluated to study the changes in voice due to different levels of cognitive loading. These features are reported in Table 4.2.

4.2.2.1 Features Based on f_0 Contour

The Yaapt (Yet another algorithm for pitch tracking) [16] package was used to extract the f_0 contour from the recordings. The algorithm uses a 35 ms analysis window for the estimation of f_0 contour. The f_0 contour was estimated by setting minimum f_0 to 60 Hz and maximum to 400 Hz which are also the default values of the algorithm. The mean and standard deviation (SD) from the f_0 contour were estimated and the ratio of SD and mean was used as a third measure to quantify the variability in f_0.

TABLE 4.2
Features Used in the Study.

LIST OF FEATURES COMPUTED

Mean fundamental frequency (f_0)

SD fundamental frequency (SD f_0)

Fundamental frequency variability (f_0 Var)

Mean Root mean square energy (rmsE)

SD Root mean square energy (SD rmsE)

Root mean square energy variability (rmsE Var)

Duration

Cepstral peak prominence (CPP)

Corrected difference between amplitudes of first and second harmonics (H1*-H2*)

Mean Harmonics to noise ratio (HNR)

Percent ratio of voiced to unvoiced frames (PR)

Normalised amplitude quotient (NAQ)

Amplitude quotient (AQ)

Closing quotient (ClQ)

Primary and secondary open quotients (OQ_1, OQ_2)

Amplitude based open quotient (OQ_a)

Quasi open quotient (QOQ)

Primary and secondary speed quotients (SQ_1, SQ_2)

4.2.2.2   Low Level Features from the Recording

The root mean square energy (rmsE) was calculated by framing each audio recording in to a non-overlapping segment of 35 ms. A custom-made MATLAB routine was employed to perform the framing and framing related feature estimation. For each of the frame rmsE was calculated using the equation 4.1.

$$rmsE = \sqrt{\frac{\sum x_i^2}{n}} \qquad (4.1)$$

Where x_i is the amplitude of the ith sample and n is the number of samples in the frame. The mean rmsE for the entire audio recording was calculated along with the features SD and variability in rmsE. The Yaapt algorithm uses spectrogram-based thresholding to determine voiced and un-voiced frame decision in the given audio recording. The decisioning of voiced and un-voiced frames further aids in the f_0 contour estimation. In this study we have tapped this voiced and un-voiced framing decision routine to evaluate a feature called the percent ratio of voiced to un-voiced frames (PR) in the audio recording.

4.2.2.3   The Glottal and Spectro-Temporal Features

Cepstral peak prominence (CPP), Corrected difference between amplitudes of first and second harmonics (H1*-H2*), Harmonics to noise ratio (HNR) were calculated using the default settings of the VoiceSauce open source software package [17]. The glottal features were extracted from the recordings using the iterative adaptive inverse filtering method (IAIF) obtained from TKK Voice Source Analysis and Parameterization (APARAT) toolbox [18]. The list of features used in the study are listed in Table 4.2.

4.3   STATISTICAL ANALYSIS

Data analysis was performed using the IBM SPSS statistical software version 28.0.1.0 (142). The features were evaluated from the recordings of controls, PD off-state and PD on-state for two levels of cognitive loading tasks viz; mild cognitive loading and enhanced cognitive loading. The Shapiro-Wilk test was applied on the data to check for normality. Since the data followed a non-normal distribution, the comparison between controls PD off-state and PD on-state were carried out using the non-parametric Mann-Whitney U test and the correlation between the features and factors like UPDRS, MoCA, age, Levodopa dosage were evaluated using the Spearman correlation test. 95% confidence interval (CI) plots were further employed to descriptively illustrate the group difference for the features. The evaluation of the effect of change in cognitive loading was performed for each study group by comparing the features of minimal cognitive loading task with that of the enhanced cognitive loading task using Wilcoxon rank test. We employed the same test to evaluate the effect of Levodopa medication in PD group by comparing the features of PD on-state and PD off-state for the two levels of cognitive loading tasks.

4.3.1   CORRELATION STUDY

4.3.1.1   Minimal Cognitive Loading Speech Task

The correlation between study factors and the speech features estimated from the recordings of controls, PD off-state and PD on-state and they were evaluated to identify the association in each group for minimal loading and enhanced loading tasks respectively. The correlation between the features and study factors for minimal cognitive loading task has been presented for three groups (control, PD off-state, on-state) in Table 4.3 followed by the enhanced cognitive loading task in Table 4.4. The correlation Table 4.3 and Table 4.4 also report the Spearman rho, p-value, 95% CI lower and upper limits for the features.

Table 4.3 accounts the correlation between study factors and the features for the minimal cognitive loading tasks for controls, PD off-state and PD on-state. For the controls, duration of the recordings, QOQ and the SD f_0 were showing weak to moderate negative correlation with the age factor whereas MoCA showed weak positive correlation with f_0 and average rmsE.

For the PD off-state, the factors Levodopa dosage, age and PD duration were showing association with the features evaluated. The Levodopa dosage showed positive weak correlation with SD f_0 and f_0 Var features. The factor age also displayed positive weak correlation with rmsE and SD rmsE. The factor PD duration rather exhibited a moderate negative correlation with the feature CPP. The only feature that exhibited a weak negative correlation with both MoCA and motor UPDRS was the AQ.

TABLE 4.3

Correlation Analysis Using Spearman's Correlation Test for the Minimal Cognitive Loading Task.

Feature-Factor	Spearman rho	p-value	Lower limit	Upper limit
Control				
Duration—Age	−0.603	0.005	−0.830	−0.205
SD f_0—Age	−0.556	0.011	−0.807	−0.137
QOQ—Age	−0.413	0.070	−0.730	0.050
f_0_MoCA	0.419	0.066	−0.043	0.733
rmsE—MoCA	0.426	0.061	−.034	0.737
rmsE—motor UPDRS	−0.460	0.042	−0.756	−0.007
PR—motor UPDRS	−0.564	0.010	−0.810	−0.148
HNR—motor UPDRS	0.568	0.009	0.154	0.812
QOQ—motor UPDRS	−0.467	0.038	−0.760	−0.016
PD off-state				
SD f_0—Ldopa Dose	0.368	0.064	−0.035	0.668
f_0 Var—Ldopa Dose	0.525	0.006	0.161	0.763
rmsE—Age	0.414	0.036	0.019	0.697
SD rmsE—Age	0.444	0.023	0.057	0.715
CPP—PD duration	−0.542	0.004	−0.773	−0.184
AQ—MoCA	−0.347	0.082	−0.655	0.058
AQ—motor UPDRS	−0.347	0.083	−0.654	0.059
PD on-state				
SD f_0—Ldopa Dose	0.420	0.037	0.017	0.705
f_0 Var—Ldopa Dose	0.441	0.027	0.044	0.718
rmsE Var—motor UPDRS	0.574	0.003	0.220	0.795
PR—MoCA	0.445	0.026	0.049	0.721
AQ—MoCA	−0.363	0.074	−0.670	0.049
PR—motor UPDRS	−0.397	0.049	−0.691	0.010
H1*-H2*—motor UPDRS	−0.338	0.099	−0.654	0.079

The PD on-state features also showed weak to moderate correlation with the study factors. The Levodopa dosage illustrated weak positive correlation with SD f_0 and f_0 Var whereas motor UPDRS had moderate positive correlation with rmsE Var and weak negative correlation with PR and H1*-H2*. The two features that displayed weak correlation with MoCA were the PR (positive correlation) and AQ (negative correlation).

4.3.1.2 Enhanced Cognitive Loading Speech Task

The correlation for the enhanced cognitive loading task is shown in Table 4.4. The control features displayed weak correlations with the factors age (SD f_0 and f_0), MoCA (f_0 Var) and motor UPDRS (H1*-H2*).

TABLE 4.4
Correlation Analysis Using Spearman's Correlation Test for the Enhanced Cognitive Loading Task.

Feature-Factor	Spearman rho	p-value	Lower limit	Upper limit
Control				
SD f_0—Age	−0.412	0.071	−0.729	0.051
f_0—Age	−0.334	0.150	−0.684	0.141
f_0 Var—MoCA	−0.378	0.101	−0.710	0.092
H1*-H2*—motor UPDRS	0.366	0.113	−0.106	0.703
PD off-state				
rmsE—Age	0.358	0.072	−0.046	0.662
SD rmsE—Age	0.457	0.019	0.072	0.723
rmsE Var—PD duration	0.335	0.094	−0.072	0.647
PR—PD duration	−0.345	0.085	−0.653	0.061
ClQ—PD duration	−0.412	0.037	−0.696	−0.017
OQ_1—PD duration	−0.398	0.044	−0.687	0.000
OQ_2—PD duration	−0.416	0.034	−0.698	−00.023
SQ_1—PD duration	0.352	0.078	−0.054	0.657
rmsE Var—Ldopa Dose	0.359	0.072	−0.045	0.662
NAQ—Ldopa Dose	−0.363	0.069	−0.664	0.041
OQ_a—Ldopa Dose	−0.424	0.031	−0.703	−0.032
PD on-state				
Duration—Age	0.384	0.058	−0.026	0.683
PR—Age	−0.368	0.070	−0.673	0.044
OQ_1—Age	−0.544	0.005	−0.778	−0.177
QOQ—Age	−0.507	0.010	−0.757	−0.128
SD f_0—Ldopa Dose	0.377	0.063	−0.034	0.679
f_0 Var—Ldopa Dose	0.374	0.066	−0.038	0.677
SD f_0—motor UPDRS	0.362	0.076	−0.051	0.669
OQ_1—motor UPDRS	−0.432	0.031	−0.713	−0.032
QOQ—motor UPDRS	−0.452	0.023	−0.725	−0.057
rmsE—PD duration	−0.355	0.082	−0.665	0.059
f_0 Var—motor UPDRS	0.341	0.095	−0.075	0.656
rmsE Var—motor UPDRS	0.494	0.012	0.111	0.750
PR—motor UPDRS	−0.487	0.014	−0.745	−0.101
CPP-motor UPDRS	0.348	0.088	−0.067	0.660
H1*-H2*-motor UPDRS	−0.404	0.045	−0.696	0.002

a) PD off-state:

In case of the PD off-state, rmsE and SD rmsE displayed a weak but positive correlation with age, rmsE Var feature had a weak positive correlation with PD duration and Levodopa dosage. The glottal features NAQ and OQ_a also shared a weak negative correlation with the Levodopa dosage. The feature PR also displayed a weak negative correlation with the factor PD duration. Glottal features were also correlated with PD off-state factors—ClQ, OQ_1 and OQ_2 showed a weak negative correlation with PD duration whereas SQ_1 had a weak positive correlation with PD duration.

b) PD on-state

As observed from Table 4.3 the PD on-state features displayed weak to moderate correlation with the study factors. The duration of the recording shared a weak positive correlation with age whereas PR shared a weak negative correlation. The features OQ_1 and QOQ shared a moderate negative correlation with the age on the other hand. SD f_0 and f_0 Var displayed a weak positive correlation with the Levodopa dosage. The features SD f_0, f_0 Var, rmsE Var, CPP shared a weak positive correlation with motor UPDRS however, OQ_1, QOQ, rmsE, PR and $H1^*$-$H2^*$ exhibited negative weak correlation with the motor UPDRS. It is observed from Table 4.3 and Table 4.4 that the features extracted from the recordings were correlated to the factors like age, MoCA, Levodopa dosage, PD duration and motor UPDRS scores.

4.3.2 EFFECT OF COGNITIVE LOADING WITHIN COHORTS

4.3.2.1 Comparison of Study Features between Minimal Cognitive Loading and Enhanced Cognitive Loading Speech Tasks for Controls

We examined the effect of increase in the cognitive loading in each of the study groups. To achieve this, the features were compared between the minimal and enhanced cognitive loading speech task using the non-parametric Wilcoxon rank test. The Table 4.5 summarises the result of comparison between features of minimal and enhanced cognitive loading tasks for controls. The features rmsE and SD rmsE were significantly different between the minimal and enhanced cognitive speech tasks for the controls. The feature CPP though not statistically significant showed a remarkable difference between the two tasks. No glottal features showed any difference between two levels of cognitive loading speech tasks for controls. Figure 4.2 shows a few features where significant difference is seen between the two levels of loading.

4.3.2.2 Comparison of Study Features between Minimal Cognitive Loading and Enhanced Cognitive Loading Speech Tasks for PD Off-State

The only feature that showed a difference between minimal and enhanced cognitive loading speech task for PD off-state was CPP. Other features with remarkable difference between two levels of cognitive loading are SD f_0, SD rmsE, ClQ, SQ_1 and HNR. The results are provided in Table 4.6. Figure 4.3 shows the difference in features SQ_1 and ClQ between minimal and enhanced cognitive loading speech tasks.

TABLE 4.5
Comparison of Features between Minimal and Enhanced Cognitive Loading Speech Tasks for Controls.

Feature	Z score	p-value
SD f_0	−0.037	0.97
f_0	−1.045	0.296
f_0 Var	−0.112	0.911
rmsE	−1.941	0.052
SD rmsE	−2.203	0.028
rmsE Var	−0.037	0.97
PR	−0.672	0.502
CPP	−1.755	0.079
H1*-H2*	−1.344	0.179
HNR	−1.344	0.179
NAQ	−0.149	0.881
AQ	−0.149	0.881
ClQ	−0.784	0.433
OQ_1	−0.709	0.478
OQ_2	−0.709	0.478
OQ_a	−0.112	0.911
QOQ	−0.709	0.478
SQ_1	−0.075	0.940
SQ_2	−1.008	0.313

TABLE 4.6
Comparison of Features between Minimal and Enhanced Cognitive Loading Speech Tasks for PD Off-State.

Feature	Z score	p-value
SD f_0	−1.562	0.118
f_0	−0.089	0.929
f_0 Var	−1.232	0.218
rmsE	−1.181	0.238
SD rmsE	−1.435	0.151
rmsE Var	−0.013	0.99
PR	−0.267	0.79
CPP	−2.222	0.026
H1*-H2*	−1.029	0.304
HNR	−1.638	0.101
NAQ	−0.648	0.517
AQ	−0.114	0.909

(Continued)

TABLE 4.6 (*Continued*)

Comparison of Features between Minimal and Enhanced Cognitive Loading Speech Tasks for PD Off-State.

Feature	Z score	p-value
ClQ	−1.486	0.137
OQ_1	−0.648	0.517
OQ_2	−1.003	0.316
OQ_a	−0.749	0.454
QOQ	−0.546	0.585
SQ_1	−1.892	0.058
SQ_2	−0.978	0.328

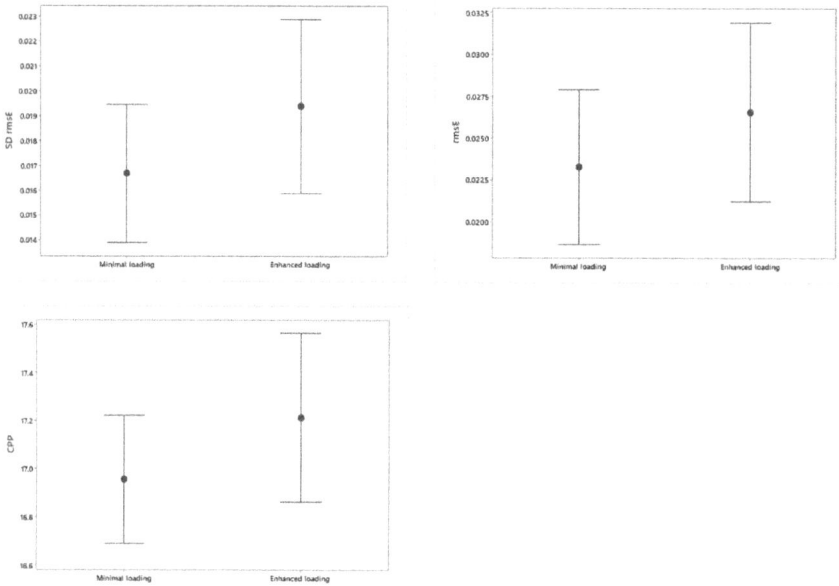

FIGURE 4.2 Comparison of features SD rmsE, CPP, rmsE between minimal and enhanced cognitive loading speech tasks for controls.

4.3.2.3 Comparison of Study Features between Minimal Cognitive Loading and Enhanced Cognitive Loading Speech Tasks for PD On-State

For the PD on-state, no features were significantly different between the two cognitive loading tasks and the results are shown in Table 4.7. The features which showed a remarkable difference between the two levels of cognitive loading tasks were SQ_1 with a p-value of 0.058 and ClQ with p-value 0.137. Figure 4.4. shows the comparison for features ClQ and SQ_1 between the minimal and enhanced loading tasks.

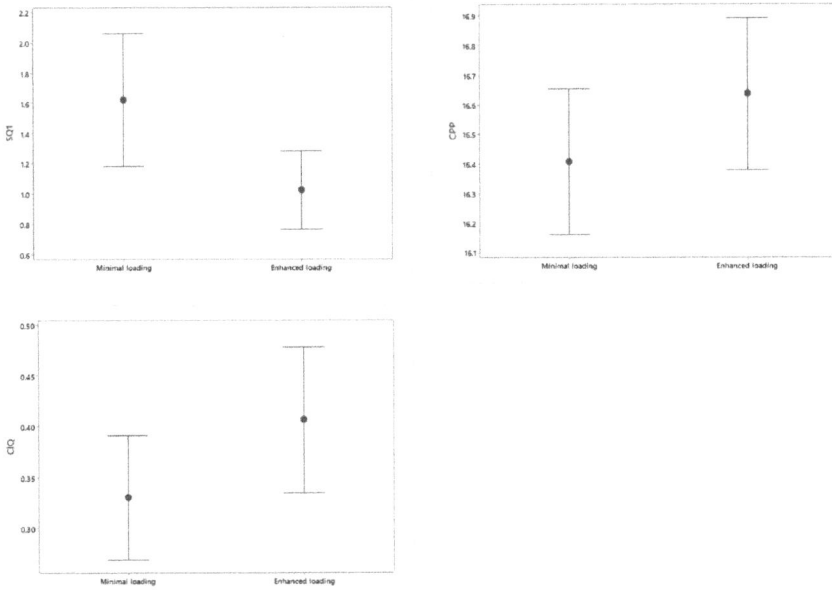

FIGURE 4.3 Comparison of features SQ_1, CPP and ClQ between minimal and enhanced cognitive loading speech tasks for PD off-state.

TABLE 4.7
Comparison of Features between Minimal and Enhanced Cognitive Loading Speech Tasks for PD On-State.

Feature	Z score	p-value
SD f_0	−0.229	0.819
f_0	−0.040	0.968
f_0 Var	−0.148	0.882
rmsE	−0.013	0.989
SD rmsE	−0.686	0.493
rmsE Var	−0.309	0.757
PR	−0.175	0.861
CPP	−.363	0.716
$H1^*-H2^*$	−0.336	0.737
HNR	−0.148	0.882
NAQ	−0.648	0.517
AQ	−0.114	0.909
ClQ	−1.486	0.137
OQ_1	−0.648	0.517
OQ_2	−1.003	0.316
OQ_a	−.749	0.454
QOQ	−0.546	0.585
SQ_1	−1.892	0.058
SQ_2	−0.978	0.328

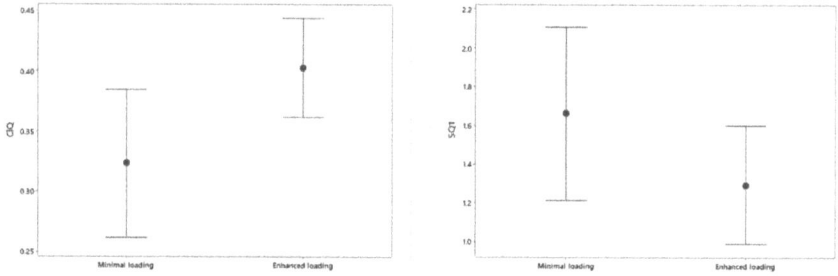

FIGURE 4.4   Comparison of features ClQ and SQ_1 between minimal and enhanced cognitive loading speech tasks for PD on-state.

4.3.3   EFFECT OF COGNITION IN SPEECH FEATURES BETWEEN CONTROL AND PD

4.3.3.1   Comparison between Control and PD Off-State for Minimal and Enhanced Cognitive Loading Tasks Using Mann-Whitney U Test

The features were compared between PD off-state and controls for minimal and enhanced cognitive loading tasks and the result is summarised in Table 4.8. For the minimal cognitive loading task (i.e., congruent word test) the features CPP and rmsE Var were significantly different between the two groups. Two other features rmsE and f_0 Var were also remarkably different between the two groups though not statistically significant (p = 0.054 and 0.057). For the enhanced cognitive loading task (i.e., in-congruent word test), CPP, duration of the recordings and rmsE were significantly different between the two groups. In addition to this though statistically not significant, the feature rmsE Var also displayed notable difference between controls and PD off-state. For both minimal and enhanced cognitive loading speech tasks, the glottal features did not exhibit any difference between controls and PD off-state. These results are summarised in Table 4.8. 95% CI plots are shown in Figure 4.5 and Figure 4.6 to illustrate the difference between the features of controls and PD off-state for minimal and enhanced cognitive load tasks.

4.3.3.2   Comparison between Control and PD On-State for Minimal and Enhanced Cognitive Loading Tasks Using Mann-Whitney U Test

The comparison of features between controls and PD on-state could indicate the effect of Levodopa medication on the vocal mechanism of the PD study group. Table 4.9 provides the summary of comparison of features between controls and PD on-state for minimal and enhanced cognitive loading tasks. In the minimal cognitive loading speech task, the features CPP, rmsE Var and PR were significantly different between the two groups. The H1*-H2* feature though statistically non-significant (p-value = 0.061), showed substantial difference between controls and PD on-state.

TABLE 4.8

Comparison of Features between Controls and PD Off-State for Minimal and Enhanced Cognitive Loading Tasks.

Features	p-value minimal cognitive loading task	p-value enhanced cognitive loading task
CPP	0.003	0.012
H1*-H2*	0.240	0.912
HNR	0.859	0.842
Duration	0.565	0.051
SD f_0	0.092	0.240
f_0	0.947	0.807
f_0 Var	0.054	0.223
rmsE	0.057	0.016
SD rmsE	0.240	0.084
rmsE Var	0.041	0.057
PR	0.063	0.249
NAQ	0.674	0.929
AQ	0.740	0.859
ClQ	0.191	0.773
OQ_1	0.438	0.965
OQ_2	0.308	0.947
OQ_a	0.690	0.610
QOQ	0.947	0.894
SQ_1	0.268	0.298
SQ_2	0.375	0.191

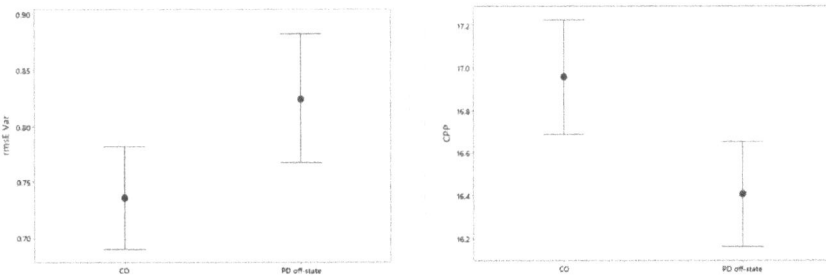

FIGURE 4.5 95% CI plot illustrating the difference between controls and PD off-state patients for the features rmsE Var and CPP in the minimal cognitive loading speech task.

95% CI plots for the features PR, CPP, H1*-H2* and ClQ between controls and PD on-state for the minimal cognitive loading task are shown in Figure 4.7. When comparing the features of controls and PD on-state for the enhanced cognitive loading speech task, we noticed significant differences in CPP, rmsE and rmsE Var of the

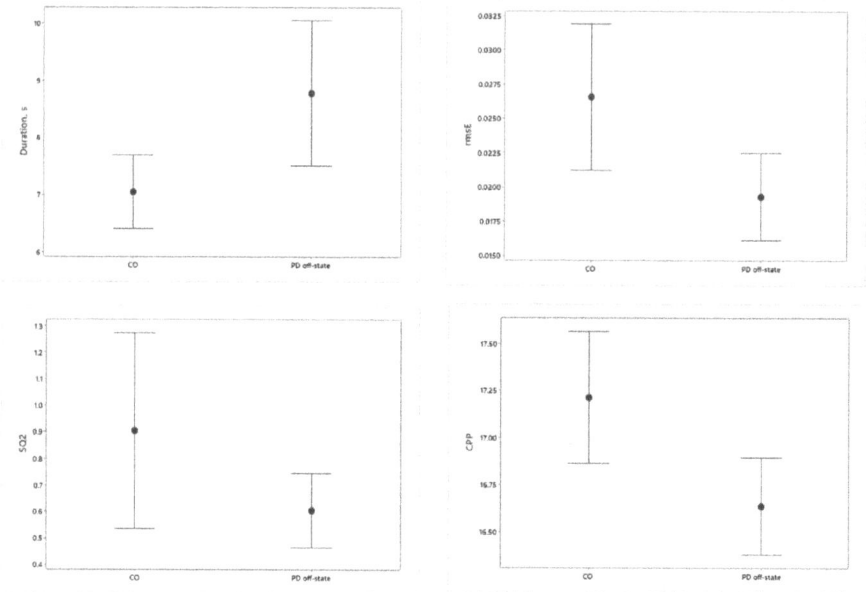

FIGURE 4.6 95% CI plot illustrating the difference between controls and PD off-state patients for the features duration, rmsE, SQ$_2$ and CPP in the enhanced cognitive loading speech task.

TABLE 4.9

Comparison of Features between Controls and PD On-State for Minimal and Enhanced Cognitive Loading Tasks.

Features	p-value minimal cognitive loading task	p-value enhanced cognitive loading task
CPP	0.016	0.001
H1*-H2*	0.061	0.178
HNR	0.681	0.150
Duration	0.698	0.071
SD f_0	0.217	0.361
f_0	0.424	0.522
f_0 Var	0.100	0.217
rmsE	0.150	0.022
SD rmsE	0.437	0.185
rmsE Var	0.047	0.047
PR	0.040	0.217
NAQ	0.648	0.537
AQ	0.784	0.784
CIQ	0.138	0.715

TABLE 4.9 (*Continued*)

Comparison of Features between Controls and PD On-State for Minimal and Enhanced Cognitive Loading Tasks.

Features	p-value minimal cognitive loading task	p-value enhanced cognitive loading task
OQ_1	0.411	0.283
OQ_2	0.263	0.150
OQ_a	0.648	0.599
QOQ	0.855	0.034
SQ_1	0.201	0.784
SQ_2	0.293	0.263

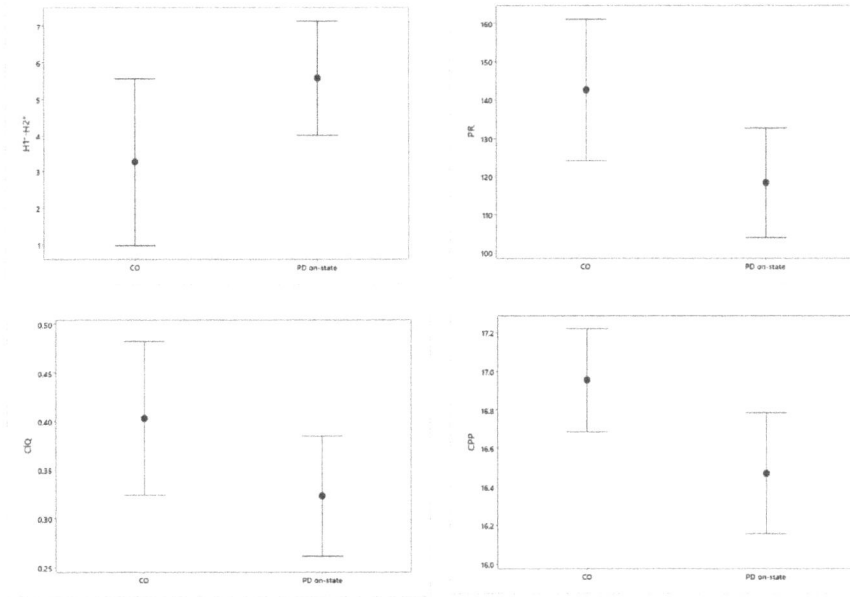

FIGURE 4.7 95% CI plot illustrating the difference between controls and PD on-state patients for the features H1*-H2*, PR, ClQ and CPP in the minimal cognitive loading speech task.

two groups. Again, the duration was showing remarkable group difference (p-value = 0.071). No glottal features differentiated the two groups in the minimal cognitive loading task whereas the feature QOQ statistically differentiated controls and PD on-state for the enhanced cognitive loading speech task. Figure 4.8 shows the 95% CI plots for the features CPP, rmsE, duration and QOQ indicating the difference between controls and PD on-state in the enhanced loading task.

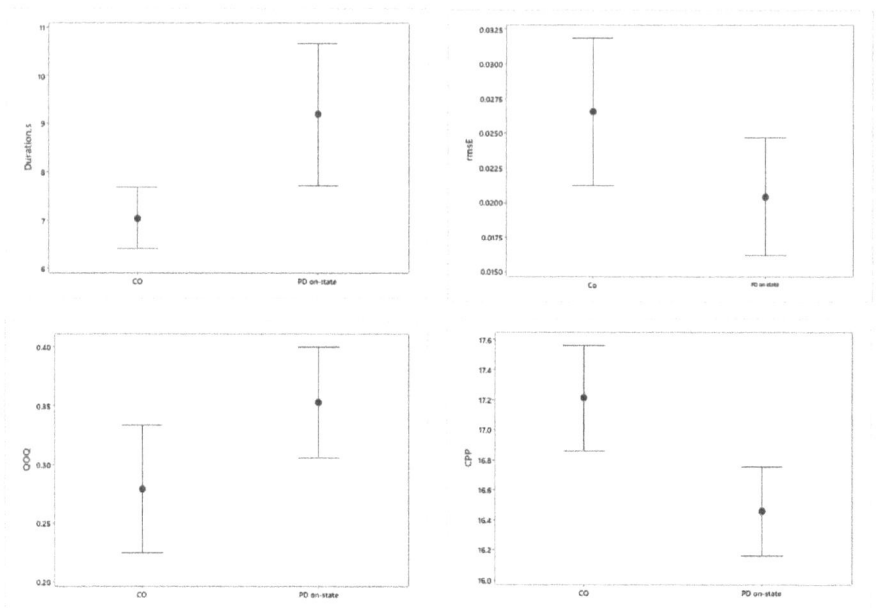

FIGURE 4.8 95% CI plot illustrating the difference between controls and PD on-state patients for the features duration, rmsE, QOQ and CPP in the enhanced cognitive loading speech task.

4.3.3.3 Comparison between PD Off-State and PD On-State for Minimal and Enhanced Cognitive Loading Tasks Using Wilcoxon Rank Test

To evaluate the effect of Levodopa medication, we performed Wilcoxon rank test between the PD off-state and PD on-state features. The results of comparison between PD off-state and on-state for minimal and enhanced cognitive loading speech tasks are summarised in Table 4.10. It was observed that there was no significant difference in the features at the minimal cognitive loading speech task. Two glottal features QOQ and SQ_2 showed statistically significant difference between PD off-state and on-state for the enhanced cognitive loading speech task. A few other features showed noticeable difference between PD off-state and on-state for the enhanced cognitive loading task like the H1*-H2*, HNR, SQ_1 and CPP. The 95% CI plots for the features SQ_2, H1*-H2* and QOQ are illustrated in Figure 4.9.

4.4 DISCUSSION

We have investigated and compared the effect of varying cognitive load in the speech of PD in the practically defined off-state and on-state of Levodopa medication and similarly aged control cohort. As a part of the investigation, we have also evaluated the effect of changing the cognitive load from a minimal level to an enhanced level in the speech of the study groups independently.

TABLE 4.10

Comparison of Features between PD Off-State and PD On-State for Minimal and Enhanced Cognitive Loading Speech Tasks Using Wilcoxon Rank Test.

Features	Z value	p-value minimal cognitive loading task	Z Value	p-value enhanced cognitive loading task
Duration	−0.821	0.412	−0.040	0.968
SD f_0	−0.982	0.326	−0.848	0.397
f_0	−1.144	0.253	−1.117	0.264
f_0 Var	−0.605	0.545	−0.740	0.459
rmsE	−1.090	0.276	−0.175	0.861
SD rmsE	−0.955	0.339	−0.632	0.527
rmsE Var	−0.013	0.989	−0.363	0.716
PR	−0.148	0.882	−0.336	0.737
CPP	−0.067	0.946	−1.332	0.183
H1*-H2*	−0.040	0.968	−1.816	0.069
HNR	−0.390	0.696	−1.371	0.17
NAQ	0.000	1	−0.417	0.677
AQ	−0.161	0.872	−0.390	0.696
ClQ	−0.201	0.841	−0.040	0.968
OQ_1	−0.282	0.778	−0.848	0.397
OQ_2	−0.443	0.658	−1.117	0.264
OQ_a	0.000	1	−0.740	0.459
QOQ	−0.040	0.968	−1.978	0.048
SQ_1	−0.563	0.573	−1.440	0.15
SQ_2	−0.201	0.841	−2.570	0.01

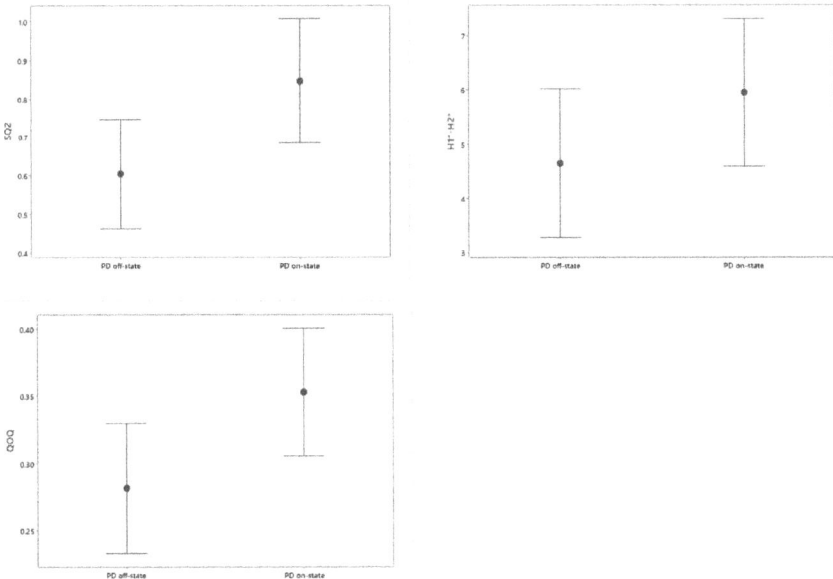

FIGURE 4.9 95% CI plot illustrating the difference between DP off-state and PD on-state patients for the features SQ_2, H1*-H2* and QOQ in the enhanced cognitive loading speech task.

TABLE 4.11
Clinical Significance of the Study Features.

Study features	Clinical significance
Mean fundamental frequency (f_0)	Perceptually associated to the vocal pitch. Difference in pitch exist between PD and controls [19, 20]
SD fundamental frequency (SD f_0)	A measure to quantify the f_0 range. Controls show higher values of SD f_0 compared to PD [19]
Fundamental frequency variability (f_0 Var)	Quantifies the variability in f_0. The feature is often associated to the monotonicity [21]
Mean Root mean square energy (rmsE)	The feature can be associated to overall energy (loudness) in the speech
Cepstral peak prominence (CPP)	A voice quality measure having strong correlation with dysphonia, breathiness and vocal weakness [22]. A lower value of this feature associated to breathiness and low periodicity [8]
Corrected difference between amplitudes of first and second harmonics (H1*-H2*)	The feature is associated to breathiness and larger the feature breathier the voice is [23]
Mean Harmonics to noise ratio (HNR)	A measure of noise present. A low HNR indicate a breathy voice [24]
Percent ratio of voiced to unvoiced frames (PR)	The feature provides the percent ratio of voiced to unvoiced frames. Higher the feature more voiced frames in the speech
Normalised amplitude quotient (NAQ)	The feature is associated to the vocal folds closing time. Larger values of the feature associated to breathiness and other vocal qualities [25]
Amplitude quotient (AQ)	The feature is associated to the vocal folds closing time. Larger values of the feature associated to breathiness and other vocal qualities [25]
Closing quotient (ClQ)	The feature is associated to the vocal folds closing time. Larger values of the feature associated to breathiness and other vocal qualities [8]
Primary and secondary open quotients (OQ_1, OQ_2)	The features are related to the open time of the vocal folds. Higher the feature value breathier the voice is [26]
Amplitude based open quotient (OQ_a)	The feature is related to the open time of the vocal folds. Higher the feature value breathier the voice is [26]
Quasi open quotient (QOQ)	The feature is related to the open time of the vocal folds. Higher the feature value breathier the voice is [26]
Primary and secondary speed quotients (SQ_1, SQ_2)	The features are associated to the symmetry of the glottal waveform. The values closer to 1 indicate breathier voice [26]

4.4.1 CLINICAL SIGNIFICANCE OF THE STUDY FEATURES

We provide the clinical significance of the study features in Table 4.11 which will facilitate the discussion of the study results.

4.4.2 CORRELATION OF FEATURES WITH STUDY

In the correlation study, we have examined the relationship between the speech features corresponding to minimal and enhanced cognitive loading tasks for PD off-state and PD on-state. The clinical relevance of all moderate correlations is summarised.

For controls we examined the association between features and study factors age, MoCA and motor UPDRS. For the minimal cognitive loading speech task, we observed moderate positive as well as negative correlation existing between duration of the recording, Sd f_0–Age (negative correlation), PR–motor UPDRS (negative correlation) and HNR–motor UPDRS (positive correlation). The negative correlation between age and SD f_0 is suggestive of a restricted f_0 variation in the control study group as age increases. There were no moderate or strong associations seen between the study features and factors for the enhanced cognitive loading speech task.

For minimal and enhanced cognitive loading speech tasks, PD off-state features had very few moderate correlations with study factors. In the minimal cognitive loading task, CPP shared a negative moderate correlation with the duration of PD disease. The feature CPP is considered as a measure of overall breathiness and dysphonia [27]. The negative moderate association between CPP and PD duration (in years) indicates possibility of breathy voice as PD duration increases. The PD off-state under the enhanced cognitive loading speech task did not exhibit any moderate to strong correlations with the speech features.

The PD on-state feature that showed moderate association with the motor UPDRS for the minimal cognitive loading task was the rms energy variability (rmsE Var). The association indicated an increase in the rms energy variability as motor UPDRS score increased.

4.4.3 EFFECT OF COGNITIVE LOADING WITHIN COHORTS

When the cognitive load was varied from minimal to enhanced level, a few speech features of controls showed remarkable difference. There was significant increase in the features CPP, rmsE and SD rmsE when the load was increased using the incongruent word test. The increase in these features suggests an increase in periodicity, energy and modulation of the voice in the control cohort.

In case of the PD off-state when the loading was increased two features Cepstral peak prominence (CPP) and primary Speed quotient (SQ_1) displayed a significant change. We were able to notice a reduction in SQ_1 and an increase in CPP features. Both these features are associated with vocal quality as illustrated in Table 4.11. The results indicate a change in periodicity of the voice as a result of increasing the cognitive loading. To reassure whether cognitive loading increase the breathiness in the PD off-state, we also examined another feature called the ClQ which again corresponds to breathiness. Figure 4.3 showed that with increase in cognitive load, ClQ also increased. It was confirmed that as cognitive load increased the voice of PD off-state became breathier (based on increase in ClQ and SQ_1).

The PD on-state features did not show statistically significant difference in the features as a result of increasing the cognitive loading. The only feature displaying noticeable difference as a result of cognitive loading was SQ_1. We observed a reduction in SQ_1 in the enhanced cognitive loading task from the range 1.2–2.1 to a range 0.9–1.6. This reduction is an indication of change in vocal quality. To confirm this, we compared the values of the feature ClQ and observed an increase in the feature for the enhanced cognitive loading task. A simultaneous increase in ClQ and reduction in SQ_1 substantiate the change in the vocal quality of PD in the defined on-state medication.

The study to evaluate the effect of increasing the cognitive loading within the study cohorts confirms that change in loading makes the voice of controls expressive and periodic whereas the loading has a disabling effect on the PD cohorts as it tends to increase the breathiness in their voice irrespective of their medication state.

4.4.4 STATISTICAL ANALYSIS

We performed Mann-Whitney U tests to compare the features of controls and PD off-state for minimal and enhanced cognitive loading. For the minimal loading task, two features (CPP, rmsE Var) significantly differentiated controls and PD off-state and three other features (f_0 Var, rmsE, PR) were remarkably different between the two groups though not statistically significant. CPP was high in controls compared to the PD off-state indicating the breathiness in PD voices. Surprisingly the variability in energy was high in PD off-state compared to controls suggesting PD off-state produced inconsistent energy throughout the task. Other features like f_0 Var, PR and rmsE were high in controls with respect to PD off-state. While comparing the controls and PF off-state for the enhanced cognitive loading task, similar results were observed. There was an increase in CPP for controls indicating a more periodic voice for controls than the PD off-state. The duration taken to complete the enhanced loading task was high for PD off-state. The rmsE was also seen as high in controls than the PD off-state participants. The results show that for minimal and enhanced cognitive loading tasks, the PD off-state voice tends to be monotonous, breathy and low in energy compared to the control cohorts.

The PD on-state features corresponding to minimal and enhanced loading were compared to controls to evaluate the effect of Levodopa medication. For the minimal cognitive loading task, we observed significant difference in CPP and PR where the features were high for controls, $H1^*$-$H2^*$ and rmsE Var were high for PD on-state. Comparing the enhanced cognitive loading task, there was significant difference in the duration taken to complete the task between controls and PD on-state (high). Features CPP and rmsE were lower in PD on-state than controls whereas QOQ and rmsE Var were higher for PD on-state. The results bring us to the conclusion that PD on-state voice is less periodic and breathier compared to controls.

In the final test we performed to evaluate the effect of medication in PD voices, we compared the features of minimal and enhanced cognitive loading tasks between PD off-state and on-state. The comparison between minimal cognitive loading task features of PD off-state and PD on-state did not show any difference. This tells us the PD voice features behave more or less the same for the minimal cognitive loading task under the effect of medication. The comparison between the features of enhanced loading task shows some difference. The features $H1^*$-$H2^*$, QOQ and SQ_1 were exhibiting significant difference between off-state and on-state. For the on-state PD the values of SQ_1 were closer to 1 compared to that of off-state. There was a significant increase in the values of QOQ and $H1^*$-$H2^*$ for on-state PD as well. These values are suggestive of an increase in breathiness in the voice of PD under the effect of medication.

4.4.5 LIMITATIONS

Analysis of single word Stroop test responses lacked sensitivity to the acoustic features of PD dysarthria and dysphonia. A sample of more advanced PD participants with obvious dysphonia could better determine how vocal impairments are influenced by cognitive loading. Cognitive deficits associated with disease progression might be a drawback of such an approach, however. Although there was some gender imbalance between control and PD groups, we assumed that this would not greatly affect the intra-individual comparisons of the important assessments—before and after cognitive loading, before and after medication in PD. We recommend that future similar studies ensure equal gender representation and sufficient statistical power for independent gender comparisons. Our results highlight the inherent difficulty in matching measurable, objective voice acoustic features with subjective, yet more descriptive perceptual characteristics, and with their physiologic correlates.

REFERENCES

[1] B. Harel, M. Cannizzaro, and P. J. Snyder, "Variability in fundamental frequency during speech in prodromal and incipient Parkinson's disease: A longitudinal case study," *Brain and Cognition*, vol. 56, no. 1, pp. 24–29, 2004.

[2] H. Liu, E. Q. Wang, L. V. Metman, and C. R. Larson, "Vocal responses to perturbations in voice auditory feedback in individuals with Parkinson's disease," *PloS ONE*, vol. 7, no. 3, p. e33629, 2012.

[3] C. Middag, J.-P. Martens, G. Van Nuffelen, and M. De Bodt, "Automated intelligibility assessment of pathological speech using phonological features," *EURASIP Journal on Advances in Signal Processing*, vol. 2009, pp. 1–9, 2009.

[4] P. Alku and E. Vilkman, "A comparison of glottal voice source quantification parameters in breathy, normal and pressed phonation of female and male speakers," *Folia Phoniatrica et Logopaedica*, vol. 48, no. 5, pp. 240–254, 1996.

[5] P. A. Kirschner, J. Sweller, F. Kirschner, and J. Zambrano R, "From cognitive load theory to collaborative cognitive load theory," *International Journal of Computer-Supported Collaborative Learning*, vol. 13, no. 2, pp. 213–233, 2018.

[6] A. Berthold and A. Jameson, "Interpreting symptoms of cognitive load in speech input," in *UM99 User Modeling*. Springer, Berlin,1999, pp. 235–244.

[7] N. Eichorn, K. Marton, R. G. Schwartz, R. D. Melara, and S. Pirutinsky, "Does working memory enhance or interfere with speech fluency in adults who do and do not stutter? Evidence from a dual-task paradigm," *Journal of Speech, Language, and Hearing Research*, vol. 59, no. 3, pp. 415–429, 2016.

[8] T. F. Yap, "Speech production under cognitive load: Effects and classification," Diss., The University of New South Wales, 2012.

[9] A. K. Namasivayam and P. Van Lieshout, "Speech motor skill and stuttering," *Journal of Motor Behavior*, vol. 43, no. 6, pp. 477–489, 2011.

[10] A. K. Ho, R. Iansek, and J. L. Bradshaw, "The effect of a concurrent task on Parkinsonian speech," *Journal of Clinical and Experimental Neuropsychology*, vol. 24, no. 1, pp. 36–47, 2002.

[11] Z. Galaz, et al., "Prosodic analysis of neutral, stress-modified and rhymed speech in patients with Parkinson's disease," *Computer Methods and Programs in Biomedicine*, vol. 127, pp. 301–317, 2016.

[12] A. J. Hughes, S. E. Daniel, L. Kilford, and A. J. Lees, "Accuracy of clinical diagnosis of idiopathic Parkinson's disease: A clinico-pathological study of 100 cases," *Journal of Neurology, Neurosurgery & Psychiatry*, vol. 55, no. 3, pp. 181–184, 1992.

[13] P. A. Kempster, S. S. O'Sullivan, J. L. Holton, T. Revesz, and A. J. Lees, "Relationships between age and late progression of Parkinson's disease: A clinico-pathological study," *Brain*, vol. 133, no. 6, pp. 1755–1762, 2010.

[14] D. Kumar, P. Kempster, S. Raghav, R. Viswanthan, P. Zham, and S. Arjunan, "Screening Parkinson's diseases using sustained phonemes," RMIT University, https://doi.org/10.25439/rmt.12618755.v1

[15] J. R. Stroop, "Studies of interference in serial verbal reactions," *Journal of Experimental Psychology*, vol. 18, no. 6, p. 643, 1935.

[16] S. A. Zahorian and H. Hu, "A spectral/temporal method for robust fundamental frequency tracking," *The Journal of the Acoustical Society of America*, vol. 123, no. 6, pp. 4559–4571, 2008.

[17] Y.-L. Shue, *The Voice Source in Speech Production: Data, Analysis and Models*. University of California, Los Angeles, 2010.

[18] M. Airas, "TKK Aparat: An environment for voice inverse filtering and parameterization," *Logopedics Phoniatrics Vocology*, vol. 33, no. 1, pp. 49–64, 2008.

[19] L. K. Bowen, G. L. Hands, S. Pradhan, and C. E. Stepp, "Effects of Parkinson's disease on fundamental frequency variability in running speech," *Journal of Medical Speech-Language Pathology*, vol. 21, no. 3, p. 235, 2013.

[20] S. Yang, et al., "The physical significance of acoustic parameters and its clinical significance of dysarthria in Parkinson's disease," *Scientific Reports*, vol. 10, no. 1, pp. 1–9, 2020.

[21] K. Tjaden, "Speech and swallowing in Parkinson's disease," *Topics in Geriatric Rehabilitation*, vol. 24, no. 2, p. 115, 2008.

[22] S. Jannetts and A. Lowit, "Cepstral analysis of hypokinetic and ataxic voices: Correlations with perceptual and other acoustic measures," *Journal of Voice*, vol. 28, no. 6, pp. 673–680, 2014.

[23] Y.-L. Shue, G. Chen, and A. Alwan, "On the interdependencies between voice quality, glottal gaps, and voice-source related acoustic measures," in *Eleventh Annual Conference of the International Speech Communication Association*, 2010. https://www.isca-speech.org/iscaweb/index.php/conferences/interspeech

[24] D. G. Childers and C. K. Lee, "Vocal quality factors: Analysis, synthesis, and perception," *the Journal of the Acoustical Society of America*, vol. 90, no. 5, pp. 2394–2410, 1991.

[25] P. Alku, T. Bäckström, and E. Vilkman, "Normalized amplitude quotient for parametrization of the glottal flow," *The Journal of the Acoustical Society of America*, vol. 112, no. 2, pp. 701–710, 2002.

[26] P. Alku and E. Vilkman, "Amplitude domain quotient for characterization of the glottal volume velocity waveform estimated by inverse filtering," *Speech Communication*, vol. 18, no. 2, pp. 131–138, 1996.

[27] O. Murton, R. Hillman, and D. Mehta, "Cepstral peak prominence values for clinical voice evaluation," *American Journal of Speech-Language Pathology*, vol. 29, no. 3, pp. 1596–1607, 2020.

Section 2

EEG—ECG Signal Processing

5 Electroencephalography and Epileptic Discharge Identification

Mohd Syakir Fathillah, Theeban Raj Shivaraja, Khalida Azudin and Kalaivani Chellappan

5.1 INTRODUCTION TO ELECTROENCEPHALOGRAPHY

5.1.1 PRINCIPLE OF ELECTROENCEPHALOGRAPHY

Electroencephalography (EEG) uses the principle of differential amplification that measures differences in electrical potential to record waves of brain activity. Potential differences are caused by summed postsynaptic potentials from pyramidal cells that create dipoles between soma and apical dendrites. A water molecule (H_2O) is a dipole, similarly a pair of equal and opposite electric charges or magnetic poles of opposite sign separated especially by a small distance known as dipole neurons are structured in the human brain. The dipoles make the major contribution to the scalp potential when neurons are activated, local currents are produced, and these currents are measured as EEG during excitations of the dendrites of many pyramidal neurons. Neurons which are radially symmetric, randomly oriented or asynchronously activated do not produce externally observable electric fields. Whereas neurons which are non-radially symmetric, spatially aligned and synchronously activated add up to produce externally observable electric fields. In 1952, Hodgkin and Huxley (H-H) introduced an electrical model as in Figure 5.1 (a) for action potential generation that records membrane current and its application to conduction and excitation in nerve which reflects the principles of EEG in brain activity recording. Figure 5.1 (b) is an electrical equivalent circuit of the linearized H-H equation.

5.1.2 THEORY OF ELECTROENCEPHALOGRAPHY

The human brain consists of billions of neurons actively interacting with each other continuously. The neurons are excitable cells with electrical properties capable in producing electrical and magnetic field. A neuron is surrounded by a charged ion e.g., Sodium (Na+), Chloride (Cl-) and potassium (K+). On the inside, it is dominated by potassium (K+) and protein (A-) ions [1]. The lipid bilayer membrane of the neuron is impermeable to these ions.

DOI: 10.1201/9781003201137-7

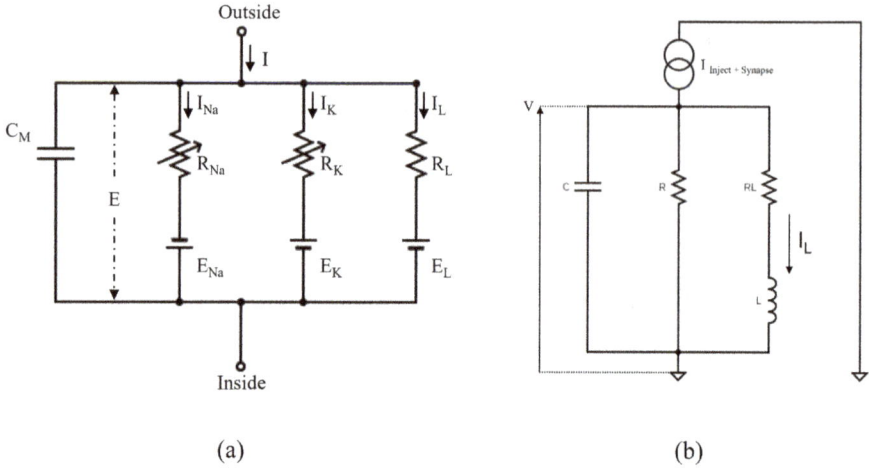

(a)   (b)

FIGURE 5.1 (a) Electrical Circuit Representing Membrane, (b) Electrical Equivalent Circuit of linearized H-H equation.

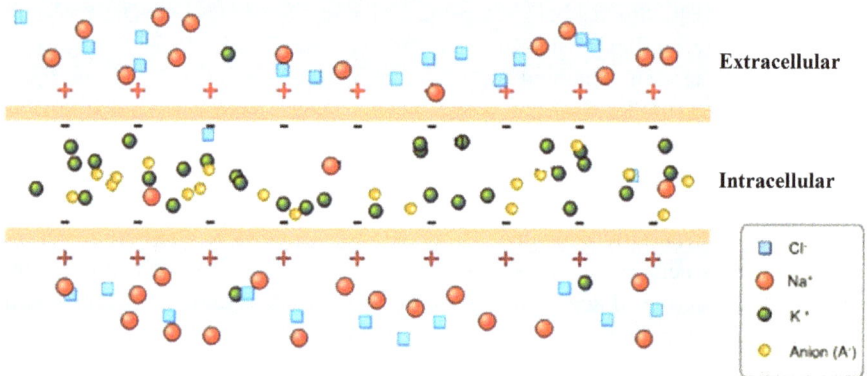

FIGURE 5.2 Illustration of ion available in extracellular (outside neuron) and intracellular (inside neuron).

Source: Fathillah M. S. 2018. Pembangunan algoritma hibrid pengesanan pelepasan cas epileptik bagi eletroensefalografi (EEG). Fakulti Kejuruteraan dan Alam Bina, UKM.

The method of entrance for the ion is via a special protein called voltage-gated channel which is specific to each ion. During the neuron resting state, the membrane potential inside the neuron is 70mV lower compared to the outside of the neuron [2]. This happens because the rate of potassium diffusion out of the cell is much more rapid than the rate of sodium diffuses inside the cell [3]. The value number may vary, depending on the neuron type and species [4].

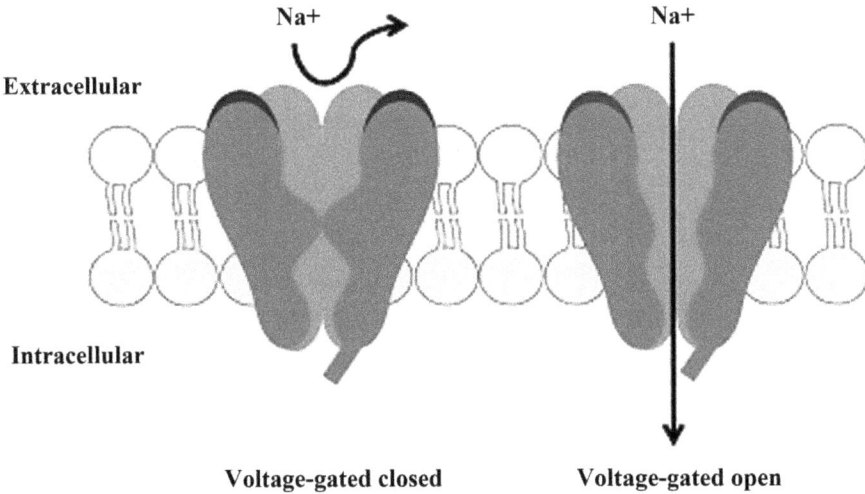

FIGURE 5.3   Illustration of voltage gate controlling the ion diffusion.

Environmental stimuli or molecules released from another neuron has the possibility to affect the membrane potential to trigger action potential. Action potential is a process by which neurons transport information from the body cell (soma) to axon terminal through axon. Action potential occurs when the membrane potential reaches the threshold of −55mV. If the neuron fails to reach threshold potential, it will go back to its resting state and no further information sending process occurs. Exceeding this threshold will cause a depolarization process to occur until it reaches membrane potential peak which is approximately +40mV. Once the peak is achieved, the repolarization process will occur. The Na+ voltage gate is closed to prevent further Na+ ion entrance and opening K+ voltage gate to release the K+ ion from the neuron. K+ channel stays open until the membrane potential drops past −70mV. This phase is called hyperpolarization. The Na+ channel will reopen when membrane potential exceeds the threshold potential. The summary of action potential process is described in Figure 5.4.

Depolarization causes a chain reaction through the axon so that it reaches the axon terminal. The ions released in the process of depolarization of presynaptic neurons cause changes in the membrane potential of postsynaptic neurons and produce a response in postsynaptic neurons until it reaches the cortical neurons.

Once the signal reaches the cortical region it is to be captured by EEG electrodes. The electrodes are placed in at least two separate locations to capture the electrical activity. The signal will undergo amplification to amplify the weak signal and filtering to remove unwanted noise. Since EEG is an analogue signal, sampling is required to convert it into a digital signal for processing or analysis.

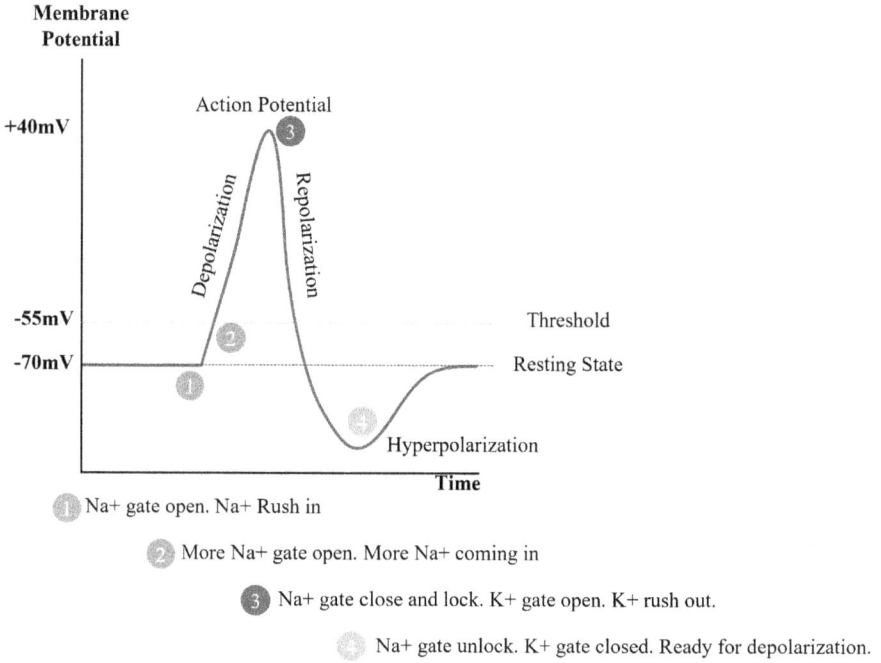

FIGURE 5.4   Action potential graph.

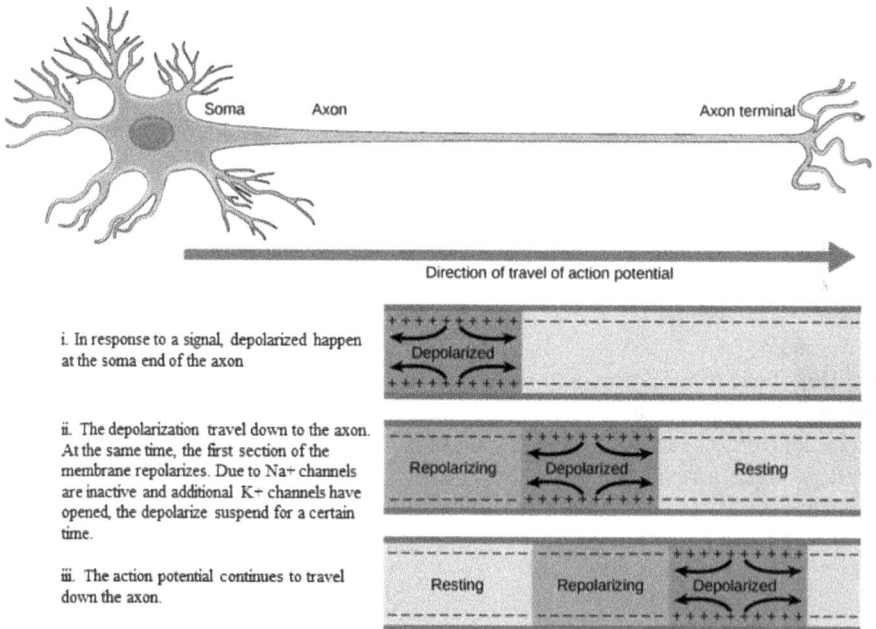

FIGURE 5.5   Travel of action potential in neuron.

5.1.3 Origin of Electroencephalography

The existence of the electrical activity of the brain (i.e., the electroencephalogram or EEG) was discovered more than a century ago by Richard Caton. In 1875, Caton performed the first known neurophysiologic recordings of animals. The advent of recording the electrical activity of human beings took another half century to occur. In the 1920s, Hans Berger, a German psychiatrist, demonstrated EEG can be recorded from the human scalp. Following the demonstration, it made a slow start before being accepted as a method of analyzing brain functions in the healthy and diseased. This acceptance came only after the demonstration by Adrian and Mathews in 1934 that the EEG, specifically, the alpha rhythm, was likely generated in the occipital lobes in humans and was not artefactual. However, until the 1970s, the neuronal sources of the alpha rhythm remained undefined, till demonstrated in a dog that the alpha rhythm is generated by a dipole layer cantered at layers IV and V of the visual cortex [5]. The mechanisms of generation and the functional significance of the EEG remained controversial for a relatively long period considering the complexity of the underlying systems of neuronal generators on one hand and the rather involved transfer of signals from the cortical surface to the scalp due to the topological and electrical properties of the volume conductor (brain, cerebrospinal fluid, skull, scalp) on the other hand [6]–[9].

5.1.4 Electroencephalography Technology

An EEG sensor is an electronic device which measures electrical activities of the brain. EEG sensors typically capture the varying electrical signals created by the activity of large groups of neurons near the surface of the brain over a period of time. They work by measuring the small fluctuations in electrical current between the skin and the sensor electrode, amplifying the signal, and filtering the irrelevant frequencies [10].

Electrodes are not part of EEG's structural machine. However, they are included in the instrumentation of EEG equipment for functional activities. All EEG machines have similar structural design. The first part is known as the jack box or electrode board, where the pin, the terminal end of the electrode, is inserted. Clinical EEG machines have jacks with anatomical descriptors. The electrode is connected with the electrode selector by a shielded multiple connector input cable. The electrode selector enables any pair of electrodes to be selected and connected to any input terminals of the amplifiers. The input panel of electrodes should include at least 23 connectors to fulfil the recommendations of the IFCN. The calibrator, which is responsible for supplying a precise preselected voltage within the range of EEG activity to every amplifier input, is another part of the input system. The amplifier is an electronic device which magnifies the potential of the brain approximately by a million. The amplified output is displayed through a computer screen or is printed.

The electrodes of an EEG device are made of a metal contact surface and a flexible insulated wire ending with a connector pin to be connected to the jack box. The reproduction of the waveform depends on the electrode impedance that is a function of the electrode resistance and capacitance. Chloride silver discs with a diameter of 4–10 mm is one of the most common reversible electrodes. Reversible electrodes have higher reproductive capacity compared to electrodes fabricated with other materials, such as platinum, silver and gold.

Montages are essential for EEG interpretation. Montage is a broad term derived from the arts that refers to a composite of different parts. A similar concept is applied in EEG, where a montage is essentially a group of derivations or channels. The number of channels in a montage varies and depends on the number of electrodes available as well as the purpose of the recording. There can be two to more than 20 channels in a montage depending on the setting. Routine EEGs are generally performed with a 10–20 system, an internationally accepted system for scalp electrode placement. Generally, there are three main purposes of montages for routine scalp EEGs. First is to display activity of the entire head to not miss any activity. Second is to compare activity on the two sides to provide lateral information. Third is to localize activity to a specific brain region if possible [11].

5.2   ELECTROENCEPHALOGRAPHY SIGNAL PROCESSING

5.2.1   ELECTROENCEPHALOGRAPHY PHYSIOLOGY AND ARTEFACTS

EEG signal is always subjected to different kinds of artefacts or noises which occur in different frequency range [12]–[14]. These artefacts are defined as unwanted components of the EEG signal that are not generated by the brain. The generated signals contaminate the original signal, causing corruption or loss of the essential information.

EEG artefacts can be separated into two categories: physiological artefact and non-physiological artefact. A physiological artefact is an artefact that is produced by the human body, e.g., respiration, cardiac activity, muscle movement, etc. A non-physiological artefact is generated by an external factor such as electrode pop, cable movement, AC electrical and electromagnetic interference.

In-depth knowledge of EEG morphology is important as some of the artefacts can mimic biomarkers for disease e.g., epileptiform abnormalities, leading to errors in diagnosis. Table 5.1 displays common artefacts in EEG with their characteristics:

TABLE 5.1

Description of Common Physiological and Non-Physiological.

#	Physiological Artefacts
1	Ocular Activity Origin: Eye Cause: Blinking, lateral eye movement Effect on EEG: Affect frontal electrode the most. Will produce high amplitude voltage peaks in low frequency.
2	Muscle Activity Origin: Muscle Cause: Electrical activity from muscle contracting such as clenching the jaw, neck and shoulder can interfere with the EEG. This signal is referred to as electromyogram (EMG). Effect on EEG: Presence of high frequency signal. The amplitude of the noise is dependent on the muscle contraction strength.

TABLE 5.1 (*Continued*)
Description of Common Physiological and Non-Physiological.

#	Physiological Artefacts
3	Cardiac Activity Origin: Heart Cause: Heart pulse. This signal is referred to as electrocardiogram (ECG). Effect on EEG: Presence of rhythmic pattern in EEG correlates with the heartbeats.
4	Perspiration Origin: Skin Cause: Sweat produced on the scalp can change the conductance value. Worst case scenario, short circuit may occur between electrodes. Effect on EEG: Presence of low frequency noise
5	Respiration Origin: Movement of chest Cause: Inhaling and exhaling may interfere with the electrodes and the scalp especially if the patient is lying on the bed. Effect on EEG: Low frequency artefact overlap with delta and theta rhythm in EEG. The artefact is synchronized with the breathing rhythm.
	Non-Physiological Artefacts
1	Electrode pop Origin: Electrode-skin contact Cause: Temporarily disconnected electrode from the scalp or spontaneous changes in electrode-skin contact Effect on EEG: Sudden high amplitude that usually affects single channel
2	AC electrical and electromagnetic interferences Origin: Power lines Cause: Electromagnetic fields from AC power source and wires can affect the EEG reading. The AC line noise is in 50Hz or 60Hz depending on the country location. Effect on EEG: Distortion of spike around 50Hz or 60Hz
3	Body movements Origin: Body movement Cause: Movement of body part such as head and hand. Other activity such as walking and running also can cause this artefact. This movement also may affect electrode-skin contact. Effect on EEG: Contaminate lower frequency of EEG

5.2.2 ELECTROENCEPHALOGRAPHY SIGNAL PRE-PROCESSING

A proper recording design and setup can eliminate most of the non-physiological noise. Physiological noise on the other hand requires a pre-processing technique to remove its presence. Failure to eliminate the noise can affect the data reliability and interpretability during analysis. EEG signal also is susceptible to variance of magnitude scale between recording or subject due to different skull thickness for each subject, subject condition and device impedance [15]. These challenges can be minimized using a pre-processing method to transform the raw EEG signal into a clean and interpretable data. Pre-processing techniques are not limited to the noise

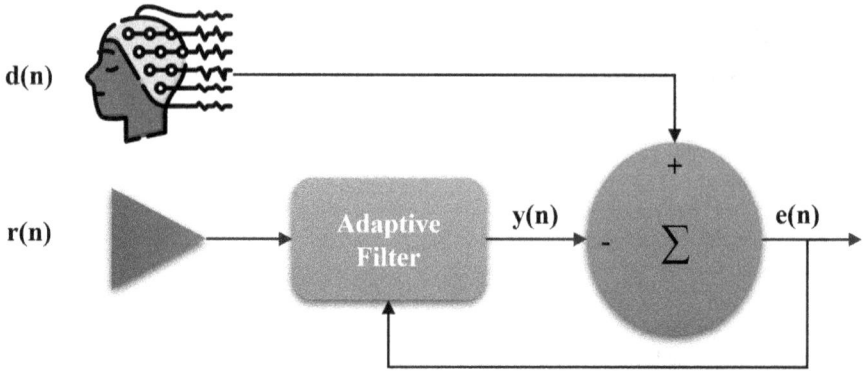

FIGURE 5.6   Adaptive filter structure.

removal. In this section, we will discuss common pre-processing techniques in noise removal, normalization and segmentation.

Noise removal techniques are separated into three categories: filtering, blind source decomposition (BSS) and source decomposition method. Bandpass filtering commonly is adopted to remove the noise outside a frequency range. A bandpass filter consists of a low-pass filter which helps to attenuate high frequency noise and the high-pass filter to reduce low frequency noises. A notch filter is usually employed along with a bandpass filter, acting as a narrow band filtering to attenuate a specific range of frequency to effectively remove noise such as an AC power line. Important criteria of filter design are model selection, model order and cut-off frequency. Another type of filter is adaptive filter that has the ability to adjust its coefficient to generate a signal mimic to the noise present that needs to be removed [16]. Figure 5.6 shows how the adaptive filter works. Coefficient of $r(n)$ will be adjusted to enable the frequency response to produce a signal similar to the unwanted noise. Combining the EEG original raw signal $d(n)$ and the $y(n)$ adapted signal will produce as noise reduced signal $e(n)$.

BSS is able to separate the mixture of signal into separate components. Independent Component Analysis (ICA) is one of the popular choices for BSS in EEG application to separate noise components from the EEG signal [17], [18].

In EEG cases, ICA assumes a set of m measured data points at a time instant (t) of the EEG as $x(t) = \left[x_1(t), x_2(t), \ldots, x_m(t) \right]^T$ The equation is a representation of the observed signal retrieved from a different source of electrode. The observed signal is a mixture of n unknown underlying source $s(t) = \left[s_1(t), s_2(t), \ldots, s_n(t) \right]^T$. Assume matrix $\mathbf{A}$ is a $n \times m$ size of the weighted sum of different source signal such that

$$x(t) = As(t) \tag{5.1}$$

The objective of ICA is to recover the original $s(t)$ from the observed signal $x(t)$ by identifying the de-mixing matrix $\mathbf{W}$ which is inverse of mixing matrix $\mathbf{A}$ such that [19]–[21]

$$\hat{s}(t) = Wx(t) \tag{5.2}$$

The ICA mechanism is illustrated as in Figure 5.7.

Wavelet denoising (WD) is a source decomposition tool to separate the noise while preserving the significant information [22], [23]. The first stage of WT is to decompose the signal using consecutive high-pass and low-pass filtering. The number of consecutive filters depends on the decomposition level. Next, the threshold of the signal is determined by using WT coefficient via median absolute deviation (MAD). MAD, δ also known as estimated noise, is calculated by using largest coefficient spectrum as follow

$$\delta_{mad} = \frac{median\{|c_0|,|c_1|,...,|c_n - 1|\}}{0.6745} \tag{5.3}$$

where $|c_0|,|c_1|,...,|c_n - 1|$ are the wavelet coefficient and value 0.6745 is used to rescale the numerator so it can be an appropriate estimator for the standard deviation for Gaussian White Noise. The noise threshold, τ is then determined by using δ and total number of samples, N

$$Threshold, \tau = \delta_{mad}\sqrt{\ln(N)} \tag{5.4}$$

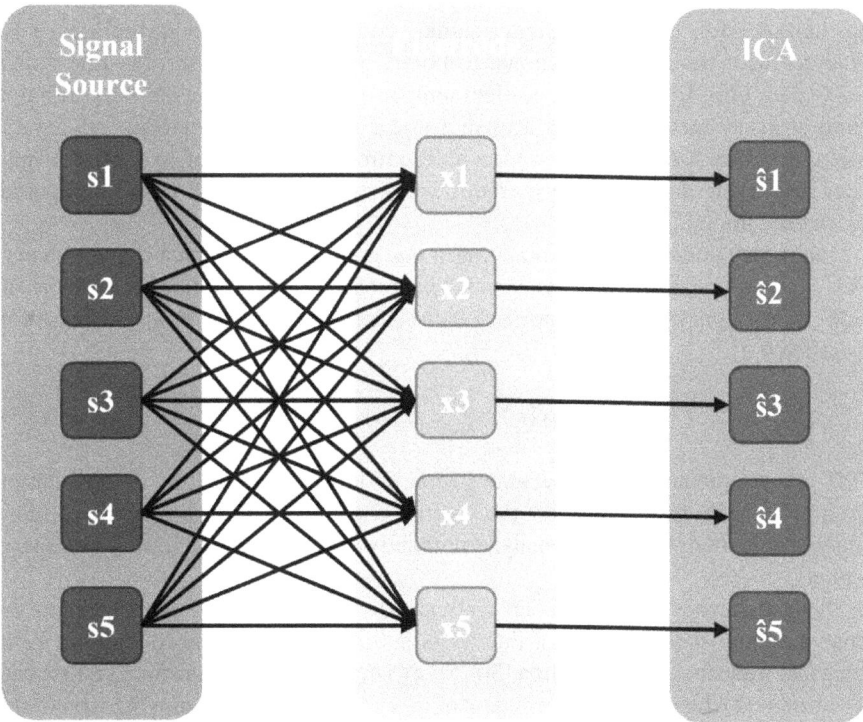

FIGURE 5.7 Fundamental of ICA process in EEG.

The final stage of wavelet denoising is reconstruction of signal which is a reverse decomposition process. Up sample technique is utilized for the coefficient where the number of levels should be the same as the decomposition level used initially.

Absolute value of EEG magnitude is affected by age, skull thickness and subject condition during the recording. Electrode quality and EEG device used also has an effect on the magnitude. To enable comparison between EEG recordings, normalization is introduced to transform the EEG into one common scale [24]. Normalization scaling will return the signal mean to 0 and ensure the standard deviation is 1. Normalization, z is denoted as the following equation, where x is the signal value, μ is the mean and σ is the signal standard deviation.

$$z = \frac{(x - \mu)}{\sigma} \tag{5.5}$$

Segmentation is required to manage lengthy EEG recording into epoch. Segmentation reduces computational cost for analysis and transforms non-stationary signal into pseudo-stationary signal [24]. Processing non-stationary signal allows a statistical feature to be extracted [25]. Epoch eases the identification of a biomarker in a specific time frame.

5.2.3 ELECTROENCEPHALOGRAPHY SIGNAL TRANSFORMATION

By nature, EEG is a time domain signal. The graph shows how brain activity changes with time. EEG is known to be 3N: non-deterministic, non-linear and noisy data [26]. EEG has an excellent temporal resolution allowing it to pinpoint time of event accurately [27]. To use a linear tool for EEG analysis, stationary assumption such as moving window is required as the cost of loss of information [28]. Time domain analysis requires lesser computational power compared to another domain.

Frequency domain is a linear approach that indicates how often a change occurs. Fast Fourier Transform (FFT) is one of the approaches to extract frequency information from a signal [29]. The approach can be derived from Fast Fourier Transform as follows:

$$x(k) = \sum_{n=0}^{N-1} x(n)^{\frac{-2\pi ink}{N}} \tag{5.6}$$

FFT reduces the number of executions from N^2 to $N \log N$, resulting in much faster processing time. FFT represents the spectral resolution with lack of time of information. It is more suitable to analyze stationary signal compared to non-stationary signal.

Time-frequency domain (TFD) method has outperformed the other domain in analyzing non-stationary signal due to its ability to provide information on both time and frequency representation [30]. Short Time Fourier Transform (STFT) is one TFD method which is the improvement of FFT that utilizes small moving window to assume the signal stationary. However, the selection of window size gives a trade-off between excellent time or frequency resolution [31].

Pseudo Wigner Ville Distribution (PWVD) is better in solving the time-frequency trade-off but has a downside due to its noise generating nature [32]. Empirical Mode Decomposition (EMD) is another approach of time-frequency domain that works by decomposing signal to intrinsic mode function (IMF) [33], [34]. IMF needs to fulfil two criteria as follows: a) have the same number of extrema and zero-crossing or differ at most by one and B) average value of envelope defined by the local maxima and minima is zero at any data location. EMD has a drawdown where it generates noise due to mode mixing problem. Partly ensemble empirical mode decomposition (PEEMD) then was invented to improve the noise-assisted method for eliminating mode mixing [35].

Wavelet Transform (WT) is one popular TFD technique applied in EEG signal [36]–[41]. There are two types of wavelet, continuous wavelet transforms (CWT) and discrete wavelet transform (DWT). The basic definition of CWT can be referred to as follows:

$$\psi\left(a,b\right)=\frac{1}{\sqrt{|a|}}\psi\left(\frac{t-b}{a}\right)$$

(5.7)

Where $\psi(t)$ is the basis function that acts as bandpass filter, a and b is defined as scale and time location respectively. The $\sqrt{|a|}$ is used for energy preservation. Consecutive high-pass and low-pass filtering is what made DWT different from CWT [42], [43]. DWT can be denoted by formula:

$$\psi\left(j,k\right)=\frac{1}{\sqrt{|2^{j}|}}\psi\left(\frac{t-k2^{j}}{2^{j}}\right)$$

(5.8)

where a and b are replaced 2^{j} and $k2^{j}$ respectively. Signal will undergo high-pass and low-pass filter to produce approximate and detail coefficient. The advantage of wavelets is they can provide accurate frequency resolution when the signal frequency is low and excellent time resolution when the frequency is high [44]. This enables wavelets to give accurate time of event when the changes happen. CWT has an advantage in analysis signal in higher resolution, but DWT offers multi resolution analysis that enables analysis of sub-band frequency and faster computational speed [45]. Challenge in wavelet application is to select the suitable basis function as it can differ in the performance, depending on the biomarker to be detected in EEG [36].

5.3   APPLICATION OF EEG IN EPILEPSY DIAGNOSIS

Epilepsy is a chronic noncommunicable disease defined as the occurrence of two or more unprovoked seizures within 24 hours [46]. It may be caused by the disturbance in brain function that may arise from the structure, abnormalities of electrical or biochemistry in brain, spinal cord or other nervous systems; however, most of the cases are idiopathic, which is no known cause [47]. The role of EEG in epilepsy diagnosis is to detect the presence of epileptic discharge as the biomarker. EEG is still used as the first step in epilepsy diagnosis due to its affordability and availability. Compared to other methods such as CT and MRI, EEG is able to provide better temporal resolution thus allowing it to track brain activity within milliseconds.

5.3.1 CHALLENGE IN EEG BASED EPILEPSY DIAGNOSIS

The conventional method of epilepsy diagnosis still relies on visual screening by a neurologist or epileptologist. This requires in-depth knowledge in understanding morphology and pattern of epileptic discharge, also the ability to differentiate them from the artefact. Often, EEG review is a tedious and time-consuming process. Plus, the data length can vary from 20 minutes to 72 hours for long-term electroencephalographic (LTE) recording. A challenge in inter-rater disagreement on the diagnosis may arise and lead to prolonged EEG diagnosis time [48].

5.3.2 EPILEPTIC DISCHARGE

Epileptic discharge is the epilepsy biomarker presence in the EEG recording. It happens when neuron generating excessive and synchronize electrical activity. Presence of epileptic discharge with physical symptoms (such as rhythmic shaking and jerking of part or whole body, up rolling eyeball and loss of conscious) is known as ictal (during seizure). Ictal state is a strong indicator that the person has had a seizure. Diagnosing epilepsy based on ictal state is a challenge since the seizure may not happen during the recording. Thus, interictal state (between seizure) is an alternative indicator for a person to have the possibility of having epilepsy. Interictal is the condition where epileptic discharge is present in the EEG, but no symptoms are shown. Interictal rarely occurs to a person who has no history of seizure [49].

5.3.3 CHARACTERISTICS AND MORPHOLOGY

Epileptic discharge can be distinguished based on its amplitude, frequency, morphology and rhythmic pattern. It is known to have higher amplitude than background signal, steep slope and specific width (20–200 milliseconds) [50]. Epileptic discharge can be separated into two categories, spike and sharp. Spike has a width of 20ms to 70ms where the sharp has a width of 70ms to 200ms [51]. Commonly the occurrence of spike and sharp will happen along with slow waves complex. Continuous spike or sharp is defined as poly-spike. Figure 5.8 shows the example of epileptic discharge.

Human brain activity is a nonlinear complex system that generate nondeterministic EEG signals. However, the occurrence of epileptic discharge will reduce the randomness of the signal, causing it to look to be repetitive [52]. The deterministic feature can be the indicator of epileptic discharge presence in EEG. A comparison example between normal, interictal and ictal is shown in Figure 5.9.

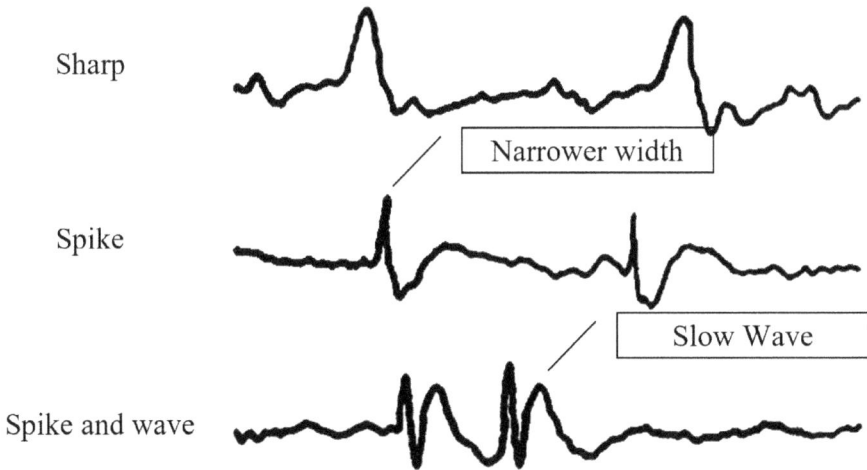

FIGURE 5.8   Example of epileptic discharge.

Source: Fathillah M. S. 2018. Pembangunan algoritma hibrid pengesanan pelepasan cas epileptik bagi eletroensefalografi (EEG). Fakulti Kejuruteraan dan Alam Bina, UKM.

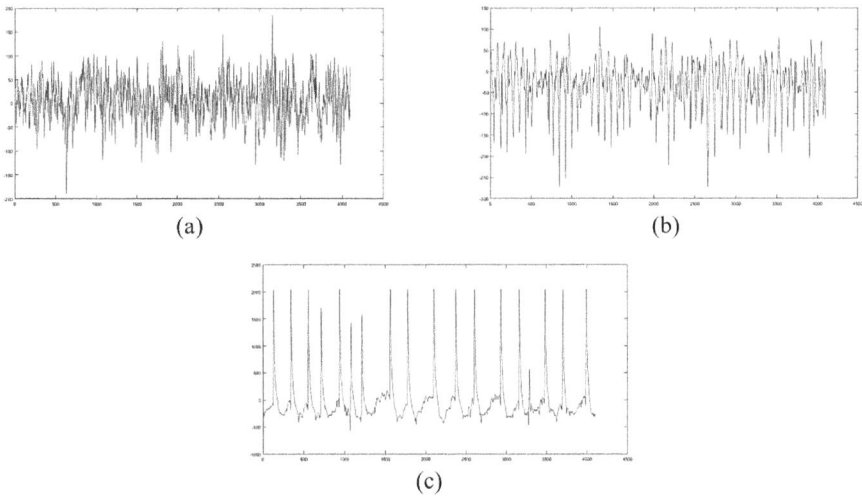

(a)

(b)

(c)

FIGURE 5.9   Comparison of rhythmic pattern between a) normal, b) interictal and c) ictal EEG.

Source: Fathillah M. S. 2018. Pembangunan algoritma hibrid pengesanan pelepasan cas epileptik bagi eletroensefalografi (EEG). Fakulti Kejuruteraan dan Alam Bina, UKM.

5.4 METHODOLOGY IN EPILEPTIC DISCHARGE DETECTION AND CLASSIFICATION

The advancement of signal processing techniques has allowed researchers to come out with an assistive computer aided system for epilepsy diagnosis. Such a system is able to expedite the diagnosis by identifying the presence and location of epileptic discharge for neurologist/epileptologist verification. A fundamental component for the algorithm development involves pre-processing and conditioning, signal transformation, feature extraction and classification. The flow of these components can be referred to in Figure 5.10. Raw EEG data first will be denoised to remove the artefacts before being normalized and segmented. Signal transformation is an option, depending on whether the signal needs to be analyzed in other domains. Feature identification and extraction enables a signal to be distinguished as a unique characteristic of a signal group. This allows classification to classify the signal based on these features into two or more groups, in this case epileptic or non-epileptic signal. In this section, we will discuss in detail the criteria in classifying epileptic and non-epileptic signals.

5.4.1 ELECTROENCEPHALOGRAPHY SIGNAL CONDITIONING

EEG signal conditioning for epilepsy signal identification is similar to any other EEG pre-processing methods. For epileptic signal it is possible to filter frequency components higher than 30Hz using lowpass filters [20].

5.4.2 ELECTROENCEPHALOGRAPHY FEATURE IDENTIFICATION

Classifying epileptic and non-epileptic signal requires understanding of characteristic of the signal. As mentioned previously, the characteristic or epileptic discharge is to have high amplitude, steep slope, specific width and synchronize pattern. Based on these characteristics, research has identified features such as energy, entropy and morphology to classify between epileptic and non-epileptic signal [53].

High amplitude and steep slope enable the energy feature to be extracted for the classification purpose. Signal energy per epoch per channel in time domain can be obtained using [54]

$$e_j(k) = \sum_{i=1}^{l} x_i^2 \qquad (5.9)$$

Where x_i represents the value of signal at i^{th} instant, l represents number of samples per epoch, k is the epoch number and j is the channel number.

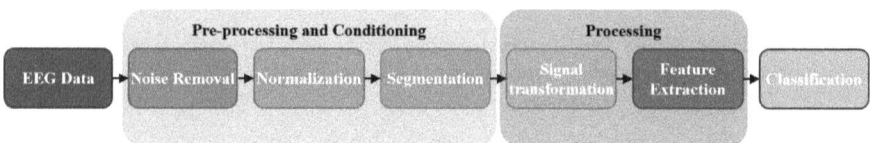

FIGURE 5.10 Epileptic discharge detection algorithm flow.

Another energy-based feature, Teager Energy Operator (TEO), is adopted due to its nonlinear properties and sensitivity to variation of signal amplitude and frequency [55]. The instantaneous energy TEO, E_n at a specific instant of time, n can be calculated as follows:

$$E_n = x_n^2 - x_{n-1}x_{n+1} = A^2 sin^2\left(\Omega\right) \tag{5.10}$$

$$E_n \approx A^2\Omega^2 \tag{5.11}$$

Assuming the signal x_n,

$$x_n = A\cos\left(\Omega n + \varphi\right) \tag{5.12}$$

where A is the amplitude and φ is the initial phase, given

$$\Omega = \frac{2\pi f}{f_s} \tag{5.13}$$

Where f and f_s are analog and sample frequency in Hz respectively.

Energy can be extracted from frequency domain by using power spectral densities as follows [56].

$$\widehat{PD}\left(f\right) = \frac{\Delta t}{N}\left|\sum_{n=0}^{N-1} x_n e^{-i2\pi fn\Delta t}\right|^2, -\frac{1}{2\Delta t} < f \leq \frac{1}{2\Delta t} \tag{5.14}$$

Given Δt is the sampling interval, N is a number of samples in a signal, x_n represent the signal f is defined as sampling frequency.

Time frequency approach via wavelet can be exploited to extract the energy feature, enabling multi resolution analysis to provide information on each sub-band [57]–[59]. Rather than having definite resolution, more flexible analysis is offered, particularly in interpretation of energy trends each frequency band. Energy calculation for each coefficient is represented in the following equation.

$$E_j = \sum_k \left|d_{j,k}\right|^2, j = 1,...,N \tag{5.15}$$

$$E_{N+1} = \sum_k \left|c_k\right|^2 \tag{5.16}$$

Where $d_{j,k}$ is wavelet coefficeint and c_k is the scaling coefficient.

Entropy analysis helps to quantify the amount of regularity and predictability of time-series data. Various entropy measurements have been adopted in epileptic discharge detection such as approximate entropy, Lempel-Ziv, Hurst exponent, Shannon entropy and Sample entropy. These techniques have different ways of extracting a deterministic feature but have the same motivation to distinguish between epileptic and non-epileptic discharge.

A classic approach is to use Shannon entropy to compute the degree of uncertainty of a signal [60]. It is defined by equation:

$$H = -\sum_{i=1}^{k} p_i \, log_2 \, p_i \tag{5.17}$$

Where p stand for probability of the unique pattern and k is defined for number of unique values in the data sample.

Approximate Entropy (ApEn) works by finding the distance between vector of two neighbouring sequences in EEG time series [61]. A large value indicates that the signal is random while a small value indicates the presence of rhythmic pattern. ApEn is inefficient on short time signal. The method then was improved by introducing Sample Entropy (SampEn). SampEn can be calculated as follows:

$$SampEn\left(m,r,N\right) = ln\left[\frac{B^m\left(r\right)}{A^m\left(r\right)}\right] \tag{5.18}$$

Where m is pattern length, N is number of data point and r is vector comparison distance.

Lempel-Ziv measure entropy by transforming a signal $S\left(n\right)$ into a binary sequence based on defined threshold, T_d. It is recommended that median of the signal is used as the threshold due to its robustness to outliers [62]. The complexity counter will increase each time a new subsequence of consecutive character is found.

$$S = s\left(1\right), s\left(2\right), \ldots, s\left(n\right) \tag{5.19}$$

$$s\left(i\right) = \{1, if \; s\left(i\right) < T_d 0, otherwise \tag{5.20}$$

Hurst exponents measure the presence or absence of long-range dependence and its extent in a signal [63], [64]. It can be defined as:

$$H = \frac{log\dfrac{R}{S}}{log T} \tag{5.21}$$

where $\dfrac{R}{S}$ is the corresponding value of rescaled range and T is the data sample duration. The Hurst exponent range is from 0 to 1. Value less than 0.5 will indicate the data is anti-correlated, which means that if the sampling point at an instantaneous time is decreased the data at the next sampling point should be increased. Alternatively, if the current sampling point is an increase, next sampling point will be a decrease. Value higher that 0.5 indicates a correlated data where an increase of current sampling point will be followed by an increase of value on the next point and vice versa.

Morphology method is adapted based on the neurologist assessment in identifying the epileptic discharge. Introduce in 1976, this method utilizes width amplitude and slope as feature [65]–[67]. Commonly, threshold is set as reference to distinguish between epileptic and non-epileptic discharge. Previous research has defined upper amplitude threshold to 575uV and 30uV for lower amplitude threshold. Width threshold is set between 45ms and 225ms. Slope threshold is set to 0.00375 × standard deviation for lower limit and 0.030 × standard deviation for upper limit [68]. In 2016, threshold was redefined by using 500uV for upper amplitude and 20uV for lower amplitude [69]. Width threshold is set to 200ms for halfwave. However, this method is not efficient since threshold for every subject is different and may mix up with artefact. Dynamic amplitude threshold using moving average was introduced to overcome the challenge [46]. The threshold Th_n for nth sample can be calculated using this equation

$$Th_n = Th_{n-1} \times \xi + f \; abs\left(\hat{z}_n - Aver_n\right) \times \left(1-\xi\right)$$ (5.22)

Where ξ is the regression factor that is 0.9967. f abs means absolute value, $Aver_n$ is the moving averange value of non-stationary component $\hat{z}_n$

5.4.3 CLASSIFICATION TECHNIQUES

Data classification is a process of organizing complicated data by its common and related categories. Classification is the process of automatically creating a model of classes from a dataset and forecast class labels based on other attributes. The general objective of classification is to be able to locate and assess the data easily when needed. EEG wave classification is used to automate the process of manually visually analyzing the EEG signals. Classification is used in EEG signal domain especially in classifying seizure signals. As discussed previously, epileptic discharge features included amplitude, frequencies amplitude, signal entropy, power and energy in frequency bands. Spatial distributions can be extracted from the segments for the classification process.

Classification algorithm may be classified into two main types: supervised and unsupervised classification. Unsupervised classification is used when the algorithm builds a model from a dataset with given input with no set of outputs. Unsupervised classifiers develop a pattern with unlabeled data. Unsupervised classification, also known as clustering, groups a dataset into common attributes or differences categories. Unsupervised classification is able to organize information and learn the pattern from previous training.

Common unsupervised classification algorithms are k-means. K-means clustering algorithm is the simplest unsupervised algorithm and it is based on centroid and distance. Its objective to distinguish a dataset into clusters based on minimization of the sum of square of points to the centroid of each cluster. Orhan used k-means to cluster wavelet coefficients of EEG segments into groups according to their features and similarities [70]. Exarchos and Tzallas used k-means clustering to classify epileptic discharge (spikes) and brief events in EEG signals [71], [72]. Wahlberg used fuzzy k-means to cluster epileptic spikes [73].

Supervised classification builds a pattern of recognition based on a dataset with inputs and known outputs. The supervised classifiers need a labelled data to classify the dataset. Supervised classification is widely used in detecting epileptic discharges from the EEG signals. The most commonly used supervised classification algorithms in the epileptic research are k-Nearest Neighbour [74], Support Vector Machine [75], Naïve Bayes [76] and decision trees [77].

k-Nearest Neighbour (kNN) is a simple, supervised classification algorithm that groups the categories based on distance Euclidean. kNN algorithm presumes that common things are close to one another. kNN algorithm is easy to implement, but it does not perform well with increase in variables.

Support Vector Machine (SVM) is a supervised classification model that categorizes dataset based on labelled train data. The algorithm searches for an optimal decision boundary, also known as hyperplane, to separate the groups [78]. SVM is widely used because of its flexibility in high-dimensional spaces and strong generalization ability [78], [79]. Kumar et al. and Zhang et al. used SVM to predict and classify seizure in the EEG signal [75], [79].

Naïve Bayes (NB) is a supervised classifier that calculates the probability of groups based on Bayesian theory with the hypothesis that each attribute is independent of each other. The benefit of NB classification is that it is simple to build, and it is suitable for multi-class classification. NB requires a small amount of training data to build the model. Sharmila and Geethanjali observed significant differences in the performance between NB and kNN classifier to detect epileptic seizure based on DWT statistical features. NB has higher accuracy and lesser computation time than kNN.

Decision tree (DT) is a supervised classification algorithm that uses a tree-like structure in which it has root node, decision node and leaf nodes. Each node represents an observation and possible outcomes of choices. Decision tree is a method for approximating discrete-valued functions. Polat and Gunes have utilized decision tree classifier to distinguish epileptic seizures and normal EEG signal [77]. This study has demonstrated that the method used has higher classification accuracy and sensitivity [77].

Previous research has established that the performance of each classifier differs from each other. Research by Acharya et al. evaluate the performance of seven different classifiers: Fuzzy Sugeno Classifier (FSC), Support Vector Machine (SVM), k-Nearest Neighbour (kNN), Probabilistic Neural Network (PNN), decision tree (DT), Gaussian Mixture Model (GMM) and Naive Bayes Classifier (NBC) [80]. The researchers observed that Fuzzy contributed the highest accuracy followed by SVM, GMM, ANN, kNN, DT and NB [80].

Another research by Sood and Booshan postulated that SVM and ANN have higher accuracy in epileptic signal classification followed by NB and Radial Base Function (RBF) [81]. It has been demonstrated that Noise-aware signal combination (NSC) has higher accuracy than SVM, kNN, ANN and NB classifiers [82]. T. Zhang in 2016 demonstrated that genetic algorithm SVM (GA-SVM) has higher accuracy compared to the normal SVM, ANN, kNN and Linear Discriminant Classifier (LDA) [83].

All the previous researchers are using the same EEG signal database from the University of Bonn. Table 5.2 demonstrated that SVM has a consistent performance in accuracy. The difference in the performance between classifiers may be based on the features and attributes chosen.

TABLE 5.2

Performance Comparison between Classifiers.

Acharya et. al. (2015)		Sood & Bhooshan (2014)		Abualsaud et. Al (2015)		Zhang & Chen (2016)	
Classifiers	Accuracy	Classifiers	Accuracy	Classifiers	Accuracy	Classifiers	Accuracy
Fuzzy	98.1%	SVM	100%	NSC	90%	GA-SVM	100%
SVM	95.9%	ANN	100%	SVM	86%	SVM	98.0%
GMM	95.9%	NB	98%	kNN	83%	ANN	98.0%
ANN	93%	RBF	96%	ANN	72%	kNN	97.3%
kNN	93%			NB	73%	LDA	96.7%
DT	88.5%						
NB	88.1%						

The supervised classification has been widely used in the research of epileptic discharge detection. This may be because the label data provide more accurate outcomes.

5.5 SUMMARY

EEG (a brain activity signal with non-invasive recording method) plays a significant role in epileptic discharge detection in clinical setting. The present techniques used in identifying the presence of epileptic discharge by classifying sharp and spike has huge potential for research expansion in increasing the accuracy, reliability and quantification. This chapter is aimed to provide guidance to explore EEG as a bio-signal that is able to capture brain activity for epileptic discharge detection and classification.

REFERENCES

[1] H. L. Attwood and W. A. Mackay, *Essentials of Neurophysiology*. Berlin: Springer, 1989.

[2] W. Tatum, A. Husain, S. Benbadis, and P. Kaplan, *Handbook of EEG Interpretation*. Berlin: Springer, 2008.

[3] A. N. Thompson, I. Kim, T. D. Panosian, T. M. Iverson, T. W. Allen, and C. M. Nimigean, "Mechanism of potassium-channel selectivity revealed by Na+ and Li+ binding sites within the KcsA pore," *Nat. Struct. Mol. Biol.*, vol. 16, no. 12, pp. 1317–1324, 2009, doi:10.1038/nsmb.1703.

[4] C. Molnar and J. Gair, "The nervous system," in *Concepts of Biology*, 1st Canadian Edition. Toronto: BCcampus, 2015, pp. 653–716.

[5] F. H. Lopes Da Silva and W. Storm Van Leeuwen, "The cortical source of the alpha rhythm," *Neurosci. Lett.*, vol. 6, no. 2–3, pp. 237–241, 1977, doi:10.1016/0304-3940(77)90024-6.

[6] C. Mulert and L. Lemieux, "EEG—fMRI: Physiological basis, technique, and applications," *EEG—FMRI Physiol. Basis, Tech. Appl.*, pp. 1–539, 2010, doi:10.1007/978-3-540-87919-0.

[7] M. Sela, et al., *Electroencephalography: An introductory Text and Atlas.* Chicago: American Epilepsy Society, 2002.

[8] J. L. Stone and J. R. Hughes, "Early history of electroencephalography and establishment of the american clinical neurophysiology society," *J. Clin. Neurophysiol.*, vol. 30, no. 1, pp. 28–44, 2013, doi:10.1097/WNP.0b013e31827edb2d.

[9] D. Hagemann, J. Hewig, C. Walter, and E. Naumann, "Skull thickness and magnitude of EEG alpha activity," *Clin. Neurophysiol.*, vol. 119, no. 6, pp. 1271–1280, 2008, doi:10.1016/j.clinph.2008.02.010.

[10] M. Soufineyestani, D. Dowling, and A. Khan, "Electroencephalography (EEG) technology applications and available devices," *Appl. Sci.*, vol. 10, no. 21, pp. 1–23, 2020, doi:10.3390/app10217453.

[11] S. R. Sinha, "EEG instrumentation," *Pract. Epilepsy*, 2018, doi:10.1891/9781617051876. 0010.

[12] J. N. Acharya and V. J. Acharya, "Overview of EEG montages and principles of localization," *J. Clin. Neurophysiol.*, vol. 36, no. 5, pp. 325–329, 2019, doi:10.1097/WNP.0000000000000538.

[13] C. Y. Jung, "A review on EEG artifacts and its different removal technique," *Asia-pacific J. Converg. Res. Interchang.*, vol. 2, no. 4, pp. 45–62, 2016, doi:10.21742/apjcri.2016.12.06.

[14] W. O. Tatum, B. A. Dworetzky, and D. L. Schomer, "Artifact and recording concepts in EEG," *J. Clin. Neurophysiol.*, vol. 28, no. 3, pp. 252–263, 2011, doi:10.1097/WNP.0b013e31821c3c93.

[15] M. Sazgar and M. G. Young, "EEG artifacts," in *Absolute Epilepsy and EEG Rotation Review.* Cham: Springer International Publishing, 2019, pp. 149–162.

[16] A. G. Correa, E. Laciar, H. D. Patiño, and M. E. Valentinuzzi, "Artifact removal from EEG signals using adaptive filters in cascade," *J. Phys. Conf. Ser.*, vol. 90, p. 012081, 2007, doi:10.1088/1742-6596/90/1/012081.

[17] J. Iriarte, et al., "Independent component analysis as a tool to eliminate artifacts in EEG: A quantitative study," *J. Clin. Neurophysiol.*, vol. 20, no. 4, pp. 249–257, 2003, doi:10.1097/00004691-200307000-00004.

[18] M. Pal, R. Roy, J. Basu, and M. S. Bepari, "Blind source separation: A review and analysis," *2013 Int. Conf. Orient. COCOSDA Held Jointly with 2013 Conf. Asian Spok. Lang. Res. Eval. O-COCOSDA/CASLRE 2013*, Gurgaon: IEEE, pp. 20–24, 2013, doi:10.1109/ICSDA.2013.6709849.

[19] S. M. M. Islam, E. Yavari, A. Rahman, V. M. Lubecke, and O. Boric-Lubecke, "Separation of respiratory signatures for multiple subjects using independent component analysis with the JADE algorithm," *Proc. Annu. Int. Conf. IEEE Eng. Med. Biol. Soc. EMBS*, vol. 2018–July, pp. 1234–1237, 2018, doi:10.1109/EMBC.2018.8512583.

[20] C. Guerrero-Mosquera, A. Malanda, and A. Navia-Vazquez, "EEG signal processing for epilepsy," *Epilepsy—Histol. Electroencephalogr. Psychol. Asp.*, vol. 3, pp. 49–74, 2012, doi:10.5772/31609.

[21] I. Rejer and P. Górski, "Independent component analysis for EEG data preprocessing—Algorithms comparison," *Lect. Notes Comput. Sci. (including Subser. Lect. Notes Artif. Intell. Lect. Notes Bioinformatics)*, vol. 8104 LNCS, pp. 108–119, 2013, doi:10.1007/978-3-642-40925-7_11.

[22] M. Mamun, M. Al-Kadi, and M. Marufuzzaman, "Effectiveness of wavelet denoising on electroencephalogram signals," *J. Appl. Res. Technol.*, vol. 11, no. 1, pp. 156–160, 2013, doi:10.1016/S1665-6423(13)71524-4.

[23] I. Garvanov, V. Jotsov, and M. Garvanova, "Data science modeling for EEG signal filtering using wavelet transforms," *2020 IEEE 10th Int. Conf. Intell. Syst. IS 2020—Proc.*, IEEE, pp. 352–357, 2020, doi:10.1109/IS48319.2020.9199843.

[24] L. Logesparan, A. J. Casson, and E. Rodriguez-Villegas, "Assessing the impact of signal normalization: Preliminary results on epileptic seizure detection," *Proc. Annu. Int. Conf. IEEE Eng. Med. Biol. Soc. EMBS*, IEEE, pp. 1439–1442, 2011, doi:10.1109/IEMBS.2011.6090356.

[25] M. Azarbad, H. Azami, S. Sanei, and A. Ebrahimzadeh, "A time-frequency approach for EEG signal segmentation," *J. AI Data Min.*, vol. 2, no. 1, pp. 63–71, 2014 [Online]. Available: http://jad.shahroodut.ac.ir/article_151_44.html%5Cnhttp://jad.shahroodut.ac.ir/pdf_151_7bf964d0d2b1ef1353e582625f70df1e.html.

[26] W. Klonowski, "Everything you wanted to ask about EEG but were afraid to get the right answer," *Nonlinear Biomed. Phys.*, vol. 3, p. 2, 2009, doi:10.1186/1753-4631-3-2.

[27] B. B. Burle, L. Spieser, C. C. Roger, L. Casini, T. Hasbroucq, and F. Vidal, "Spatial and temporal resolutions of EEG: Is it really black and white? A scalp current density view," *Int. J. Psychophysiol.*, vol. 97, no. 3, pp. 210–220, 2015, doi:10.1016/j.ijpsycho.2015.05.004.

[28] V. K. Harpale and V. K. Bairagi, "Time and frequency domain analysis of EEG signals for seizure detection: A review," *Int. Conf. Microelectron. Comput. Commun. MicroCom 2016*, 2016, https://www.semanticscholar.org/.

[29] V. K. Mehla, A. Singhal, P. Singh, and R. B. Pachori, "An efficient method for identification of epileptic seizures from EEG signals using Fourier analysis," *Phys. Eng. Sci. Med.*, vol. 44, no. 2, pp. 443–456, 2021, doi:10.1007/s13246-021-00995-3.

[30] H. Hassanpour, M. Mesbah, and B. Boashash, "Time-frequency feature extraction of newborn EEC seizure using SVD-based techniques," *EURASIP J. Appl. Signal Proc.*, vol. 2004, no. 16, pp. 2544–2554, 2004, doi:10.1155/S1110865704406167.

[31] V. Giurgiutiu and L. Yu, "Comparison of short-time fourier transform and wavelet transform of transient and tone burst wave propagation signals for structural health monitoring," *4th Int. Work. Struct. Heal. Monit.*, pp. 1–9, 2003. https://www.semanticscholar.org/

[32] C. J. Hope and D. J. Furlong, "Time-frequency distributions for timbre morphing: The wigner distribution versus the STFT," *Proc. SBCMIV 4th Symp. Bras. Comput. Music.*, 1997, http://www.ciaranhope.com/paper1.html.

[33] V. Bajaj and R. B. Pachori, "Classification of seizure and non-seizure EEG signals using empirical mode decomposition," *IEEE Trans. Inf. Technol. Biomed.*, vol. 16, no. 6, pp. 1135–1142, 2012, doi:10.1109/TITB.2011.2181403.

[34] R. J. Martis, et al., "Application of empirical mode decomposition (Emd) for automated detection of epilepsy using eeg signals," *Int. J. Neural Syst.*, vol. 22, no. 06, p. 1250027, 2012, doi:10.1142/S012906571250027X.

[35] J. Zheng, J. Cheng, and Y. Yang, "Partly ensemble empirical mode decomposition: An improved noise-assisted method for eliminating mode mixing," *Signal Processing*, vol. 96, no. PART B, pp. 362–374, 2014, doi:10.1016/j.sigpro.2013.09.013.

[36] O. Faust, U. R. Acharya, H. Adeli, and A. Adeli, "Wavelet-based EEG processing for computer-aided seizure detection and epilepsy diagnosis," *Seizure*, vol. 26, pp. 56–64, 2015, doi:10.1016/j.seizure.2015.01.012.

[37] H. Goelz, R. D. Jones, and P. J. Bones, "Continuous wavelet transform for the detection and classification of epileptiform activity in the EEG," *Annu. Int. Conf. IEEE Eng. Med. Biol.—Proc.*, vol. 2, p. 941, 1999, doi:10.1109/iembs.1999.804095.

[38] J. Gutiérrez, R. Alcántara, and V. Medina, "Analysis and localization of epileptic events using wavelet packets," *Med. Eng. Phys.*, vol. 23, no. 9, pp. 623–631, Nov. 2001, doi:10.1016/S1350-4533(01)00096-0.

[39] H. S. Liu, T. Zhang, and F. S. Yang, "A multistage, multimethod approach for automatic detection and classification of epileptiform EEG," *IEEE Trans. Biomed. Eng.*, vol. 49, no. 12, pp. 1557–1566, 2002, doi:10.1109/TBME.2002.805477.

[40] M. S. Fathillah, K. Chellappan, R. Jaafar, R. Remli, and W. A. W. Zaidi, "Time-frequency analysis in ictal and interictal seizure epilepsy patients using electroencephalogram," *J. Theor. Appl. Inf. Technol.*, vol. 96, no. 11, pp. 3433–3443, 2018.

[41] A. Rosado and A. Rosa, "Automatic detection of epileptiform discharges in the EEG," in *IEEE EMBS Bioengin.*, IEEE, 2011, no. February, pp. 1–9.

[42] P. E. T. Jorgensen and M.-S. Song, "Comparison of discrete and continuous wavelet transforms," *arXiv Prepr. arXiv0705.0150*, pp. 1–30, 2013, doi:10.1007/978-1-4614-1800-9_34.

[43] R. H. Chowdhury, M. B. I. Reaz, M. A. B. M. Ali, A. A. A. Bakar, K. Chellappan, and T. G. Chang, "Surface electromyography signal processing and classification techniques," *Sensors (Basel)*, vol. 13, no. 9, pp. 12431–12466, 2013, doi:10.3390/s130912431.

[44] C. Torrence and G. P. Compo, "A practical guide to wavelet analysis," *Bull. Am. Meteorol. Soc.*, vol. 79, no. 1, pp. 61–78, 1998, doi:10.1175/1520-0477(1998)079<0061:APGTWA >2.0.CO;2.

[45] H. S. Liu, T. Zhang, and F. S. Yang, "A multistage, multimethod approach for automatic detection and classification of epileptiform EEG," *Ieee Trans. Biomed. Eng.*, vol. 49, no. 12, 2, pp. 1557–1566, 2002, doi:10.1109/TBME.2002.805477.

[46] R. S. Fisher, et al., "ILAE official report: A practical clinical definition of epilepsy," *Epilepsia*, vol. 55, no. 4, pp. 475–482, 2014, doi:10.1111/epi.12550.

[47] S. T. Herman, "Classification of epileptic seizures," *Contin. Lifelong Learn. Neurol.*, vol. 13, no. 4, pp. 13–47, 2007, doi:10.1212/01.CON.0000284533.66143.12.

[48] J. J. Halford, et al., "Inter-rater agreement on identification of electrographic seizures and periodic discharges in ICU EEG recordings," *Clin. Neurophysiol.*, vol. 126, no. 9, pp. 1661–1669, 2015, doi:10.1016/j.clinph.2014.11.008.

[49] L. J. Hirsch and H. A. Haider, "Electroencephalography (EEG) in the diagnosis of seizures and epilepsy," *UpToDate*, no. graph 1, pp. 1–31, 2015.

[50] R. S. Fisher, H. E. Scharfman, and M. DeCurtis, "How can we identify ictal and interictal abnormal activity?" *Adv. Exp. Med. Biol.*, vol. 813, 2014, pp. 3–23.

[51] S. Noachtar, C. Binnie, J. Ebersole, F. Mauguière, A. Sakamoto, and B. Westmoreland, "A glossary of terms most commonly used by clinical electroencephalographers and proposal for the report form for the EEG findings. The international federation of clinical neurophysiology," *Electroencephalogr. Clin. Neurophysiol.*, vol. 52, pp. 21–41, 1999, doi:10.1055/s-2003-812583.

[52] P. Li, C. Karmakar, C. Yan, M. Palaniswami, and C. Liu, "Classification of 5-S epileptic EEG recordings using distribution entropy and sample entropy," *Front. Physiol.*, vol. 7, pp. 1–9, 2016, doi:10.3389/fphys.2016.00136.

[53] H. Azami, H. Hassanpour, J. Escudero, and S. Sanei, "An intelligent approach for variable size segmentation of non-stationary signals," *J. Adv. Res.*, vol. 6, no. 5, pp. 687–698, 2015, doi:10.1016/j.jare.2014.03.004.

[54] P. P. Muhammed Shanir, Y. U. Khan, and O. Farooq, "Time domain analysis of EEG for automatic seizure detection," *Emerg. Trends Electr. Electron. Eng.*, pp. 1–5, 2015.

[55] C. Kamath, "A new approach to detect epileptic seizures in electroencephalograms using teager energy," *ISRN Biomed. Eng.*, vol. 2013, no. i, pp. 1–14, 2013, doi:10.1155/2013/358108.

[56] H. R. Al Ghayab, Y. Li, S. Siuly, and S. Abdulla, "Epileptic EEG signal classification using optimum allocation based power spectral density estimation," *IET Signal Process.*, vol. 12, no. 6, pp. 738–747, 2018, doi:10.1049/iet-spr.2017.0140.

[57] L. Guo, D. Rivero, J. A. Seoane, and A. Pazos, "Classification of EEG signals using relative wavelet energy and artificial neural networks," *Proc. first ACM/SIGEVO Summit Genet. Evol. Comput.—GEC '09*, no. July 2015, ACM, p. 177, 2009, doi:10.1145/1543834.1543860.

[58] M. S. Fathillah, R. Jaafar, K. Chellappan, R. Remli, and W. A. W. Zainal, "*Interictal Epileptic Discharge EEG Detection Based on Wavelet and Multiresolution Analysis*," IEEE, 2017, doi:10.1109/ICSEngT.2017.8123435.

[59] Y. Paul, "Various epileptic seizure detection techniques using biomedical signals: a review," *Brain Informatics*, vol. 5, no. 2, 2018, doi:10.1186/s40708-018-0084-z.

[60] C. E. Shannon, "A mathematical theory of communication," *Bell Syst. Tech. J.*, vol. 27, no. July 1928, pp. 379–423, 1948, doi:10.1145/584091.584093.

[61] M. S. Fathillah, R. Jaafar, K. Chellappan, R. Remli, and W. A. W. Zainal, "Multiresolution analysis on nonlinear complexity measurement of EEG signal for epileptic discharge monitoring," *Malaysian J. Fundam. Appl. Sci.*, vol. 14, no. 2, pp. 219–225, 2018.

[62] M. Aboy, R. Hornero, D. Abásolo, and D. Álvarez, "Interpretation of the Lempel-Ziv complexity measure in the context of biomedical signal analysis," *IEEE Trans. Biomed. Eng.*, vol. 53, no. 11, pp. 2282–2288, 2006, doi:10.1109/TBME.2006.883696.

[63] S. Geng, W. Zhou, Q. Yuan, D. Cai, and Y. Zeng, "{EEG} non-linear feature extraction using correlation dimension and Hurst exponent," *Neurol. Res.*, vol. 33, no. 9, pp. 908–912, 2011, doi:10.1179/1743132811Y.0000000041.

[64] A. El-Kishky, "Assessing entropy and fractal dimensions as discriminants of seizures in EEG time series," *2012 11th Int. Conf. Inf. Sci. Signal Process. their Appl. ISSPA 2012*, IEEE, pp. 92–96, 2012, doi:10.1109/ISSPA.2012.6310687.

[65] Y. C. Liu, C. C. K. Lin, J. J. Tsai, and Y. N. Sun, "Model-based spike detection of epileptic EEG data," *Sensors (Basel)*, vol. 13, no. 9, pp. 12536–12547, 2013, doi:10.3390/s130912536.

[66] Z. Ji, T. Sugi, S. Goto, X. Wang, and M. Nakamura, "Multi-channel template extraction for automatic EEG spike detection," in *2011 IEEE/ICME Int. Conf. Compl. Med. Eng., CME 2011*, IEEE, pp. 179–184, 2011, doi:10.1109/ICCME.2011.5876728.

[67] J. Gotman and P. Gloor, "Automatic recognition and quantification of interictal epileptic activity in the human scalp EEG," *Electroencephalogr. Clin. Neurophysiol.*, vol. 41, no. 5, pp. 513–529, 1976, doi:10.1016/0013-4694(76)90063-8.

[68] W. R. S. Webber, B. Litt, K. Wilson, and R. P. Lesser, "Practical detection of epileptiform discharges (EDs) in the EEG using an artificial neural network: A comparison of raw and parameterized EEG data," *Electroencephalogr. Clin. Neurophysiol.*, vol. 91, no. 3, pp. 194–204, 1994, doi:10.1016/0013-4694(94)90069-8.

[69] E. I. Zacharaki, I. Mporas, K. Garganis, and V. Megalooikonomou, "Spike pattern recognition by supervised classification in low dimensional embedding space," *Brain Informatics*, vol. 3, no. 2, pp. 73–83, 2016, doi:10.1007/s40708-016-0044-4.

[70] U. Orhan, M. Hekim, and M. Ozer, "EEG signals classification using the K -means clustering and a multilayer perceptron neural network model," *Expert Syst. Appl.*, vol. 38, no. 10, pp. 13475–13481, 2011, doi:10.1016/j.eswa.2011.04.149.

[71] T. P. T. P. Exarchos, A. T. A. T. Tzallas, D. I. D. I. Fotiadis, S. Konitsiotis, and S. Giannopoulos, "EEG transient event detection and classification using association rules," *IEEE Trans. Inf. Technol. Biomed.*, vol. 10, no. 3, pp. 451–457, 2006, doi:10.1109/TITB.2006.872067.

[72] A. T. Tzallas, P. S. Karvelis, C. D. Katsis, D. I. Fotiadis, S. Giannopoulos, and S. Konitsiotis, "A method for classification of transient events in EEG recordings: Application to epilepsy diagnosis," *Methods Inf. Med.*, vol. 45, no. 6, pp. 610–621, 2006, doi:06060610 [pii].

[73] P. Wahlberg and G. Lantz, "Methods for robust clustering of epileptic EEG spikes," *IEEE Trans. Biomed. Eng.*, vol. 47, no. 7, pp. 857–868, 2000, doi:10.1109/10.846679.

[74] A. Mohamed, K. B. Shaban, and A. Mohamed, "Effective seizure detection through the fusion of single-feature enhanced-k-NN classifiers of EEG signals," in *2013 ISSNIP*

Biosignals and Biorobotics Conference: Biosignals and Robotics for Better and Safer Living (BRC), IEEE, pp. 1–6, 2013, doi:10.1109/BRC.2013.6487534.

[75] Y. Kumar, M. L. Dewal, and R. S. Anand, "Epileptic seizure detection using DWT based fuzzy approximate entropy and support vector machine," *Neurocomputing*, vol. 133, pp. 271–279, 2014, doi:10.1016/j.neucom.2013.11.009.

[76] A. Sharmila and P. Geethanjali, "DWT based detection of epileptic seizure from EEG signals uInaive bayes and k-NN classifiers," *IEEE Access*, vol. 4, pp. 7716–7727, 2016, doi:10.1109/ACCESS.2016.2585661.

[77] K. Polat and S. Gü Nes, "Classification of epileptiform EEG using a hybrid system based on decision tree classifier and fast Fourier transform." *Appl. Math. Comput.*, vol. 187, no. 2, pp. 1017–1026, 2007, doi:10.1016/j.amc.2006.09.022.

[78] V. Vapnik and O. Chapelle, "Bounds on error expectation for support vector machines," *Neural Comput.*, vol. 12, no. 9, pp. 2013–2036, 2000, doi:10.1162/089976600300015042.

[79] Z. Zhang, Y. Zhou, Z. Y. Chen, X. H. Tian, S. H. Du, and R. M. Huang, "Approximate entropy and support vector machines for electroencephalogram signal classification," *Neural Regen. Res.*, vol. 8, no. 20, pp. 1844–1852, 2013, doi:10.3969/J. ISSN.1673-5374.2013.20.003.

[80] U. R. Acharya, F. Molinari, S. V. Sree, S. Chattopadhyay, K. H. Ng, and J. S. Suri, "Automated diagnosis of epileptic EEG using entropies," *Biomed. Signal Process. Control.*, vol. 7, no. 4, pp. 401–408, 2012, doi:10.1016/j.bspc.2011.07.007.

[81] M. Sood and S. V Bhooshan, "Automatic processing of EEG signals for seizure detection using soft computing techniques," *2014 Recent Adv. Innov. Eng.*, p. IEEE; Poornima Coll Engn, 2014, doi:10.1109/ICRAIE.2014.6909180.

[82] K. Abualsaud, M. Mahmuddin, M. Saleh, and A. Mohamed, "Ensemble classifier for epileptic seizure detection for imperfect EEG data," *Sci. World J.*, vol. 2015, 2015, doi:10.1155/2015/945689.

[83] M. Li, W. Chen, and T. Zhang, "Classification of epilepsy EEG signals using DWT-based envelope analysis and neural network ensemble," *Biomed. Signal Process. Control*, vol. 31, pp. 357–365, 2017, doi:10.1016/J.BSPC.2016.09.008.

6 A Novel End-to-End Secure System for Automatic Classification of Cardiac Arrhythmia

Narendra K. C., Pradyumna G. R.
and Roopa B. Hegde

6.1 INTRODUCTION

Cardiovascular diseases (CVDs) are a significant cause of death globally, with an estimate of affecting 17.9 million lives each year [1]. CVDs are disorders related to heart and blood vessels, including coronary heart disease, cerebrovascular disease, rheumatic heart disease and other conditions. More than four out of five CVD deaths are due to heart attacks and strokes, and one-third of these deaths occur prematurely in people under 70 years of age. Developing countries like India have one of the highest burdens of CVD worldwide [2]. Cardiac arrhythmias and their long-term effects are the primary cause of CVDs that are overlooked in fatal issues.

Arrhythmia is a medical condition that occurs due to irregular heart rate. This irregular heart rhythm can be identified by a process called electrocardiography (ECG). In this process, electrodes are placed on the chest to detect the heart's electrical activities. This produces a tiny waveform called an ECG signal, and the normal ECG signal is shown in Figure 6.1. Any change in the pattern of signal is considered abnormal. Physicians assess the signals for identification of any irregular way. This process is laborious, and the assessment result is subjective based on experience. Also, a lot of training is required for the accurate detection of abnormal ECG patterns. Detection of heart arrhythmia in the medical field is essential for timely diagnosis and treatment.

In the past few years, many algorithms have been proposed for automatic annotation and classification abnormalities in ECG signals using the deep learning (DL) approach. The challenge in developing automated ECG classification is to manage the irregularities in the signals, which is very important to detect the patient status. Hence, an accurate automated system is highly desirable. DL approach using convolution neural network (CNN) has shown promising results, and hence it is a striking success in classification tasks. Recently, a deep network trained using 12-lead ECG recordings of MIT-BIH database outperformed manual identification with an F1-Score of 80% in classifying six types of abnormalities [4]. Further, an accuracy

DOI: 10.1201/9781003201137-8

TABLE 6.1
Abbreviations.

Explication	Abbreviation
Electrocardiogram	ECG
Intrinsic mode functions	IMF
Cardiovascular disease	CVD
Deep learning	DL
Convolution neural network	CNN
Recurrent neural network	RNN
Long short-term memory	LSTM
Normal beat	N
Left bundle branch block beat	L
Right bundle branch block beat	R
Atrial premature beat	A
Premature ventricular contraction	V
Isolated QRS-like artifact	I
Empirical mode decomposition	EMD
One-time password	OTP
Diffie Hellman key exchange algorithm	DHKEA
Message authentication code	MAC

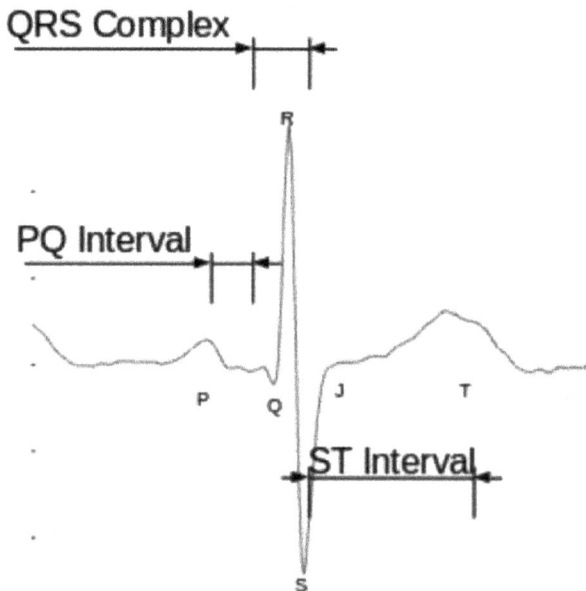

FIGURE 6.1   Normal ECG signal [3].

of 97% was reported [5] for the classification of 17 types of arrhythmias listed in the MIT-BIH dataset by combining tower graph-based features election and CNN methods. To address the irregularities present in the ECG signal, a robust and anti-noise capable 1D CNN consisting of 12 layers was developed [6]. Also, five micro-types of heartbeat taken from MIT- BIH Arrhythmia database achieved an average classification accuracy of around 97%. The occurrence of noise is unavoidable while recording ECG signals. Noise cancellation without loss of the necessary information is an essential preprocessing step for developing automated classification systems. A combination of recurrent neural networks (RNNs) with bidirectional long short-term memory (LSTM) model was built [7] for noise cancellation and ECG signal classification. The model was trained and tested using the QT database and achieved an average accuracy of 99.4% for the variety of four types of ECG signals. However, the aforementioned automated ECG classifier networks were tested on a single dataset. A practically usable system needs to be tested using various datasets to evaluate the performance for all possible variations. Recently, the performance of deep CNN was assessed for four ECG databases, namely MIT-BIH arrhythmia database, American Heart Association (AHA) database, QT database and common standards for electro-cardiography (CSE) database [8].

Researchers demonstrated that higher accuracy could be achieved by developing a deep network. Such an efficient system will improve the decision making and clinical practice even in rural healthcare centers. Patients from remote areas face difficulty accessing primary health care facilities for their sustainability due to a lack of facilities. Moreover, in remote villages, expert persons are not available who can observe the symptoms accurately. Hence, an online secured transmission technique is needed for transmitting various clinical information to physicians.

To use telemedicine facilities, security and privacy of personal physiological data is a significant concern, and it should be protected while being transmitted in public communication channels. Hence, there is a need for secure data transmission to ensure confidentiality. This issue was addressed by many research groups [9–15]. To improve the security of private cypher codes, a study showed that chaotic pseudorandom number generators within specific control parameters could dynamically produce unordered sequence numbers to set the secret keys for a regular remote key update [9]. The model was evaluated using 100 ECG fragments of the MIT-BIH database and obtained a peak signal-to-noise ratio (SNR) of about 35.26 ± 3.77 dB. Internet of things (IoT) based health monitoring systems aid remote health assessment when clubbed with secure data transmission. A lightweight ECG signal strength analysis was developed for real-time automatic classification [10]. The system was evaluated using MIT-BIH and PhysioNet challenges databases. Data security and privacy is an extensive practice in tertiary hospitals. An algorithm to securely encrypt the patient's data and transmit it to the authorized doctor in hospital premises was proposed [11] using wireless devices. Metaheuristic salp swarm-based secure ECG transmission was presented [12] to meet the challenges of medical data security and privacy. An encryption key was generated from the weight vector of the fittest salp, and decryption was achieved in 8.8 seconds. Nowadays, medical facilities are being provided to the rural population through remote healthcare centers. These centers send medical data to tertiary centers for accurate diagnosis. Hence, data security and privacy play an essential role. To address this issue, a lightweight cellular automaton-based encryption algorithm was proposed [13] for the secure transmission of ECG data using the MIT-BIH arrhythmia database. The telehealth system plays a vital role in providing primary medical care in rural regions. An attempt towards this was carried out [14] by developing a telehealth system for the secure transmission of various physiological data classification of the physiological data. This system achieved an overall classification accuracy of 99.56%. Real-time health monitoring and checking devices are commonly used in the sports field to evaluate the health of athletes. However, the transmission of athletes' data demands privacy and security as it contains personal information. The data block scrambling method was presented [15] to secure ECG data of tennis players. This method encodes the data without key management.

In this work, we proposed a novel end-to-end secure system for automated classification of six types of cardiac arrhythmia, namely normal beat (N), left bundle branch block beat (L), right bundle branch block beat (R), atrial premature beat (A), premature ventricular contraction (V) and isolated QRS-like artifact (I). Further, a lightweight cryptographic scheme is developed that aids in the remote diagnosis of the ECG signals. The proposed two-key authentication scheme ensures the confidentiality of the signals between the intended users.

6.2   METHODOLOGY

ECG data is noisy, and it is hard to detect arrhythmic events in ECG signals. Convolutional neural networks are widely used to identify arrhythmia in the signals automatically. Towards denoising the signals, nonlinear and non-stationary signal analysis methods are employed. The enhancement algorithms proposed in the literature contribute to improving the system performance. In this work, we propose an innovative denoising algorithm and verify its utility in enhancing the system's accuracy, F1-score, precision and recall of the proposed method. In this work, we consider two different databases and two different architectures. The study is carried out to verify the viability of the temporal enhancement algorithm, namely the empirical mode decomposition.

6.2.1   DATASET

The data used in this study is as reported in [16]. The S12L-ECG dataset consists of data records from various subjects across various ages and genders. The signal is acquired with a sampling rate of 400 Hz, and the data is recorded for 7s. The ECG records are taken from all 12 leads and are considered for decision-making for detecting arrhythmia. Annotations performed by various cardiologists take the ground truth for the conditions. The details of the database can be found in https://doi.org/10.5281/zenodo.3625006.

The second database used in this study is the MIT-BIH database that contains records of 47 subjects, and the data was sampled at 360 Hz. The annotations are provided for 23 records. This part of the database can be used for training the network. The details of the network are provided in the next section.

6.2.2 PROPOSED METHOD

In this work, the effect of denoising ECG signals is evaluated on the system performance. The proposed algorithm for denoising employs empirical mode decomposition (EMD) [16]. EMD is a nonlinear and non-stationary signal analysis method that provides nonlinear and non-stationary components derived from the signal termed as intrinsic mode functions (IMFs). In this work, ten IMFs are generated on both datasets, and the IMFs are selected by computing the Hurst exponent. If the IMF is more significant than a specified threshold, then the IMF is set to reconstruct the signal. If the IMF is lesser than the selected threshold, then the IMF is not considered in the rebuilding of the movement. This substantially improves the quality of the signal and will also improve the system performance. The proposed methodology is shown in Figure 6.2. The raw data is encrypted using a cryptographic method before transmitting, and it is decrypted at the receiving end. Denoising is performed on the decoded data, followed by the design of a deep network for classifying arrhythmia.

The enhanced data is provided as input to a deep neural network that consists of two convolutional 1D layers that operate on 4096 samples of the ECG record with 64 filters. The second layer takes the input of 4096 and has 16 filter elements. This layer is followed by a max pooling layer with scaling factor 2 and a ReLU unit. Residual blocks then follow this. These residual blocks are used as libraries that are defined in. This network is trained using the S12L-ECG database. The training phase of the network was done using the backpropagation method for error correction. The network was trained for 40 epochs of data, and 85k parameters were introduced. Adam optimization was used while preparing the network.

Another network consisting of ten layers is designed and trained using the MIT-BIH database. Each frame of ECG data consists of 320 samples. The input layer of the network has 64 filters that perform 1-D convolutions. The second layer is a pooling layer that will scale down by 2, followed by a convolution layer of 1D. This is followed by a batch normalization layer followed by convolution layer, max pooling layer and batch normalization layer followed by flattening layers and dense layers that are fully connected. The final layer provides the output class of the arrhythmia that the frame of data belongs to. The network employs the backpropagation algorithm with an Adam optimizer for training the network.

The DL approach can classify ECG data into one of the six types. However, practically usable real-time health monitoring systems, either in tertiary centers or rural healthcare centers, demand data security and privacy. Data security can be achieved

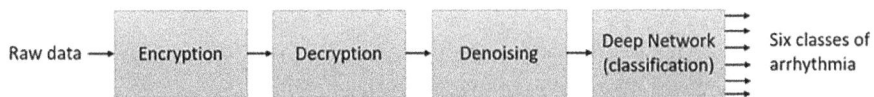

Raw data → Encryption → Decryption → Denoising → Deep Network (classification) → Six classes of arrhythmia

FIGURE 6.2   The proposed workflow.

ECG Data divided into 8 blocks

1	2	3	4	5	6	7	8

Scrambled blocks for the OTP 57328146

5	7	3	2	8	1	4	6

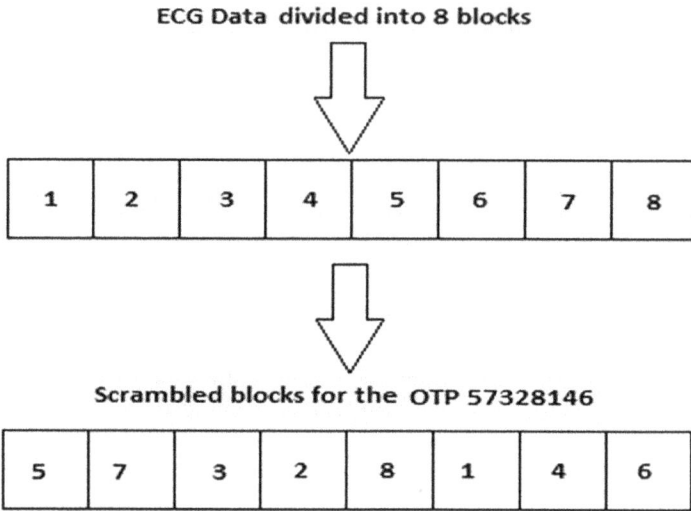

FIGURE 6.3 OTP based data scrambling.

by encoding the ECG data while transmitting and decoding the same at the receiving end. A novel two key lightweight scheme is used in the proposed work while transmitting the data to the other end. The proposed lightweight scheme uses a one-time password (OTP) based critical generation method. A two-step encryption technique is used in this method, with the key size limited to 32 bits. In the first step, once the source and destination are identified, a unique eight-digit OTP is generated, and the generated OTP will be shared between the source and destination. This OTP acts as a base for the encryption technique. The ECG signal to be delivered to the destination is divided into eight blocks of data, and each block is scrambled based on the order of the digits present in the unique OTP, as shown in Figure 6.3. In the second step, the Diffie Hellman key exchange algorithm (DHKEA) is used for exchanging the key between the source and destination. The obtained key from the DHKEA is used to perform encryption using a standard symmetric-key algorithm (Simplified Data Encryption Standard (SDES)). Finally, a message authentication code (MAC) is added to the encrypted block to maintain the authenticity of the data. The working of encryption and decryption is as shown in Table 6.2.

6.3 RESULTS AND DISCUSSIONS

This section provides the results of the experimentation that was conducted. Confusion matrices of both the networks with and without denoising are shown in Figure 6.4 and Figure 6.5. The results indicate a significant improvement in the system performance when the denoised data is used.

The confusion matrices indicate that the EMD based denoising has significantly improved the system performance. The improvement can be attributed to the

TABLE 6.2

Encryption and Decryption Stages.

Encryption Stage	Decryption Stage
Get the ECG data in the binary form	Store the received OTP
Divide the ECG data into eight blocks	Store the received key K1
Generate OTP, share it with the receiver	Generate MAC for the received encrypted block
Scramble the ECG data blocks based on the OTP	Verify the MAC
Generate a symmetric key K1 using DHKEA, share it with the receiver	Decrypt the data using SDES and K1
Using the key K1, perform encryption using SDES	Divide the decrypted data into eight blocks
Generate MAC for the encrypted block, attach to encrypted data and transmit	Arrange the output based on the OTP

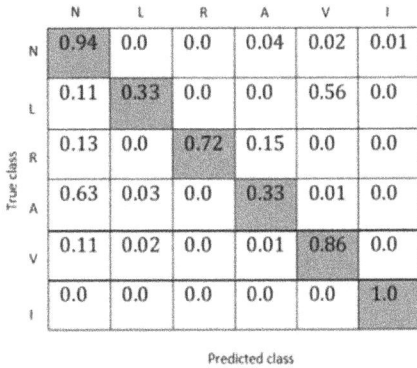

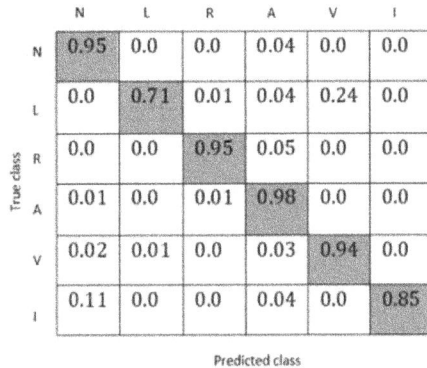

FIGURE 6.4 Confusion matrix for the S12L-ECG lead without denoising (left) and EMD denoising (right) after training for 40 epochs.

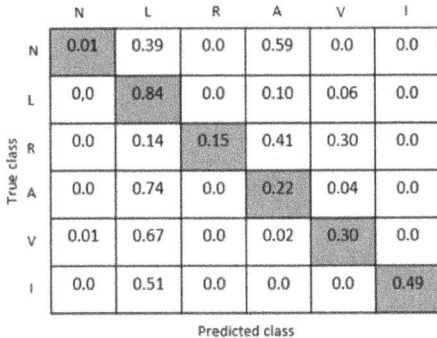

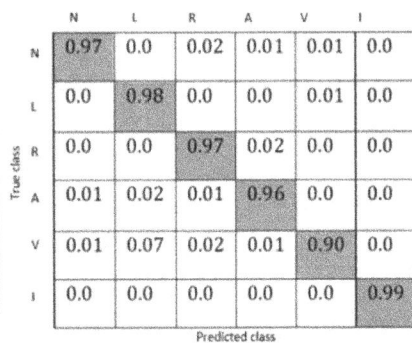

FIGURE 6.5 Confusion matrix for the MIT-BIH database without denoising (left) and EMD denoising (right) after training 40 epochs.

reduction in the components common to the arrhythmia classes. As observed in Figure 6.4, out of the six classes the R and the L are the most confused classes. This could be due to the short time nonlinearities observed in the signal common to both types. Particularly in the case of the left bundle branch, the most confused class is premature ventricular contraction. A probable reason for this would be similar temporal characteristics in the ECG signal. This confusion is significantly reduced in the confusion matrix, obtained by providing denoised signals to the network.

In the second study, the MIT-BIH database is used for training the network. The confusion matrix indicates when the raw ECG signal is presented to the network. The network performance is less than the network performance when the denoised signal is provided as input to the network. The improvement in the system performance is improved significantly. The probable reason for this improvement is that the IMFs selected for the reconstruction of the two lead ECG data drastically alleviate the effect of the other IMFs that hamper the quality of the ECG signal.

The results in Table 6.3 indicate that the network trained on the MIT-BIH ECG database is better than the network trained on the S12L-ECG dataset.

The average accuracy of the deep network mode on MIT-BIH data outperforms the network trained on the S12L-ECG web. In both the networks, the denoised data is drastically improving the accuracy and F1-Score performance. This improvement can be attributed to the choice of the IMFs used in reconstructing the ECG signal.

The working of encryption and decryption was tested using the proposed secure algorithm. The OTP based scrambled block of data was encrypted using the SDES symmetric key algorithm and sent to the destination after adding the MAC. On the receiver side, the received information was tested for message authenticity and decrypted using the key generated by DHKEA. Since the decoded bits were in the scrambled form, the unique OTP was used again to retrieve the original data. In the existing systems, the size of the keys varies from 512 bits to 2048 bits, which is not suitable for IoT models. Also, an increase in bits results in high computational cost. Hence, the critical size was limited to 32 bits in the proposed lightweight encryption method. This method provides an option for secure transfer of ECG signals with less computation cost and less computation time.

TABLE 6.3

Accuracy, Precision, Recall and F1-Score of the Classifiers Used in This Work.

Network/Database	Accuracy	Precision	Recall	F1-Score
Network1- S12L-ECG-without denoising	89.48	0.69	0.76	0.68
Network1- S12L-ECG-with denoising	97.06	0.91	0.925	0.91
Network 2—MIT-BIH-without denoising	62.79	0.33	0.43	0.30
Network 2—MIT-BIH-with denoising	98.78	0.963	0.965	0.963

6.4 CONCLUSION

This work focuses on developing a secure end-to-end scheme that encrypts, decrypts and performs denoising of ECG signals to identify cardiac arrhythmias. In this work, we propose a novel end-to-end secure system for classifying cardiac arrhythmia. Towards this end, two deep learning architectures are explored and the performance comparison of the networks is made. The network performance is evaluated on the raw data and denoised data. The results indicate a significant improvement in the system's accuracy, precision, recall and F1-score. The improvement in the performance is attributed to the selection of the intrinsic mode functions (IMF) of the signal on which empirical mode decomposition is performed.

Further, a lightweight cryptographic scheme is developed that aids in the remote diagnosis of the ECG signals. The proposed two-key authentication scheme ensured the confidentiality of the signals between the intended users. The annotated ECG could be made available for further diagnosis.

Extensive experiments carried out on the MIT-BIH arrhythmia databases indicate significant improvement in the system performance in the classification of arrhythmia. The proposed novel lightweight cryptography scheme is also analyzed by creating test cases and reported to provide sufficient security and confidentiality of the ECG data.

REFERENCES

1. Ruan Y, Guo Y, Zheng Y, Huang Z, Sun S, Kowal P, Shi Y, Wu F., Cardiovascular disease (CVD) and associated risk factors among older adults in six low-and middle-income countries: results from SAGE Wave 1. *BMC Public Health*, 18(1): 1–13, 2018. doi: 10.1186/s12889-018-5653-9. PMID: 29925336; PMCID: PMC6011508. www.who.int/health-topics/cardiovascular-diseases#tab=tab_1, Accessed on 20/11/2021.

2. Huffman MD, Prabhakaran D, Osmond C, Fall CH, Tandon N, Lakshmy R, Ramji S, Khalil A, Gera T, Prabhakaran P, Biswas SK, Reddy KS, Bhargava SK, Sachdev HS., Incidence of cardiovascular risk factors in an Indian urban cohort result from the New Delhi birth cohort. *Journal of American College of Cardiology*, 57(17): 1765–74, 2012. doi: 10.1016/j.jacc.2010.09.083. PMID: 21511113; PMCID: PMC3408699. www.ncbi.nlm.nih.gov/pmc/articles/PMC3408699/, Accessed on 20/11/2021.

3. Karlen W, Raman S, Ansermino JM, Dumont GA., Multiparameter respiratory rate estimation from the photoplethysmogram. *IEEE Transactions on Biomedical Engineering*, 60(7):1946–53, 2013. doi: 10.1109/TBME.2013.2246160. PMID: 23399950. www.alivecor.com/education/ecg.html, Accessed on 12/10/2021.

4. Antônio H. Ribeiro, Manoel Horta Ribeiro, Gabriela M. M. Paixão, Derick M. Oliveira, Paulo R. Gomes, Jéssica A. Canazart, Milton P. S. Ferreira, Carl R. Andersson, Peter W. Macfarlane, Wagner Meira Jr., Thomas B. Schön, Antonio Luiz P. Ribeiro, Automatic diagnosis of the 12-lead ECG using a deep neural network. *Nature Communications*, 11:1760, 2020. https://doi.org/10.1038/s41467-020-15432-4.

5. A. Subasi, S. Dogan, T. Tuncer, A novel automated tower graph-based ECG signal classification method with hexadecimal local adaptive binary pattern and deep learning. *Journal of Ambient Intelligence and Humanized Computing*, 14:711–725, 2021. https://doi.org/10.1007/s12652-021-03324-4.

6. Mengze Wu, Yongdi Lu, Wenli Yang, Shen Young Wong, A study on arrhythmia via ECG signal classification using the convolutional neural network. *Frontiers in Computational Neuroscience*, 14:106, 2021.

7. Siti Nurmaini, Alexander Edo Todas, Annisa Darmawahyuni, Muhammad Naufal Rachmatullah, Jannes Effendi, Firdaus Firdaus, Bambang Tutuko, Electrocardiogram signal classification for automated delineation using bidirectional long short-term memory. *Informatics in Medicine Unlocked*, 22:100507, 2021.

8. Yang Meng, Guoxin Liang, Mei Yue, Deep learning-based arrhythmia detection in electrocardiograph. *Scientific Programming*, 7:Article ID 9926769, 2021. https://doi.org/10.1155/2021/9926769.

9. Chia-Hung Lin, Jian-Xing Wu, Pi-Yun Chen, Chien-Ming Li, Neng-Sheng Pai, Chao-Lin Kuo, Symmetric cryptography with a chaotic map and a multilayer ML network for physiological signal infosecurity: Case study in electrocardiogram. *IEEE Access*, 9:26451–26467, 2021, https://doi.org/10.1109/ACCESS.2021.3057586.

10. Guangyu Xu, IoT-assisted ECG monitoring framework with secure data transmission for health care applications, *IEEE Access*, 8:74586–74594, 2020. https://doi.org/10.1109/ACCESS.2017.

11. Kalaivanan Sugumar, Balaji Ganesh S., Encrypted ECG data transmission in the wireless medium for health care applications. *Authorea*, 2020. https://doi.org/10.22541/au.159060745.54140641.

12. Joydeep Dey, Arindam Sarkar, Sunil Karforma, Bappaditya Chowdhury, Metaheuristic secured transmission in Telecare Medical Information System (TMIS) in the face of post-COVID-19. *Journal of Ambient Intelligence and Humanized Computing*, 14:6623–6644, 2021. https://doi.org/10.1007/s12652-021-03531-z.

13. S. Harinee, Anand Mahendran, Secure ECG signal transmission for smart healthcare. *International Journal of Performability Engineering*, 17:8:711–721, 2021.

14. Hocine Hamil, Zahia Zidelmal, Mohamed Salah Azzaz, Samir Sakhi, Redouane Kaibou, Salem Djilali, Djaffar Ould Abdeslam, Design of a secured telehealth system based on multiple biosignals diagnosis and classification for IoT application. *Expert Systems*, 39:4:1–27, 2021. https://doi.org/10.1111/exsy.12765.

15. Bo Yang, Bojin Cheng, Yixuan Liu, Lijun Wang, DL-enabled block scrambling algorithm for securing telemedicine data of table tennis players. *Springer Professional*, 1–14, 2021.

16. Zão Leonardo, Rosângela Coelho, Patrick Flandrin, Speech enhancement with EMD and hurst-based mode selection, IEEE/ACM transactions on audio. *Speech, and Language Processing*, 22:5:899–911, 2014.

7 Machine Learning for Detection and Classification of Motor Imagery in Electroencephalographic Signals

Juliana C. Gomes, Vanessa Marques,
Caio de Brito, Yasmin Nascimento,
Gabriel Miranda, Nathália Córdula,
Camila Fragoso, Arianne Torcarte,
Maíra A. Santana, Giselle Moreno and
Wellington Pinheiro dos Santos

7.1 INTRODUCTION

The independence of patients with neurological trauma or motor disabilities is still very limited. These individuals often have a high degree of awareness and intellectual capacity, but with partial or complete physical, emotional, social and vocational dependence. With this in mind, several studies have sought to expand the interaction of these individuals with the external environment in a more autonomous way (Rashid, Sulaiman, Abdul Majeed, Musa, Ab Nasir, Bari, and Khatun, 2020; Cantillo-Negrete, Carino-Escobar, Carrillo-Mora, Rodriguez-Barragan, Hernandez-Arenas, Quinzaños-Fresnedo, Hernandez-Sanchez, Galicia-Alvarado, Miguel-Puga, and Arias-Carrion, 2021; Wang and Bezerianos, 2017). Thus, by monitoring brain activity, these scientists want to create an alternative to the natural communication between the Central Nervous System, the region where the neural processes for muscle contraction originate, and the muscles (Corbet, Iturrate, Pereira, Perdikis, and Millán, 2018).

In many of these cases, despite the motor impairment, brain signals are generated in regions of the cortex when movements are imagined. Signs of this type can be monitored by the electroencephalography (EEG) technique, which can be non-invasive and portable. That is, by distributing electrodes over the subject's scalp, it is possible to obtain signals from certain areas of the cortex activated during a specific

DOI: 10.1201/9781003201137-9

task (de Oliveira, de Santana, Andrade, Gomes, Rodrigues, and dos Santos, 2020; Ieracitano, Mammone, Bramanti, Hussain, and Morabito, 2019). Thus, through these brain-machine interfaces it is possible to assess brain activity, in addition to controlling assistive technologies such as wheelchairs, neuroprostheses, communication or entertainment devices, such as music composition and digital games (Cantillo-Negrete, Carino-Escobar, Carrillo-Mora, Elias-Vinas, and Gutierrez-Martinez, 2018; Do, Wang, King, Chun, and Nenadic, 2013).

In this sense, the translation of these brain stimuli is useful and promising in the control of brain-machine interfaces, but it is still a challenge for many researchers. One of the approaches adopted in order to translate brain signals is Motor Imagery by Electroencephalography (IM-EEG). IM-EEG is the acquisition of brain activity signals associated with imagined movements. Therefore, IM-EEG is an intentional mental rehearsal of a motor behavior, without an associated external movement or stimulus. When imagining the movement, the subject reactivates the kinesthetic memory of a previously performed movement, giving the impression that it is being performed again. Its strategy is based on sensorimotor rhythms. When analyzing brain signals in the frequency ranges of 8–12Hz and 12–30Hz, there is a reduction in their amplitudes when there is intention or execution of a motor task, called Event-Related Desynchronization (ERD). On the other hand, after the imagination or execution of the movement, there is an Event-Related Synchronization (ERS). Thus, by observing the ERDs and ERSs, it is possible to control a BCI (Machida and Tanaka, 2018; Ono, Wada, Kurata, and Seki, 2018; Padfield, Zabalza, Zhao, Masero, and Ren, 2019; Tariq, Trivailo, and Simic, 2017).

IM-EEG has great complexity due to the variability of intensities and frequencies of signals produced by each individual. The use of machine learning becomes fundamental so that these imagined movements can be detected and classified. These learning machines need to be able to learn quickly, since the phenomenon of neuroplasticity causes great variation in EEG signals. This makes the task of adapting assistive technologies to users customized for each subject (Burianová, Marstaller, Rich, Williams, Savage, Ryan, and Sowman, 2020).

Given the preceding, this work will study the classification of motor imagery using simple and fast classifiers, such as decision trees, Support Vector Machines and Bayesian networks. For this, signal pre-processing techniques will be tested, including windowing, bandpass filters, wavelet decomposition and application of statistical thresholds. The methodology will be tested using the BCI Competition IV database (Leeb, Brunner, Müller-Putz, Schlögl, and Pfurtscheller, 2008).

This chapter is organized as follows: in section 2, works related to the classification of motor imagery based on EEG signals are critically reviewed. Then, in section 3 the proposal of this work is made, in addition to explaining the theoretical concepts necessary to understand this work. In section 4 experimental results will be shown. Finally, in section 5 the results and possible future work will be discussed.

7.2   RELATED WORKS

Machine learning has been widely used in the classification and identification of motor imagery based on EEG signals. However, there are still limitations that have not been overcome even with different approaches present in the literature.

The EEG signal is widely used when dealing with BCI, mainly motor imagery. In this type of study, classical classifiers use attributes related to event-related synchronization and desynchronization (ERS and ERD, respectively) in mu and beta rhythms. In these classic approaches, the model is trained and evaluated using the individual's own signals. In deep learning, this approach produces an insufficient amount of data for a single subject.

In the work of Zhang, Robinson, Lee, and Guan (2021), they proposed a method to work around this problem using transfer learning. Transfer learning allows the use of deep learning in small databases, through pre-training in a large and controlled database. Although effective, this method does not dispense with the use of a large database, which continues to be a limitation of unsupervised learning.

Thereby, Zhang et al. (2021) tried to adapt a deep CNN model. The aim of the authors was to take the attributes extracted from convolutional filters and adapt them to a subject not yet introduced to the model. The database consists of EEG signals from 54 healthy subjects performing binary motor imagery tasks. The signals were recorded using 62 Ag/AgCl electrodes at a sampling rate of 1000Hz. Then, the signals were undersampled by a factor of 4, and an eight-order type 1 Chebyshev filter was applied. The classification step was subdivided into subject-specific, subject-independent and subject-adaptive. The first one is to train the model only with the signals of the subject to be evaluated. The second is to train the model with the signals of all subjects except the one to be evaluated. And finally, the training is done with the signals of all the subjects and part of the signals of those who want to be evaluated.

In the subject-adaptive approach, the authors varied the amount of signals coming from the target subject (from 10% to 100% in steps of 10%). In addition, five different schemes were tested for the subject-adaptive approach. In this case, they varied the amount of deep CNN training parameters and the learning rate.

The subject-specific and subject-independent methods had accuracy of 63.54% ± 14.25% and 84.19% ± 9.98%, respectively. For a learning rate equal to 1, the best accuracies were all present in the scheme with fewer training parameters. Overall, the study achieved good results, with accuracy values above 80%. In other state-of-the-art studies on the same dataset, mean accuracy values were 74/15% ± 15.83%.

Similarly, Amin, Alsulaiman, Muhammad, Mekhtiche, and Hossain (2019) proposed the use of a new method of deep learning, based on the use of various combinations of CNN models. Initially, they also used transfer learning.

In a second moment, the work highlighted the fact that the depth and quantity of filters present in the CNN algorithms must vary for a better classification. In this sense, the authors proposed a method with the fusion of four layers of CNN, related to the different frequency ranges of brain waves captured by the EEG. Thus, after learning transference, the data were then concatenated and sent as input to a multilayer perceptron neural network or to an autocoder, to perform the fusion. In sequence, the data were segmented into two-second windows. Finally, they took another step of one-on-one training using CNN methods. At the end of the architecture, a softmax function was added to the system.

Thus, good results were obtained, with an increase in the system's accuracy by 10% for cases of cross-training, where learning and testing were done with different patients. However, it is noteworthy that the database used had only nine patients, which may limit the study results. Finally, it was still possible to overcome another limitation of the deep learning method, related to the identification of attributes that are relevant to the algorithm. For this, the researchers developed correlation maps that indicate the spatial distribution of the attributes used for different EEG wave frequencies.

The acquisition of signals of motor imagery depends on several long sessions. Moreover, any subtle movements, such as blinking or contracting a muscle, interfere with the quality of the signals. This scenario makes it difficult to obtain data for training and consequently results in a decrease in the accuracy of the classifier. In this sense, Dai, Zhou, Huang, and Wang (2020) presented an alternative through a hybrid scale convolutional neural network (HS-CNNs) as a classifier and a data augmentation method for training.

In this work, two databases (2a and 2b from BCI Competition 2008) were used, both with nine participants. In database 2a, two collection sessions were carried out, with 288 trials each, using 22 electrodes. It consists of four classes, including right hand, left hand, tongue and foot movements. On the other hand, in base 2b, five sessions were carried out per participant, the first two with 122 trials and the last three with 160. Only the last three sections used visual feedback. Data acquisition was made from three electrodes (C3, CZ and C4), with a sampling rate of 250Hz. Signals from both databases were processed with bandpass filters of 4–7Hz, 8–13Hz and 13–32Hz. Then, the authors chose three kernels of different sizes (1×45, 1×65 and 1×85) for time domain attribute extraction. In the following, each of these results went through a vertical kernel for spatial attribute extraction and all attributes were merged and concatenated into one dimension. Finally, the vector served as input to a fully connected layer.

As accuracy in neural networks depends directly on the amount of data provided for training, the authors also proposed a method of generating artificial data for training based on real data. This new method can be divided into three steps: 1) signal segmentation, 2) time domain recombination and 3) frequency domain exchange. In order to evaluate the database augmentation method, the authors compared it with two popular methods in the literature. Thus, the average accuracy for the windowing and white noise addition methods were 80.1% and 86.1%, respectively. However, the hybrid classifier presented a performance of 85.6% and 87.6% for the classifier without and with data enhancement, respectively. Overall, the proposed method resulted in an increase of 23.25% in average accuracy for database 2a and 19.7% for database 2b.

The main idea of Zhao, Zhang, Zhu, You, Kuang, and Sun (2019)'s work is to find electroencephalogram (EEG) representation methods that can obtain optimized temporal and spatial characteristics. The authors' strategy was to represent three-dimensional signal windows, formed from sequences of vectors in two dimensions that preserve the spatial conformation. From there, they used a 3D CNN as a classifier. The 3D CNN has three branches with different sizes of receptive fields. Each of these branches is made up of three convolutional layers and three dense layers, including the shared layer. Thus, the combination of the three branches serves to acquire the attributes of motor imagery. Finally, these three networks are combined, serving as input to a softmax.

In order to validate the proposed methodology, Zhao et al. (2019) also used the 2a database. The signals were pre-processed with bandpass filters between 0.5 and 100Hz, and a 50Hz notch filter. In addition, they tested and investigated the effect of electrode selection on classification.

The results of experiments with single-branch networks showed that it is not possible to choose just one for all subjects in the dataset, as each one stood out in the performance of some subject. These observations point to the need for subject-specific networks. On the other hand, when using the Multi-branch 3D CNN, the results were superior in all cases.

The results also revealed that each of the four electrode conformations in the experiment demonstrated a relatively lower accuracy than the full model (using the 22 electrodes). Despite this, some configurations showed comparable performance. On the other hand, the training and testing time with all electrodes is longer. Thus, an optimized choice is needed in terms of time and classification performance.

Another study, also motivated by the limitations of the use of motor imagery associated with BCI, was proposed by Zhu, Li, Li, Yao, Zhang, and Xu (2019). The need for a high amount of time for individual training sessions becomes one of the limitations addressed in the study. As the training data obtained by the individuals vary among themselves, it is necessary that each new individual carries out a new training session. However, in many cases this becomes unfeasible and submits the individuals to exhaustive trials. Thus, Zhu et al. proposes a method of training based on deep learning that uses data from other individuals for a new classification through a separate channel convolutional network (SCCN).

The proposed model used Common spatial patterns (CSP) as a method for feature extraction. CSP is widely used in problems where EEG signals are used in BCI systems. However, the information is kept in the domain of time. In the EEG signals, the study applied a filter in the range of 8–30Hz, which comprises alpha and beta brain rhythms. The authors used two databases. The first contains EEG data obtained from 25 individuals. In this first dataset, 15 Ag/AgCl electrodes were used at a sampling frequency of 1000Hz. The analyzed classes were right hand and left hand. The second database used was the BCI Competition IV-2b, mentioned earlier.

The steps taken to implement SCCN were the conversion of the input signal to CSP, the encoding of these signals for each channel and classification. The results were compared with k-nearest neighbors (KNN), logistics regression (LR), machine support vector (SVM) and linear discriminant analysis (LDA) classifiers. The results showed an improvement in accuracy of up to 13% and 15% for the first and second base used, respectively, in relation to the other algorithms. However, the study does not have a good amount of data, and there is also a difficulty in understanding the logic behind CNN's feature extraction.

7.3 MATERIALS AND METHODS

7.3.1 PROPOSED METHOD

The present work aims to perform the classification of binary motor imagery using simple and fast classifiers. In this case, we use well-known and established classifiers such as decision trees, Support Vector Machines and Bayesian networks. However, given the complexity of the problem, different pre-processing techniques were tested. The purpose of this step is to remove artifacts such as eye movements and electrical network interference.

Accordingly, the signals were initially segmented into 1s sliding windows with an overlap of 0.5s. Then, each signal window was evaluated using two approaches: setup 1 and setup 2. Setup 1 consists of using raw EEG signals, without any type of processing. Setup 2, contrariwise, consists of the application of three processing steps:

(1) Filtering with a 5th order Butterworth bandpass filter. The filter selects the alpha and beta frequency ranges (8–32Hz);
(2) Decomposition into 10 levels of wavelets;
(3) Application of a statistical threshold on each wavelet component based on the ATAR algorithm proposed by Bajaj, Carrión, Bellotti, Berta, and De Gloria (2020).

Subsequently, we extracted numerical attributes from setups 1 and 2 (both in the time and frequency domains). The attributes were then selected using the Evolutionary Search Optimization method. For the classification step, we used a subject-specific approach. That is, only the signals from the subject to be evaluated during the training process of the model were used.

Finally, in addition to the experiments performed with the collected EEG signals, experiments were also performed with augmented databases with synthetic instances. This was done with the aim of resembling scenarios in which more data were available per patient and analyzing the impact of such increase in motor imagery sections on classification performance. Accordingly, we utilized the SMOTE (*Synthetic Minority Oversampling Technique*) method to generate the augmented bases.

The methodology of this work is summarized in the diagram from Figure 7.1.

7.3.2 DATABASE

The database used consists of EEG signals obtained from nine volunteers. The subjects were 24.7 ± 3.3 years old, being a group formed by men and women. The volunteers had normal or corrected vision and were right-handed. The data were collected through an electroencephalogram (EEG) during the sessions. At the time of data acquisition, the individuals remained seated in front of a monitor, on which the instructions to be followed were given (Leeb, Lee, Keinrath, Scherer, Bischof, and Pfurtscheller, 2007; Leeb et al., 2008).

Data collection was performed using an EEG cap (Easycap, Germany) with electrode distribution following the extended version of the 10–20 system. Thus, three electrodes identified as C3, Cz and C4, with a sampling frequency of 250Hz, were used to acquire the signals. They were filtered with a bandpass filter, with a frequency range of 0.5Hz to 100Hz, and a notch filter of 50Hz. In addition, three channels were also used to capture electrooculography (EOG) signals. They intended to

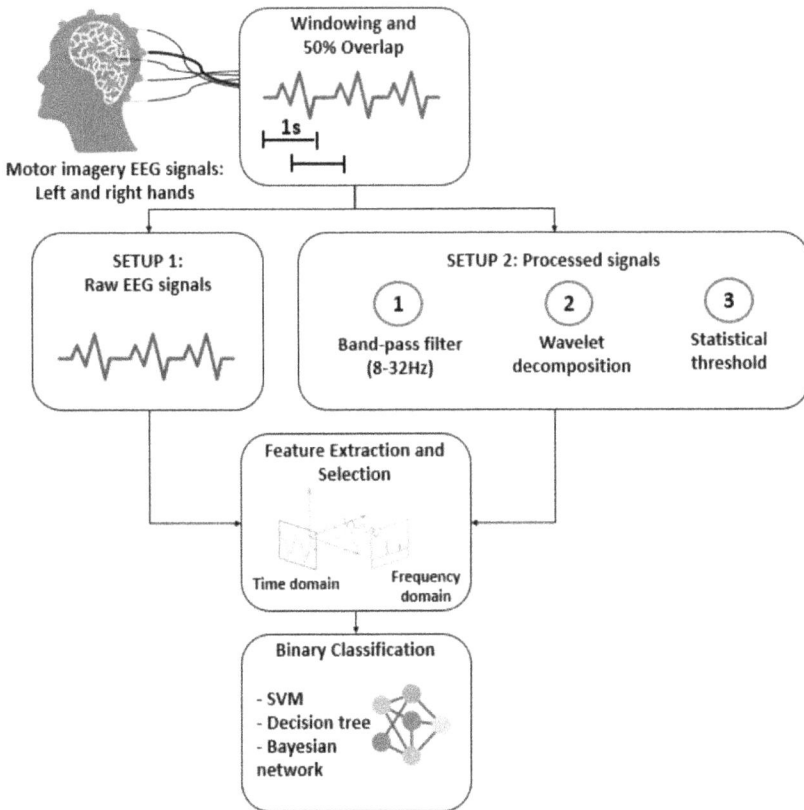

FIGURE 7.1 General methodology of this work: firstly, the EEG signals were segmented into sliding windows. Two approaches were tested: 1) usage of raw signals; 2) usage of processed signals. In setup 2 the signals were filtered by a bandpass filter (832Hz). Then, a wavelet decomposition and a statistical threshold were applied to each wavelet component. Afterwards, we extracted numerical attributes in both setups. Finally, the attributes were selected by the Evolutionary Search method as the last step before proceeding to the classification process.

analyze the influence of these signals on the EEG signal captured throughout the sessions (Leeb et al., 2007, 2008).

The EEG data were collected through the course of five sessions for each volunteer. At the beginning of each session, EOG signals were captured. For approximately 5 minutes before each session, the volunteer was instructed to (1) remain 2 minutes looking at a fixed point on the monitor, with their eyes open (2) close their eyes for 1 minute; and (3) move their eyes for 1 minute. Signage for the start and end of each step was made using an audible alert (Leeb et al., 2008).

Subsequently, the session was started to obtain the EEG data corresponding to the motor imagery (MI) tasks. The classes analyzed in this database were left hand (class 1) and right hand (class 2). Five sessions were held, the first two with visual feedback and the last three without visual feedback. Each of the sessions consisted of six rounds. In these rounds, there were 10 attempts (trials) for the left hand class and 10 attempts for the right hand. That is, in each of the six rounds there were 20 attempts, corresponding to 120 attempts per session (Leeb et al., 2008).

In the first two sessions, done without visual feedback, the following rounds were performed: (1) a fixed cross was shown on the monitor, followed by an audible alert to prepare for the next step; (2) an arrow pointing to the left or right, depending on the aimed class, was displayed on the monitor for 1.25 seconds; and (3) for 4 seconds, the volunteer imagined the hand movement corresponding to the visual stimulus given in the previous step. Between the end and start of each round, breaks of 1 and 1.5 seconds were taken. The volunteers performed the first two sessions (without visual feedback) over a two week period, on two different days (Leeb et al., 2008).

In the subsequent three sessions, with visual feedback, the rounds included the following steps: (1) a gray face was shown centered on the monitor and after 2 seconds, an audible alert was emitted; (2) in the range of 3 to 7.5 seconds, the clue corresponding to the class was displayed; and (3) based on this information, the volunteer should move their face to the right or left, according to the clue shown. If the imagined movement was in accordance with the clue, the face displayed on the screen turned green (correct direction) or red (wrong direction) otherwise. When the period ended, the monitor went blank and there was a 1- to 2-second pause between attempts (Leeb et al., 2007, 2008).

Figure 7.2 simplifies the protocol used to acquire the signals.

7.3.3 SIGNAL PROCESSING WITH WAVELETS
AND STATISTICAL THRESHOLD

Physiological signals are commonly contaminated by various artifacts. Those artifacts can be physiological (eye or muscle movements, heartbeat) or non-physiological (electrical network, interference from other medical equipment). Recent studies observed promising results in the use of wavelets for processing biomedical signs (Bozhokin, Suslova, and Tarakanov, 2019; Rampal, Maciel, and Hirsch, 2018; Stepanov, 2017).

Most biomedical signals, including the EEG, are non-stationary. This means that the properties of the signal vary over time. With that in mind, the wavelet transformation can be an alternative for locating events of interest in EEG signals. The wavelet transformation (WT) exploits low frequency signals which are spread over time. At the same time, the WT also exploit rapidly occurring high frequency signals (Bajaj et al., 2020).

An analysis with wavelets can be done in a continuous or discrete manner. In this chapter we will focus on the Discrete Wavelet Transform (DWT). At DWT, the signal is decomposed into two: 1) low-pass signal; 2) high-pass signal. Then a DWT decomposes the low-pass signal recursively using the same wavelet function. This is done as many times as necessary and the signal is decomposed into levels (Rioul

MOTOR IMAGERY SESSION

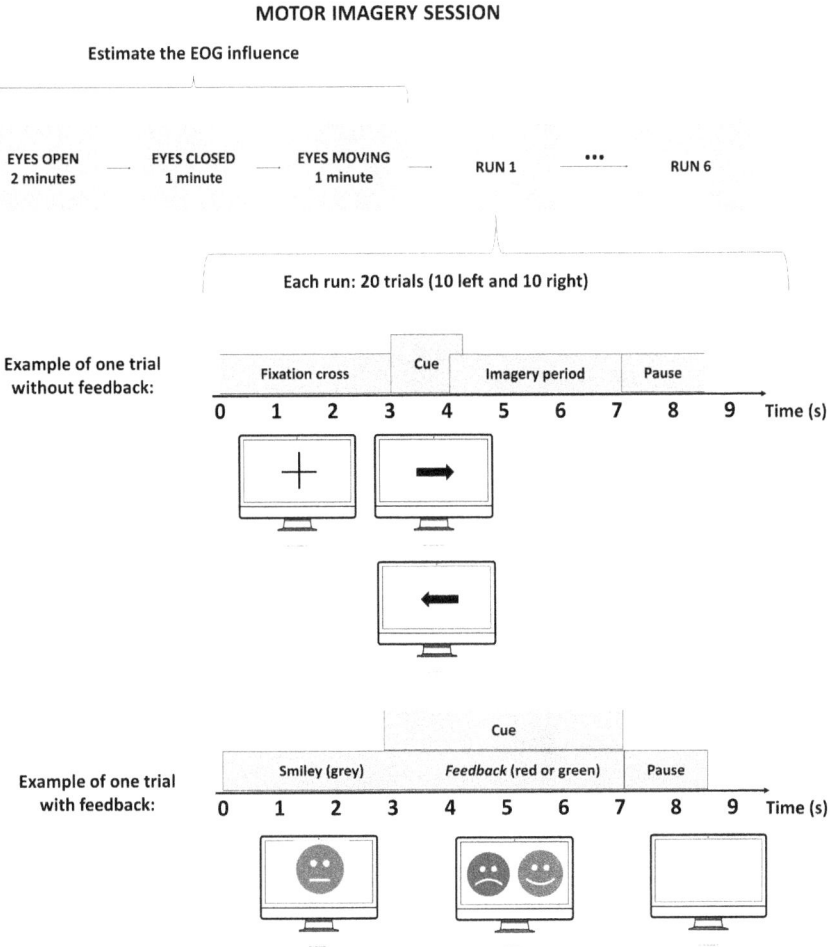

FIGURE 7.2   Acquisition protocol from BCI Competition 2008 database 2b.

and Duhamel, 1992). Figure 7.3 exemplifies a wavelet decomposition with DWT. In Figure 7.3, the LP block is composed by a low-pass filter $f(n)$ and a downsampler block. The HP block is composed by a $g(n)$ high-pass filter and a downsampler block.

Besides the wavelet decomposition, this work also applied a statistical threshold for removing artifacts in wavelet components. The formulated algorithm was the Automatic and Adjustable Artifact Removal (ATAR), proposed by Bajaj et al. (2020).

7.3.4   FEATURE EXTRACTION

When we have a bioelectrical signal over time, we can extract several statistical attributes from this signal. Among those we can include the mean value of the signal,

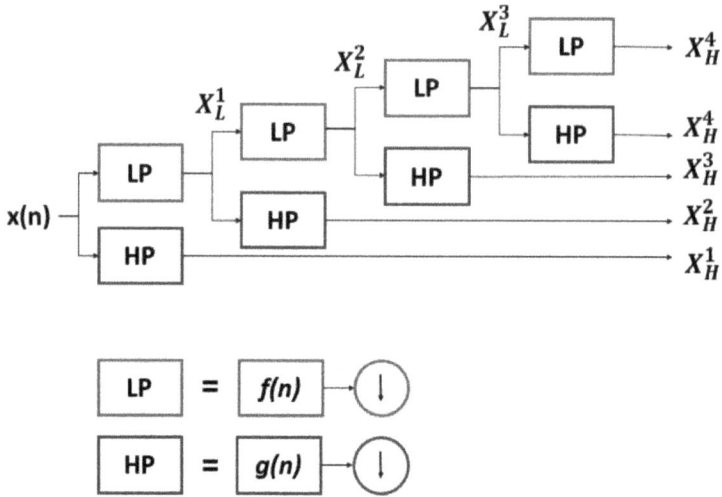

FIGURE 7.3 Discrete wavelet decomposition: the signal is decomposed into two: low-pass signal and high-pass signal. DWT successively decomposes the low-pass signal.

the variance, the standard deviation, for instance. We can also transform the signal from the time domain to the frequency domain and extract attributes such as the average frequency of the signal, average power and frequency variance. In this way, we extracted 34 numerical attributes in both time and frequency domains (Espinola, Gomes, Pereira, and dos Santos, 2021a, 2021b). All attributes as well as their mathematical expressions can be seen in the Table 7.1. We used the GNU/Octave environment to perform feature extraction.

7.3.5 CLASSIFICATION

7.3.5.1 Decision Trees

To understand the concept of decision trees, it is necessary to understand some general concepts. The first of these is the concept of a tree. In computing, a tree means a data structure with several storage components that generate information, which we call nodes. These nodes will be key points for obtaining certain commands that will generate responses for a certain action. Therefore, decision trees are trees that store sets of information in their nodes. This information will define a useful answer for a certain study or classification (Barbosa, Gomes, de Santana, de Lima, Calado, Bertoldo Júnior, Albuquerque, de Souza, de Araújo, Mattos Júnior et al., 2021).

We will give a practical example. When starting a scientific study there will usually be a key question, which can also be called a root node. This question will guide the professional to a certain field of study. From this root node, further information and subset nodes related to other nodes may emerge. At the end of the process, the decision of the path to take will lead to the decision of the scientific study to which the professional will be assigned. In the same way, machine

TABLE 7.1
List of Numerical Attributes and Their Mathematical Expressions.

Attribute	Mathematical expression	Attribute	
Mean (μ)	$\mu = \dfrac{1}{N}\sum\limits_{n=1}^{N} x_n$	Zero crossings	$ZC = \sum\limits_{n=1}^{N-1} [sgn(x_n \times x_{n+1}) \cap \|x_n - x_{n+1}\| \geq threshold]$ $sgn(x) = \{ \begin{smallmatrix} 1, if\ x \geq threshold \\ 0, \quad otherwise \end{smallmatrix}$
Variance	$var = \dfrac{1}{N-1}\sum\limits_{n=1}^{N} (x_n - \mu)^2$	Slope Sign Change	$SSC = \sum\limits_{n=1}^{N-1} [f(x_n - x_{n-1}) \times (x_n - x_{n+1})]]$ $f(x) = \{ \begin{smallmatrix} 1, if\ x \geq threshold \\ 0, \quad otherwise \end{smallmatrix}$
Standard deviation (σ)	$\sigma = \sqrt{\dfrac{1}{N-1}\sum\limits_{n=1}^{N} \|x_n - \mu\|^2}$	Hjorth parameter activity	$Hjorth_{activity} = \dfrac{1}{N-1}\sum\limits_{n=1}^{N} (x_n - \mu)^2$
Root mean square	$RMS = \sqrt{\dfrac{\sum_{n=1}^{N} (x_n)^2}{N}}$	Hjorth parameter mobility	$Hjorth_{mobility} = \dfrac{\sqrt{var\left(\frac{d\,x(t)}{dt}\right)}}{var(x(t))}$
Average Amplitude Change	$AAC = \dfrac{1}{N}(\sum\limits_{n=1}^{N} \|\dfrac{d\,x(t)}{dt}\|)$	Hjorth parameter complexity	$Hjorth_{complexity} = \dfrac{Hjorth_{mobility}\left(\frac{d\,x(t)}{dt}\right)}{Hjorth_{mobility}(x(t))}$
Difference Absolute Deviation	$DASDV = \sqrt{\dfrac{1}{N}\sum\limits_{n=1}^{N} \left(\dfrac{d\,x(t)}{dt}\right)^2}$	Mean frequency	$MNF = \dfrac{\sum_{j=1}^{M} f_j P_j}{\sum_{j=1}^{M} P_j}$ Where f_j, P_j are the frequencies and power of the spectrum, respectively, and M is the length of the frequencies
Integrated Absolute Value	$IAV = \sum\limits_{n=1}^{N} x_n$	Median frequency	$MDF = \dfrac{1}{2}\sum\limits_{j=1}^{M} P_j$
Logarithm Detector	$LOGD = e^{\left(\frac{1}{N}\sum_{n=1}^{N} \log(\|x_n\|)\right)}$	Mean power	$MNP = \sum\limits_{j=1}^{M} \dfrac{P_j}{M}$
Simple Square Integral	$SSI = \sum\limits_{n=1}^{N} x_n^2$	Peak frequency	$PKF = \max(P_j)$
Mean Absolute Value	$MAV = \dfrac{1}{N}\sum\limits_{n=1}^{N} \|x_n\|$	Power Spectrum ratio	$PSR = \dfrac{PKF}{\sum_{j=1}^{M} P_j}$
Mean Logarithm Kernel	$MLOGK = \dfrac{1}{N}\|\sum\limits_{n=1}^{N} x_n\|$	Total Power	$TP = \sum\limits_{j=1}^{M} P_j$
Skewness (s)	$s = \dfrac{\frac{1}{N}\sum_{n=1}^{N}(x_n - \mu)^3}{\sigma^3}$	First Spectral Moment	$SM1 = \sum\limits_{j=1}^{M} f_j P_j$
Kurtosis	$kurt = \dfrac{\frac{1}{N}\sum_{n=1}^{N}(x_n - \mu)^4}{\sigma^4}$	Second Spectral Moment	$SM2 = \sum\limits_{j=1}^{M} f_j^2 P_j$
Maximum Amplitude	$MAX = \max(x_n)$	Third Spectral Moment	$SM3 = \sum\limits_{j=1}^{M} f_j^3 P_j$
Third Moment	$M3 = \|\dfrac{1}{N}\sum\limits_{n=1}^{N} (x_n)^3\|$	Variance of Central Frequency	$VCF = \dfrac{SM2}{TP} - \left(\dfrac{SM1}{TP}\right)^2$
Fourth Moment	$M4 = \|\dfrac{1}{N}\sum\limits_{n=1}^{N} (x_n)^4\|$	Waveform length	$WL = \sum\limits_{n=1}^{N-1} \|x_{n+1} - x_n\|$
Fifth Moment	$M5 = \|\dfrac{1}{N}\sum\limits_{n=1}^{N} (x_n)^5\|$	Shannon Entropy	$S = \sum\limits_{i} s_i^2 \log(s_i^2)$

learning uses the concept of decision trees through algorithms that will conduct information in a non-parametric and supervised way. This means that this dataset will provide the necessary information for the machine to understand the decision tree.

Building a decision tree will require dividing a dataset using attributes. This attribute contains primary data about each dataset. From these, other information will be sectioned to delimit an initial answer set, which will successively consummate the final answer in the algorithm used. There are several algorithms that are used in machine learning to build a tree, such as CART, ID3 and C4.5. In these algorithms, the information is processed and sectioned in certain cut points that will configure a better delimitation of attributes for the basis of the decision. For the calculation of cut points, all information is tested to calculate possible data gains, as shown in Equation 7.1:

$$Gain(S, A) = Entropy - \sum_{\upsilon} \frac{|S_{\upsilon}|}{|S|} Entropy(S_{\upsilon}), \quad \upsilon \in A \tag{7.1}$$

where S represents the entropy of the system and A the attributes used in the calculation. The systems entropy arises through Equation 7.2:

$$Entropy = \sum p_i \times \log_2 p_i \tag{7.2}$$

where p_i represents the probabilities of an event in a class i. Finally, at each iteration of the algorithm, the information gains are calculated successively and the decision tree in that system gets an answer consistent with the calculations.

7.3.5.2   Support Vector Machines

Support Vector Machines (SVM) are supervised machines learning algorithms responsible for creating a classification or regression of the data by implementing a boundary that best separates the elements. The algorithm calculates support vectors in order to find the separation line between the classes, which we can call the hyperplane. These support vectors are the coordinates that are closest to the line. The distance between these points is called the margin. Thus, the algorithm seeks to calculate the best hyperplane, with the lowest classification error and also the highest margin.

The SVM has a parameter C that will control the size of the margin separating these two classes. In this sense, the higher the value adopted for C, the smaller the margins will be and, consequently, the classifier will be more tolerant to errors. The opposite happens when this parameter is reduced.

7.3.5.3   Bayesian Methods

A Bayesian network is a directed acyclic graph (DAG). This type of graph has nodes and arcs. In the context of the classification task, each node represents a base attribute, and each arc represents a probabilistic dependency between nodes (Pearl, 1988). The function of the Bayes Net algorithm is to calculate the posterior probability distribution of the classification node, given the probability distribution of all the other nodes (attributes) (Cheng and Greiner, 1999; Gomes, Barbosa, Santana, Bandeira, Valença, de Souza, Ismael, and dos Santos, 2020).

The Naive Bayes classifier is also derived from Bayesian networks. However, a strong assumption is made: all attributes are probabilistically independent of each other, given a class distribution C. That is: $P(X_i|X_n,C) = P(X_i|C)$ for all values of X_i, X_n and C, when $P(C) > 0$ (Friedman, Geiger, and Goldszmidt, 1997). By making this assumption, the calculation of the probabilistic distribution is simplified immensely, increasing speed and decreasing computational cost. Although this assumption is unrealistic, the algorithm shows satisfactory performance, especially when the attributes have low correlation between each other (Rish, 2001).

7.3.6   DATA AUGMENTATION: SMOTE

SMOTE is a famous method of data augmentation. The algorithm idea is to accomplish an interpolation between nearby instances, both in the same class. Therefore, by introducing new examples, the SMOTE assists the classifier generalization capacity (Fernández, Garcia, Herrera, and Chawla, 2018). As a result of its popularity, many works have applied the method, which added other techniques and/or proposed modifications (Demidova and Klyueva, 2017; Douzas, Bacao, and Last, 2018; Jiang, Lu, and Xia, 2016).

The SMOTE idea is: the algorithm generates synthetic instances based on data similarity, instead of replicating the samples. At first, the SMOTE selects a sample of the class that it wants to increase. We will call this sample s_1. Then, it selects k neighbors closest to the sample s_1 using the KNN algorithm. In this case, it is considered a euclidean distance. Thus, one of the k-nearest samples is randomly selected. We will call this second sample s_2. Finally, a synthetic sample s_{new} is calculated from the samples s_1 and s_2. This one is generated according to the Equation 7.3. This means that the synthetic sample created is spatially placed between the samples that generated it (s_1 and s_2). In conclusion, the sample s_{new} is added to the database (Chawla, Bowyer, Hall, and Kegelmeyer, 2002).

$$s_{new} = (s_1 + s_2) \times n, \quad n \in [0,1] \tag{7.3}$$

7.3.7   EXPERIMENTS SETUP

In this work, the codes for pre-processing and attribute extraction of the signals were implemented in GNU/Octave.[1] The steps of attributes selection and classification were tried on *software* Weka.[2]

In the classification stage the following setups have been tried: SVM with polynomial kernel with exponent 1, 2 and 3, random tree, Naive Bayes, Bayes Network and Random Forests with the tree number ranging from 10 to 100, considering steps of 10. Each configuration was run 30 times using cross validation of 10 folds.

For increased bases with synthetic instances, we apply an increase of 150%. That is, bases with 2034 instances now have 5084 instances, since the raise is applied in each class separately. It is important to highlight that the two closest neighbors were considered for the KNN algorithm application.

7.3.8 METRICS

Accuracy is a global average of hits that the classifier obtained in relation to the total number of predictions. It is considered one of the simplest methods for evaluating results. During a classification, we can obtain different results, such as: a false positive (FP), false negative (FN), true positive (VP) and true negative (VN). From this we can calculate the value of the accuracy, as shown in Equation 7.4 (Gomes, Masood, Silva, da Cruz Ferreira, Júnior, dos Santos Rocha, de Oliveira, da Silva, Fernandes, and dos Santos, 2021).

$$Accuracy = \frac{TP + TN}{TP + TN + FP + FN} \tag{7.4}$$

where TP is the true positives, TN is the true negatives, FP is the false positives and FN the false negatives.

Accuracy alone may not be a good evaluation method, since we may have unbalanced data and validate a classifier even though it has a high value of false negatives and/or positives.

The kappa coefficient, on the other hand, indicates how well the data agree with the classification made through qualitative variables. The kappa coefficient is defined by Cohen (1960) as the proportion of concordant terms after the concordance attributed to chance is removed from consideration (Chmura Kraemer, Periyakoil, and Noda, 2002; Cohen, 1960, 1968). It has been successfully applied in several health applications (Barbosa et al., 2021; Cruz, Cruz, and Santos, 2018; da Silva, de Santana, de Lima, de Andrade, de Souza, de Almeida, da Silva, de Lima, and dos Santos, 2021; de Lima, da Silva-Filho, and dos Santos, 2014; de Lima, da Silva-Filho, and Dos Santos, 2016; Macedo, Santana, dos Santos, Menezes, and Bastos-Filho, 2021; Pereira, Santana, Gomes, de Freitas Barbosa, Valença, de Lima, and dos Santos, 2021; Santana, Pereira, Silva, Lima, Sousa, Arruda, Lima, Silva, and Santos, 2018). The kappa index can be calculated using Equation 7.5.

$$\kappa = \frac{\rho_o - \rho_e}{1 - \rho_e} \tag{7.5}$$

where ρ_o is observed agreement, or accuracy, and p_e is the expected agreement, defined as the following:

$$\rho_e = \frac{(TP + FP)(TP + FN) + (FN + TN)(FP + TN)}{(TP + FP + FN + TN)^2} \tag{7.6}$$

Thus, the value of the κ coefficient can vary from −1 to 1. The closer the value is to 1, the higher the probability of agreement between the judges. On the other hand, the closer to zero the value of κ is, the higher the probability of being a random agreement. Finally, when κ is negative, there is an indication that there is no agreement between the data (Barbosa et al., 2021).

7.4 RESULTS

In order to perform the experiments with the two proposed configurations, one of the subjects that belonged to the dataset was selected. In this paper, subject 8 was chosen since it presents good results in several studies. Therefore, it is likely that it is a more assertive participant, which means that he probably managed to follow the patterns indicated in the sessions of motor imagery better than the others. Accuracy results and kappa index for this participant are shown in Tables 7.2, 7.3, 7.4 and 7.5.

TABLE 7.2
Accuracy Values in the Classification of Motor Imagery with EEG Signals for Subject 8.

Setup		Accuracy - SUBJECT 8															
		RF 10	RF 20	RF 30	RF 40	RF 50	RF 60	RF 70	RF 80	RF 90	RF 100	Naive B	Bayes	SVM E1	SVM E2	SVM E3	RT
Setup 1	Mean	66.88	68.54	69.19	69.57	69.90	69.90	70.07	70.16	70.26	70.22	53.04	63.99	65.55	66.86	68.06	62.60
	Std	2.93	3.02	2.97	2.99	3.08	3.03	3.02	3.13	3.16	3.12	2.31	3.07	3.17	3.12	2.91	3.15
Setup 2	Mean	65.28	65.81	66.19	66.25	66.48	66.55	66.65	66.77	66.66	66.70	63.73	65.13	65.74	67.20	67.66	63.72
	Std	3.04	3.17	2.94	3.05	3.12	3.05	3.04	3.18	3.01	2.99	3.17	3.31	3.30	3.17	3.05	3.26

TABLE 7.3
Accuracy Values in the Classification of Motor Imagery with the Database Expanded with SMOTE for Subject 8.

Setup		Accuracy - SUBJECT 8 (SMOTE)															
		RF 10	RF 20	RF 30	RF 40	RF 50	RF 60	RF 70	RF 80	RF 90	RF 100	Naive B	Bayes	SVM E1	SVM E2	SVM E3	RT
Setup 1	Mean	81.52	83.97	84.86	85.29	85.61	85.83	85.95	86.04	86.19	86.26	63.36	66.72	67.17	68.86	70.14	63.32
	Std	1.70	1.57	1.52	1.48	1.41	1.48	1.44	1.43	1.41	1.39	1.89	1.88	1.91	1.82	1.85	1.85
Setup 2	Mean	81.73	84.09	84.96	85.46	85.73	85.96	86.15	86.30	86.39	86.46	63.36	66.72	67.17	68.86	70.14	63.32
	Std	1.68	1.65	1.64	1.56	1.57	1.56	1.60	1.54	1.48	1.49	1.89	1.88	1.91	1.82	1.85	1.85

TABLE 7.4
Kappa Index Values in the Classification of Motor Imagery with EEG Signals for Subject 8.

Setup		Kappa Index - SUJECT 8															
		RF 10	RF 20	RF 30	RF 40	RF 50	RF 60	RF 70	RF 80	RF 90	RF 100	Naive B	Bayes	SVM E1	SVM E2	SVM E3	RT
Setup 1	Mean	0.34	0.37	0.38	0.39	0.40	0.40	0.40	0.40	0.41	0.40	0.06	0.28	0.31	0.34	0.36	0.25
	Std	0.06	0.06	0.06	0.06	0.06	0.06	0.06	0.06	0.06	0.06	0.06	0.05	0.06	0.06	0.06	0.06
Setup 2	Mean	0.31	0.32	0.32	0.33	0.33	0.33	0.33	0.34	0.33	0.33	0.27	0.30	0.31	0.34	0.35	0.27
	Std	0.06	0.06	0.06	0.06	0.06	0.06	0.06	0.06	0.06	0.06	0.06	0.07	0.07	0.06	0.06	0.07

TABLE 7.5
Kappa Index Values in the Classification of Motor Imagery with the Database Expanded with SMOTE for Subject 8.

Setup		Kappa Index - SUJECT 9 (SMOTE)															
		RF 10	RF 20	RF 30	RF 40	RF 50	RF 60	RF 70	RF 80	RF 90	RF 100	Naive B	Bayes	SVM E1	SVM E2	SVM E3	RT
Setup 1	Mean	0.63	0.68	0.70	0.71	0.71	0.72	0.72	0.72	0.72	0.73	0.05	0.30	0.35	0.40	0.44	0.27
	Std	0.03	0.03	0.03	0.03	0.03	0.03	0.03	0.03	0.03	0.03	0.03	0.04	0.03	0.03	0.03	0.04
Setup 2	Mean	0.63	0.68	0.70	0.71	0.71	0.72	0.72	0.73	0.73	0.73	0.27	0.33	0.34	0.38	0.40	0.27
	Std	0.03	0.03	0.03	0.03	0.03	0.03	0.03	0.03	0.03	0.03	0.04	0.04	0.04	0.04	0.04	0.04

Table 7.2 presents the accuracy results with the original database, that is, without the addition of synthetic instances. For setup 1, we observed that the best classification performance was obtained using Random Forest with 90 trees, which presented an accuracy average of 70.26 ± 3.16. On the other hand, the worst classifier was Naive Bayes, with mean accuracy of 53.04 ± 2.31. However, setup 2 (which had been through signal processing) showed better results with SVM with exponent 3, with a mean accuracy of 67.66 ± 3.05. However, considering the standard deviation, all settings obtained similar results.

Table 7.3 presents the results after the application of the SMOTE method. In this case, it is possible to observe that the classification performance was superior to the previous case, using only the collected data. In addition, with the expanded database, the classifier that stood out was the Random Forest with 100 trees, which presented an average accuracy of 86.26 ± 1.39 and 86.46 ± 1.49 for setups 1 and 2, respectively. Tables 7.4 and 7.5 present the kappa index values, following the previous trend. Thus, setup 1 obtained a mean kappa of 0.41 ± 0.06 and 0.73 ± 0.03 in the original and expanded bases, respectively. In both cases, Random Forests stood out.

In addition to the tables, the boxplots in Figures 7.4 and 7.5 show a comparison of the best classifiers for each configuration. Again, it is possible to see that the results

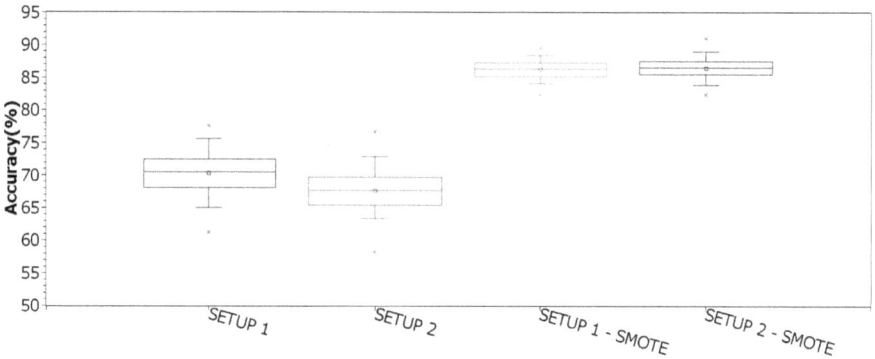

FIGURE 7.4   Accuracy: Classification performance for Subject 8.

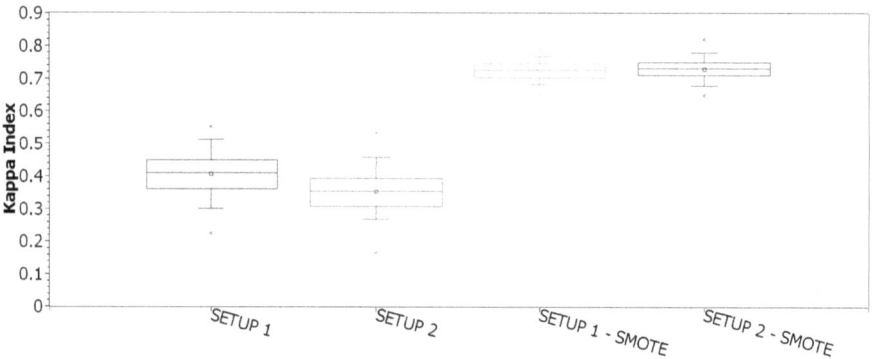

FIGURE 7.5   Kappa index: Classification performance for Subject 8.

with the enlarged database are better. Also, setup 1 is subtly superior to setup 2, considering the accuracy and the kappa index.

Since setup 1 proved to be superior for subject 8, it was chosen to carry out experiments with the other subjects. Thus, Table 7.6 shows the accuracy and kappa index values found for each subject, using Random Forest with 90 trees. In turn, Table 7.7 presents the results for each subject using Naive Bayes as a classifier. Hence, similarly to subject 8, the other subjects performed better with decision trees. Furthermore,

TABLE 7.6

Classification Performance for the Nine Subjects Using Setup 1 and Random Forest as a Classifier.

Subjects	Random Forest (90 trees) - Setup 1	
	Accuracy	Kappa Index
Subject 1	62.65 ± 3.34	0.25 ± 0.07
Subject 2	62.41 ± 3.32	0.25 ± 0.07
Subject 3	52.62 ± 3.32	0.05 ± 0.07
Subject 4	76.66 ± 2.88	0.53 ± 0.06
Subject 5	67.58 ± 3.03	0.35 ± 0.06
Subject 6	55.96 ± 3.23	0.12 ± 0.06
Subject 7	65.69 ± 3.11	0.31 ± 0.06
Subject 8	70.26 ± 3.16	0.41 ± 0.06
Subject 9	71.40 ± 3.08	0.43 ± 0.06

TABLE 7.7

Classification Performance for the Nine Subjects Using Setup 1 and Naive Bayes as a Classifier.

Subjects	Naïve Bayes - Setup 1	
	Accuracy	Kappa Index
Subject 1	58.20 ± 3.61	0.16 ± 0.07
Subject 2	52.60 ± 3.04	0.05 ± 0.06
Subject 3	50.34 ± 1.30	0.01 ± 0.03
Subject 4	71.66 ± 3.13	0.43 ± 0.06
Subject 5	57.30 ± 3.57	0.15 ± 0.07
Subject 6	50.27 ± 1.38	0.01 ± 0.03
Subject 7	56.61 ± 2.86	0.13 ± 0.06
Subject 8	53.04 ± 2.31	0.06 ± 0.05
Subject 9	59.86 ± 3.37	0.20 ± 0.07

Tables 7.6 and 7.7 show that subjects 4, 8 and 9 presented a superior classification performance compared to the other subjects, with an average accuracy of 76.66 ± 2.88, 70.26 ± 3.16 and 71.40 ± 3.08, respectively.

Finally, all subjects were also evaluated with setup 1, using the database expanded with SMOTE. Tables 7.8 and 7.9 present these results, in which it is possible to notice a considerable increase in both metrics of evaluation.

TABLE 7.8

Classification Performance for the Nine Subjects Using Setup 1 and the Enlarged Database with SMOTE.

Subjects	Random Forest (90 trees) - Setup 1 SMOTE	
	Accuracy	Kappa Index
Subject 1	81.91 ± 1.83	0.64 ± 0.04
Subject 2	84.29 ± 1.64	0.69 ± 0.03
Subject 3	75.75 ± 1.79	0.51 ± 0.04
Subject 4	86.85 ± 1.42	0.74 ± 0.03
Subject 5	81.40 ± 1.76	0.63 ± 0.04
Subject 6	74.32 ± 1.78	0.48 ± 0.04
Subject 7	83.21 ± 1.66	0.66 ± 0.03
Subject 8	86.26 ± 1.39	0.72 ± 0.03
Subject 9	83.34± 1.63	0.67 ± 0.03

TABLE 7.9

Classification Performance for the Nine Subjects Using Setup 1 and the Enlarged Database with SMOTE.

Subjects	Naïve Bayes - Setup 1 SMOTE	
	Accuracy	Kappa Index
Subject 1	60.67 ± 2.25	0.21 ± 0.05
Subject 2	54.06 ± 2.17	0.08 ± 0.04
Subject 3	50.56 ± 0.76	0.01 ± 0.02
Subject 4	72.28 ± 1.89	0.45 ± 0.04
Subject 5	63.70 ± 2.11	0.27 ± 0.04
Subject 6	50.11 ± 0.93	0.00 ± 0.02
Subject 7	57.64 ± 1.84	0.15 ± 0.04
Subject 8	63.36 ± 1.89	0.05 ± 0.03
Subject 9	62.09 ± 2.17	0.24 ± 0.04

7.5 DISCUSSION

When analyzing the results presented in the previous section, we noticed a highlight for setup 1 (raw data). This indicates that filtering and artifact removal methods have likely stripped important information from the signals. Nevertheless, future experiments can apply a greater number of levels in wavelet decomposition, since at that time only 10 levels were tested. In addition, other types of filtering can be evaluated, using softer filters and in other frequency ranges. It is possible that these changes will bring interesting results.

Furthermore, by using the SMOTE method to extend the database, more experiments were designed. With the significant improvement in results compared to the original database, we observed that the results would be better if more trials were collected from each subject. Considering subject 2, for instance, there was an increase of 62.41 ± 3.32 to 84.29 ± 1.64 of average accuracy. Despite the improved results in scenarios with more data, we understand that in assistive rehabilitation technology scenarios, small signal acquisition is less tiring and tedious. In addition, by increasing the collection time of sessions, there is a risk that the volunteer becomes more dispersed and less efficient in motor imagery. Thus, it is necessary to find a balance between collection time and classification performance.

7.6 CONCLUSION

In this chapter we propose the classification of EEG-based motor imagery using simple and fast classifiers. Thus, we tested decision trees, SVMs and Bayesian networks in the binary classification of the base 2b data (BCI Competition 2008). However, considering the nonstationary nature of the signal, we applied signal processing techniques aiming at removing artifacts. In this case, bandpass filtering, wavelet decomposition and statistical thresholding of the wavelet components were applied. Thus, the processed signals were compared with the collected signals.

In addition to processing the signals, synthetic instances were created using the SMOTE method. The intent with the increased base was to simulate scenarios where more data was collected from patients. Thus, each class was increased by 150%.

The results showed that the processed signals performed slightly worse than the original data. This indicates that other processing configurations should be tested in the future (e.g., selection of other frequency ranges and decomposition into wavelets with different levels). Furthermore, the results with the extended basis were considerably better. Therefore, this work suggests that the data provided by the base is not sufficient for generalization. This can be explained by the large variability of signals between subjects. Thus, when training a machine with only data from the target subject, the number of signals used is quite limited.

Finally, in future work other classifiers can be tested. We suggest using deep learning with transfer learning. We believe that this methodology will allow better attribute extraction while maintaining speed, which is important for real-time applications such as motor imagery.

NOTES

1 *Software* free available on: www.gnu.org/software/octave/index
2 *Software* free available on: www.cs.waikato.ac.nz/ml/weka/

REFERENCES

Amin, S. U., M. Alsulaiman, G. Muhammad, M. A. Mekhtiche, and M. S. Hossain. Deep learning for eeg motor imagery classification based on multi-layer cnns feature fusion. *Future Generation Computer Systems*, 101:542–554, 2019.

Bajaj, N., J. R. Carrión, F. Bellotti, R. Berta, and A. De Gloria. Automatic and tunable algorithm for eeg artifact removal using wavelet decomposition with applications in predictive modeling during auditory tasks. *Biomedical Signal Processing and Control*, 55:101624, 2020.

Barbosa, V. A. d. F., J. C. Gomes, M. A. de Santana, C. L. de Lima, R. B. Calado, C. R. Bertoldo Júnior, J. E. d. A. Albuquerque, R. G. de Souza, R. J. E. de Araújo, L. A. R. Mattos Júnior, et al. Covid-19 rapid test by combining a random forest-based web system and blood tests. *Journal of Biomolecular Structure and Dynamics*, 1–20, 2021.

Bozhokin, S. V., I. B. Suslova, and D. Tarakanov. Elimination of boundary effects at the numerical implementation of continuous wavelet transform to nonstationary biomedical signals. In *BIOSIGNALS*, pages 21–32, 2019, https://www.scitepress.org/HomePage.aspx.

Burianová, H., L. Marstaller, A. N. Rich, M. A. Williams, G. Savage, M. Ryan, and P. F. Sowman. Motor neuroplasticity: A meg-fmri study of motor imagery and execution in healthy ageing. *Neuropsychologia*, 146:107539, 2020.

Cantillo-Negrete, J., R. I. Carino-Escobar, P. Carrillo-Mora, D. Elias-Vinas, and J. Gutierrez-Martinez. Motor imagery-based brain-computer interface coupled to a robotic hand orthosis aimed for neurorehabilitation of stroke patients. *Journal of Healthcare Engineering*, 2018, 2018.

Cantillo-Negrete, J., R. I. Carino-Escobar, P. Carrillo-Mora, M. A. Rodriguez-Barragan, C. Hernandez-Arenas, J. Quinzaños-Fresnedo, I. R. Hernandez-Sanchez, M. A. Galicia-Alvarado, A. Miguel-Puga, and O. Arias-Carrion. Brain-computer interface coupled to a robotic hand orthosis for stroke patients neurorehabilitation: A crossover feasibility study. *Frontiers in Human Neuroscience*, 15:293, 2021.

Chawla, N. V., K. W. Bowyer, L. O. Hall, and W. P. Kegelmeyer. Smote: Synthetic minority oversampling technique. *Journal of Artificial Intelligence Research*, 16:321–357, 2002.

Cheng, J. and R. Greiner. Comparing bayesian network classifiers. In *Proceedings of the Fifteenth Conference on Uncertainty in Artificial Intelligence*, UAI'99, San Francisco, CA: Morgan Kaufmann Publishers Inc, 1999, page 101108. ISBN 1558606149.

Chmura Kraemer, H., V. S. Periyakoil, and A. Noda. Kappa coefficients in medical research. *Statistics in Medicine*, 21(14):2109–2129, 2002.

Cohen, J. A coefficient of agreement for nominal scales. *Educational and Psychological Measurement*, 20(1):37–46, 1960.

Cohen, J. Weighted kappa: nominal scale agreement provision for scaled disagreement or partial credit. *Psychological bulletin*, 70(4):213, 1968.

Corbet, T., I. Iturrate, M. Pereira, S. Perdikis, and J. D. R. Millán. Sensory threshold neuromuscular electrical stimulation fosters motor imagery performance. *Neuroimage*, 176:268–276, 2018.

Cruz, T., T. Cruz, and W. Santos. Detection and classification of mammary lesions using artificial neural networks and morphological wavelets. *IEEE Latin America Transactions*, 16(3):926–932, 2018.

Dai, G., J. Zhou, J. Huang, and N. Wang. HS-CNN: A CNN with hybrid convolution scale for eeg motor imagery classification. *Journal of Neural Engineering*, 17(1):016025, 2020.

da Silva, A. L. R., M. A. de Santana, C. L. de Lima, J. F. S. de Andrade, T. K. S. de Souza, M. B. J. de Almeida, W. W. A. da Silva, R. d. C. F. de Lima, and W. P. dos Santos. Features selection study for breast cancer diagnosis using thermographic images, genetic algorithms, and particle swarm optimization. *International Journal of Artificial Intelligence and Machine Learning (IJAIML)*, 11(2):1–18, 2021.

de Lima, S. M., A. G. da Silva-Filho, and W. P. Dos Santos. A methodology for classification of lesions in mammographies using zernike moments, elm and svm neural networks in a multi-kernel approach. In *2014 IEEE International Conference on Systems, Man, and Cybernetics (SMC)*, pages 988–991. IEEE, 2014.

de Lima, S. M., A. G. da Silva-Filho, and W. P. Dos Santos. Detection and classification of masses in mammographic images in a multi-kernel approach. *Computer Methods and Programs in Biomedicine*, 134:11–29, 2016.

Demidova, L. and I. Klyueva. Svm classification: Optimization with the smote algorithm for the class imbalance problem. In *2017 6th Mediterranean Conference on Embedded Computing (MECO)*, pages 1–4. IEEE, 2017.

de Oliveira, A. P. S., M. A. de Santana, M. K. S. Andrade, J. C. Gomes, M. C. Rodrigues, and W. P. dos Santos. Early diagnosis of parkinsons disease using eeg, machine learning and partial directed coherence. *Research on Biomedical Engineering*, 36(3):311–331, 2020.

Do, A. H., P. T. Wang, C. E. King, S. N. Chun, and Z. Nenadic. Brain-computer interface controlled robotic gait orthosis. *Journal of Neuroengineering and Rehabilitation*, 10(1):1–9, 2013.

Douzas, G., F. Bacao, and F. Last. Improving imbalanced learning through a heuristic oversampling method based on k-means and smote. *Information Sciences*, 465:1–20, 2018.

Espinola, C. W., J. C. Gomes, J. M. S. Pereira, and W. P. dos Santos. Detection of major depressive disorder using vocal acoustic analysis and machine learningan exploratory study. *Research on Biomedical Engineering*, 37(1):53–64, 2021a.

Espinola, C. W., J. C. Gomes, J. M. S. Pereira, and W. P. dos Santos. Vocal acoustic analysis and machine learning for the identification of schizophrenia. *Research on Biomedical Engineering*, 37(1):33–46, 2021b.

Fernández, A., S. Garcia, F. Herrera, and N. V. Chawla. Smote for learning from imbalanced data: progress and challenges, marking the 15-year anniversary. *Journal of Artificial Intelligence Research*, 61:863–905, 2018.

Friedman, N., D. Geiger, and M. Goldszmidt. Bayesian network classifiers. *Machine Learning*, 29(23):131163, 1997. ISSN 0885–6125. https://doi.org/10.1023/A:1007465528199.

Gomes, J. C., V. A. d. F. Barbosa, M. A. Santana, J. Bandeira, M. J. S. Valença, R. E. de Souza, A. M. Ismael, and W. P. dos Santos. Ikonos: An intelligent tool to support diagnosis of Covid-19 by texture analysis of x-ray images. *Research on Biomedical Engineering*, 1–14, 2020.

Gomes, J. C., A. I. Masood, L. H. d. S. Silva, J. R. B. da Cruz Ferreira, A. A. F. Júnior, A. L. dos Santos Rocha, L. C. P. de Oliveira, N. R. C. da Silva, B. J. T. Fernandes, and W. P. dos Santos. Covid-19 diagnosis by combining rt-pcr and pseudo-convolutional machines to characterize virus sequences. *Scientific Reports*, 11(1):1–28, 2021.

Ieracitano, C., N. Mammone, A. Bramanti, A. Hussain, and F. C. Morabito. A convolutional neural network approach for classification of dementia stages based on 2d-spectral representation of eeg recordings. *Neurocomputing*, 323:96–107, 2019.

Jiang, K., J. Lu, and K. Xia. A novel algorithm for imbalance data classification based on genetic algorithm improved smote. *Arabian Journal for Science and Engineering*, 41(8):3255–3266, 2016.

Leeb, R., C. Brunner, G. Müller-Putz, A. Schlögl, and G. Pfurtscheller. BCI competition 2008—graz data set B. *Graz University of Technology, Austria*, 1–6, 2008.

Leeb, R., F. Lee, C. Keinrath, R. Scherer, H. Bischof, and G. Pfurtscheller. Brain—computer communication: motivation, aim, and impact of exploring a virtual apartment. *IEEE Transactions on Neural Systems and Rehabilitation Engineering*, 15(4):473–482, 2007.

Macedo, M., M. Santana, W. P. dos Santos, R. Menezes, and C. Bastos-Filho. Breast cancer diagnosis using thermal image analysis: A data-driven approach based on swarm intelligence and supervised learning for optimized feature selection. *Applied Soft Computing*, 2021:107533, 2021.

Machida, R. and H. Tanaka. Visualization of erd/ers on leg motor imagery. In *International Symposium on Affective Science and Engineering ISASE2018*, pages 1–6. Japan Society of Kansei Engineering, Osaka, 2018.

Ono, Y., K. Wada, M. Kurata, and N. Seki. Enhancement of motor-imagery ability via combined action observation and motor-imagery training with proprioceptive neurofeedback. *Neuropsychologia*, 114:134–142, 2018.

Padfield, N., J. Zabalza, H. Zhao, V. Masero, and J. Ren. Eeg-based brain-computer interfaces using motor-imagery: Techniques and challenges. *Sensors*, 19(6):1423, 2019.

Pearl, J. *Probabilistic Reasoning in Intelligent Systems: Networks of Plausible Inference*. Morgan Kaufmann Publishers Inc., San Francisco, CA, 1988. ISBN 1558604790.

Pereira, J., M. A. Santana, J. C. Gomes, V. A. de Freitas Barbosa, M. J. S. Valença, S. M. L. de Lima, and W. P. dos Santos. Feature selection based on dialectics to support breast cancer diagnosis using thermographic images. *Research on Biomedical Engineering*, 37(3):485–506, 2021.

Rampal, N., C. B. Maciel, and L. J. Hirsch. Electroencephalography and artifact in the intensive careunit. In *Atlas of Artifacts in Clinical Neurophysiology*, page 59. Springer, Berlin, 2018.

Rashid, M., N. Sulaiman, P. P. Abdul Majeed, A. Musa, R. M. Ab Nasir, A. F. Bari, and S. Khatun. Current status, challenges and possible solutions of EEG-based brain-computer interface: A comprehensive review. *Frontiers in Neurorobotics*, 25:1–35, 2020.

Rioul, O. and P. Duhamel. Fast algorithms for discrete and continuous wavelet transforms. *IEEE Transactions on Information Theory*, 38(2):569–586, 1992.

Rish, I. An empirical study of the naive bayes classifier. In *IJCAI 2001 Workshop on Empirical Methods in Artificial Intelligence*, volume 3, pages 41–46. IBM, New York, 2001.

Santana, M. A. d., J. M. S. Pereira, F. L. d. Silva, N. M. d. Lima, F. N. d. Sousa, G. M. S. d. Arruda, R. d. C. F. d. Lima, W. W. A. d. Silva, and W. P. d. Santos. Breast cancer diagnosis based on mammary thermography and extreme learning machines. *Research on Biomedical Engineering*, 34:45–53, 2018.

Stepanov, A. B. Wavelet analysis of compressed biomedical signals. In *2017 20th Conference of Open Innovations Association (FRUCT)*, pages 434–440. IEEE, 2017.

Tariq, M., P. M. Trivailo, and M. Simic. Detection of knee motor imagery by mu erd/ers quantification for bci based neurorehabilitation applications. In *2017 11th Asian Control Conference (ASCC)*, pages 2215–2219. IEEE, 2017.

Wang, H. and A. Bezerianos. Brain-controlled wheelchair controlled by sustained and brief motor imagery bcis. *Electronics Letters*, 53(17):1178–1180, 2017.

Zhang, K., N. Robinson, S.-W. Lee, and C. Guan. Adaptive transfer learning for eeg motor imagery classification with deep convolutional neural network. *Neural Network*. ISSN 0893–6080. www.sciencedirect.com/science/article/pii/S0893608020304305.

Zhao, X., H. Zhang, G. Zhu, F. You, S. Kuang, and L. Sun. A multi-branch 3d convolutional neural network for eeg-based motor imagery classification. *IEEE Transactions on Neural Systems and Rehabilitation Engineering*, 27(10):2164–2177, 2019.

Zhu, X., P. Li, C. Li, D. Yao, R. Zhang, and P. Xu. Separated channel convolutional neural network to realize the training free motor imagery bci systems. *Biomedical Signal Processing and Control*, 49:396–403, 2019.

8 Emotion Recognition from Electroencephalographic and Peripheral Physiological Signals Using Artificial Intelligence with Explicit Features

Maíra A. Santana, Juliana C. Gomes,
Arianne S. Torcate, Flávio S. Fonseca,
Amanda Suarez, Gabriel M. Souza,
Giselle M. M. Moreno and
Wellington Pinheiro dos Santos

8.1 INTRODUCTION

Emotions are crucial to the social construction of the human being. They are present in countless everyday situations, guiding our actions, tastes, desires and memories. Emotions can be understood as involuntary physiological responses. They are visibly distinguishable and shaped throughout life, according to the experiences of each individual (Bomfim et al., 2019). It is important to understand that emotion is not just a variable that can be quickly identified.

On the contrary, emotion has distinct relationships with other elements, such as sensations, facial expressions, voice, and body movement (de Oliveira and Jaques, 2013). According to Ferreira and Torro-Alves (2016), emotions are essential for the regulation of social interaction, as they guide our preferences, motivations, choices and decision-making, in addition to being indispensable for establishing good verbal and non-verbal communication (Chaturvedi et al., 2021; Dorneles et al., 2020).

Each human being has a unique and peculiar way to express their emotions, which can be through heartbeat, temperature change, tone of voice and others.

DOI: 10.1201/9781003201137-10

Naturally, each individual is born with the ability to perceive, understand and recognize different emotions, both in themselves and in others. Throughout life we also develop the ability to express them in everyday social situations (Miguel et al., 2013; Woyciekoski and Hutz, 2009).

From a neurological point of view, human emotions activate a series of affective-cognitive brain structures. This activity can be observed through non-invasive electroencephalographic signals (EEG) by placing surface electrodes over the scalp (Izard, 1977, 1991). Studies in the literature confirm that in recent years EEG signals have received a lot of attention by researchers. It is mostly due to EEG's ability to provide accurate, simpler, cheaper and more portable solutions than other methods used to acquire brain information (Alarcao and Fonseca, 2017; Andrade et al., 2020; de Oliveira et al., 2020; Gupta et al., 2018; Santana et al., 2020a).

Despite having its origins in the mid-twentieth century, the artificial intelligence (AI) field has gained notable visibility only during the last decade. Many advances in the area have been made since then, developing multiple applications in different areas of interest (Abdallah et al., 2020; Chen et al., 2019; Commowick et al., 2018; Cruz et al., 2018; de Freitas Barbosa et al., 2021; dos Santos et al., 2008; Gomes et al., 2020; Islam et al., 2021b; Ong, 2021; Pereira et al., 2020a, 2020b, 2020c; Portela et al., 2009; Rodrigues et al., 2019; Santana et al., 2018, 2020b; Torcate et al., 2020; Zawacki-Richter et al., 2019; Zhao et al., 2020). In this context, the use of AI associated with emotion recognition is also an increasing area of study given the high complexity and subjectivity of this task (Atkinson and Campos, 2016; Islam et al., 2021b; Mohammadi et al., 2017; Ong, 2021; Santana et al., 2020a; Zheng et al., 2017).

There is still much to be explored in the automatic emotion recognition field. Therefore, this study aims to contribute to the improvement of computational techniques for recognition of affective states and the enrichment of the literature on the subject. In this chapter we propose an approach based on classic intelligent methods (i.e. Random Forest, Support Vector Machines (SVM) and Extreme Learning Machines (ELM)) to classify emotions in EEG and peripheral physiological signals from the HCI Tagging Database (Soleymani et al., 2011). Peripheral physiological signals were respiration rate, Electrocardiogram (ECG), Galvanic Skin Response (GSR) and temperature. All signals are represented by a set of explicit time-frequency-statistical features which showed to be suitable and relevant to differentiate the six existing emotions' categories.

In the early sections of the chapter we will present some of the most recent and relevant studies regarding emotion recognition from physiological signals. Their main challenges and limitations are also discussed. Then, the proposed method will be described, followed by the results and discussions. Finally, our conclusions will be exposed pointing out some relevant issues and findings.

8.2   RELATED STUDIES

In the work by Song et al. (2020), they performed emotion recognition from EEG signals and other physiological signals from the MAHNOB-HCI database. In order to minimize the loss function through backpropagation, two distinct models of Convolutional Neural Networks (CNN) were tested. They use the 32-channel EEG signal as input to a first 2D-CNN model that does feature extraction. Then a

1-hidden-layer Multilayer Perceptron (MLP) does the emotion rating. The second model is based on a 1D-CNN and was used to process physiological signals, also called peripherals here, this model also performs feature extraction from those signals. This input has six channels, three for ECG, one for GSR, one for respiration and one for volunteer temperature. Finally, these attributes are concatenated by a MLP in the same way as in the previous model. The authors state that in general, CNNs performed better than other methods in terms of feature extraction. For EEG, the 2D-CNN model reached an accuracy of 61.5% and 58% in excitation and valence respectively being 9.1% and 1% better than the SVM. For peripheral signals, the gains were 11.81% in excitation and 10.7% in valence.

With a similar approach, in 2021, Sepúlveda et al. (2021) published a paper performing the recognition of emotions from Electrocardiogram (ECG) signals. He starts the article by highlighting the importance of physiological signals as a viable and non-invasive way of recognizing emotions. Human-machine interfaces based on cardiac signals would be a practical alternative to systems based on EEG or facial recognition, especially for daily use. Therefore, the authors propose the use of the Wavelet transform to improve performance and analysis in this type of approach. As a result, the proposed model obtained an accuracy of 88.8% in the valence dimension, 90.2% in the arousal and 95.3% in the two-dimensional approach. Their achieved results were better than those already found in previous studies, validating the model as a very useful alternative for wearable devices.

The research carried out by Hasanzadeh et al. (2021) aimed to propose a fuzzy logic-based parallel cascade model to predict continuous emotional assessment. For this, emotions experienced by individuals were identified from their EEG signals. These emotions were induced by seven songs that evoked emotions categorized into melancholic, happy and neutral at different levels of valence and arousal. A total of 15 volunteers participated in the study. They aged between 21 and 30 years old, being three men and 12 women. They used wavelet analysis to extract the time-varying power spectrum of EEG signals in theta, alpha and beta frequency bands. In order to choose the EEG characteristics that best related to emotion, Mutual Information (MI) was applied. To evaluate the proposed model (FPC), the authors used Linear Regression (LR), Support Vector Regression (SVR) and Long Term Memory Recurrent Neural Network (LSTM-RNN) with 4-fold cross validation. Their findings are promising, since FPC obtained the smallest root mean squared error (RMSE) in relation to the other models, which was equivalent to 0.082. However, it is important to consider that the study was carried out with a small sample size, both of participants and of musical excerpts. Therefore, these results may not be generalizable for approaches that use larger numbers of individuals or use different music genres.

Islam et al. (2021a) developed a low complexity emotion recognition method from EEG signals based on audiovisual stimuli and using deep machine learning with Convolutional Neural Network (CNN). Initially, the authors used the DEAP dataset and converted the one dimensional EEG data into images characterized by Pearson's Correlation Coefficient (PCC). Subsequently, the images were inserted into the CNN so that emotions could be identified, following the classification rule through logistic regression. The results achieved by the model during the classification of emotions stand out in terms of precision obtained in valence (78.22%) and arousal (74.92%). A differential of this study that should be taken into account is their strategy of using converted PCC images. This approach resulted in a process that had a positively lower cost and computational complexity when compared to others in literature.

The research carried out by Kang et al. (2021) proposes a learning model of stacking sets of brain networks for recognizing emotions in deaf subjects, based on EEG signals. To perform the study, 15 deaf people aged between 17 and 24 years were recruited to participate in the experiments. EEG data were collected while the subjects watched emotional stimuli induced by 15 videos, responsible for evoking a single target emotion, which could be positive, neutral or negative. The Phase Lock Value (PLV) was used to calculate the correlation between the EEG channels, and later construction of the brain network using double thresholds. The spatial characteristics of the brain network were extracted using local differentiation and global integration. The stacking set learning framework was used to classify the merged features. In order to analyze the proposed model, experiments were performed in the SEED database and then the model was applied to data from deaf participants. The results demonstrate that the proposed model has the potential to classify three types of emotions with an accuracy of 0.984 (std: 0.005). In addition, the authors investigated and highlighted regions of the brain that are important for the recognition of emotions in deaf people, such as the global interchannel relationships between the frontal, temporal, parietal and occipital lobes.

In the work of Zhao et al. (2018), they use a new technology for recognizing emotions from the reflection of radio-frequency waves in the human body. The equipment, EQ-Radio, emits and receives RF signals modulated by the individual's body movements. It then analyzes the intercepted signal and synthesizes the associated physiological signals such as heartbeat and breathing. The equipment analyzes and uses features extracted from these signals to identify the emotions experienced by the subject. For this, an approach with SVM classifiers is applied to distinguish between four classes of emotions: anger, sadness, joy and pleasure. The study uses 30 volunteers and compares EQ-Radio performance with a model using conventional ECG monitors. They found that the accuracy in measuring beats matches the ones of common ECG monitoring equipment. They point out that for the same experiment, images were recorded to compare performance with image-based models as well. The accuracy in rating emotions reached 87% for the new device, compared to an accuracy of 88.2% using conventional ECG monitors. As for the model using images, the accuracy was of 39.5%. They found that the emotions experienced are not always apparent on the face, and therefore the ECG signal is of great help in this emotion recognition. The new technology is a promising alternative to conventional and inconvenient methods of monitoring ECG with wires connected to the subject's body (Zhao et al., 2018).

In the study of Sorinasa et al. (2020), they are interested in real-time recognition between positive and negative emotions. They presented 14 video clips for audio-visual stimulation to 24 volunteers. Seven clips were related to negative emotions while the other seven clips were of positive emotions. The clips were played in a randomized and counterbalanced way for each individual. Between each clip 30 seconds of black screen was presented, while the volunteers rated the previous clip on a scale of 1 to 9 for two dimensions: valence and arousal. They acquired the EEG signals using Ag-AgCl electrodes with the international 10/10 system. The signal was recorded with a sampling rate of 1kHz to be later filtered by high-pass and low-pass filter of 0.5Hz and 45Hz, respectively. For the extraction of signal features, they used a non-parametric method based on multi-resolution analysis of the wavelet transform. They tested different signal window sizes, ranging from 1s to 12s. Some tests were applied to assess the statistical differences between the windowing sizes and to identify its optimal size. In the classification process, they tested different algorithms: LDA and QDA discriminant analysis methods, linear and non-linear kernels with SVM, k-Nearest Neighbors (KNN), Gaussian Naive Bayes algorithm (GNB), Gradient Boosting (GB) and Random Forest (RF). They used the recursive feature elimination (RFE) method to perform feature selection. They use the F1 Score as the main metric, reaching values of up to 0.997 for the best classifiers (QDA and KNN). Although the authors obtained very good results, the number of clips and individuals analyzed was relatively small, and further studies with a larger volume of data (Sorinasa et al., 2020) are preferable.

The paper developed by Liu et al. (2020) studies the emotion recognition through the collection of physiological signals from the observer. For this, the work presents data resulting from the experiment with clips of the AFEW database, to stimulate the six primary emotions and the neutral state in those who watch them. This database consists in a collection of physiological data of individuals who react to videos of acted emotional expressions. The AFEW database covers a range of signals such as EDA, SKT, PPG, beat intervals and heart rate signals. In the article, for the emotion recognition problem, they tested a basic three linear layers network, of which two are hidden and one corresponds to the output layer. Then, the accuracy of the method in the classification of the seven emotions was tested, including data from one or multiple participants. Later, the model was used to classify the database from the valence and arousal levels. In this sense, emotions were divided into two classes, low/high arousal or low/high valence. As a result of the experiments, it was possible to see that the positive/negative valence information acts to increase the accuracy of the classification problem. Therefore, the authors created a two-step network, in which low or high arousal information was added to the input data for the recognition of the seven emotional states, resulting in an improvement of up to 4.9% compared to the basic network.

In 2020, Doma and Pirouz (2020) developed a research with the goal of comparing different classical machine learning algorithms in the classification of emotional data of cerebral (EEG) and peripheral signals. The authors used the DEAP database, which contains multimodal signals. The authors tested the performance of techniques such as SVM and Decision Trees, with or without PCA (Principal Component Analysis), in the problem of classifying four categories: valence, excitement, dominance and taste. As a result, the work points out that for each of the four classes of binary training, a different classification model demonstrates better results. This proves that different techniques must be used to identify different emotional states.

In the article, A Globally Generalized Emotion Recognition System Involving Different Physiological Signals, Ali et al. (2018) present two main models to overcome the problem of recognizing subject-dependent emotions. In general, their architecture consists of extracting relevant features, followed by automatic calibration of features and classification based on a Cellular Neural Networks (CNN) model. For system training, they also used MAHNOBHCI database in addition to a database developed by the group. Therefore, two experimental approaches were performed: using the same data for training and testing; and using different databases for the two phases. The authors tested four classifiers: RBSVM, Naive Bayes, KNN and Artificial Neural Network. Finally, the study showed promising results for the use of Cellular Neural Networks, when compared to other models, in addition to demonstrating an increase in performance with the use of calibration. Thus, as final considerations, the authors find that their method was a possible solution for the diversity of sensor and equipment brands to acquire signals. It may also be an alternative for emotion recognition in different subjects.

In Table 8.1 we present the main information from these related studies, such as the type of signals they used, the number of classes, the computation techniques applied and the main findings. In the last line of this table is our method, showing that our approach is well contextualized in the state-of-the-art.

8.3 MATERIALS AND METHODS

The method proposed in this study consists of the following six main steps:

(1) Signal database acquisition;
(2) Signal pre-processing;
(3) Signal feature extraction;
(4) Development of four knowledge bases;
(5) Classification;
(6) Statistical analysis of results.

Figure 8.1 illustrates these steps. In the first stage, based on the work carried out by Soleymani et al. (2011), we have the representation of the database composed of signals of the volunteers' physiological parameters. From the 47 channels available in the database, we removed nine channels that were not used in the acquisition process. In other words, only 38 channels were selected, which had different physiological parameters. In the second step, we submitted the signals to a pre-processing stage. We performed signal windowing using a 5-second window, with a 1-second overlap. In the third step, we extracted 35 features from each signal window. Then we organized this data as an ARFF (Attribute Relation File Format) file. From this file we generated three other knowledge bases, thus resulting in four ARFFs. The first ARFF refers to the complete dataset, after signal pre-processing and feature extraction, with 1330 attributes and 15907 instances. The second knowledge base was obtained by performing a feature selection process using the Best First method, thus reducing the number of features to 25. The third ARFF is the full base after using the Gaussian Distribution Based Balance resampling method to reduce the dataset to around 10% of the initial number of instances. The fourth and last ARFF refers to the use of both resampling and feature selection methods and is composed of 1596 instances each represented by 25 features. In the fifth step, we performed the classification using the Random Forest, SVM and ELM classification algorithms. All

TABLE 8.1
Summary of Related Works.

Work	Database	Signals	Classes	Algorithms	Main results
Song et al. (2020)	MAHNOB-HCI (public)	EEG, ECG, GSR, respiration rate, and temperature	2	CNN	Accuracy of 61% for classification from EEG signals, and 73% using the other physiological signals.
Sepúlveda et al. (2021)	AMIGOS (public)	ECG	2	LDA, decision trees, Naive Bayes, KNN, and SVM	Maximum accuracy of 95.3% for classification in terms of arousal and valence using decision trees.
Hasanzadeh et al. (2021)	Private	EEG	3	Fuzzy Parallel Cascades (FPC) model	The proposed model performed better than others from literature, achieving the lowest RMSE of 0.082.
Islam et al. (2021a)	DEAP (public)	EEG	2	CNN	Maximum accuracy of 78.22% in the classification in terms of arousal and valence.
Kang et al. (2021)	Private and SEED (public)	EEG	3	CNN	Maximum accuracy of 98.4%.
Zhao et al. (2018)	Private	Heart rate and respiration rate from radio frequency signals	4	SVM	Classification with 87% of accuracy.
Sorinasa et al. (2020)	Private	EEG	2	LDA, QDA, SVM, KNN, Naive Bayes, Gradient Boosting, and Random Forest	QDA and KNN performed better than the other methods, with maximum accuracy of 99.70%.
Liu et al. (2020)	PAFEW (private)	electrodermal activity (EDA)	7	A three linear layers network with two hidden layers and an output layer	68.66% of accuracy in identifying arousal, and 72.72% for valence.
Doma and Pirouz (2020)	DEAP (public)	EEG	4	SVM, KNN, LDA, Logistic Regression, and Decision Trees	SVM performed better than the other methods, with an F1-score of 84.73% and 98.01% recall. In this condition they also used principal component analysis (PCA) for dimensionality reduction, which improved the results.
Ali et al. (2018)	Private and MAHNOB-HCI (public)	ECG, EDA, and temperature	4	Cellular Neural Network (CNN)	They proposed an approach to provide a novel subject-independent emotion recognition system. Their method achieved an accuracy between 80% and 89% when using the same physiological sensors for both training and testing. The authors also found 71.05% accuracy when the physiological sensors in training are different from those used in testing.
Our approach	MAHNOB-HCI (public)	EEG, ECG, GSR, temperature, and respiration rate	6	SVM, ELM, and Random Forest	In addition to classification, we carried out a study of feature selection and resampling. Random Forest models achieved the best overall performance, with accuracies up to 100% and low data dispersion.

(1) Signal database acquisition;
(2) Signal pre-processing;
(3) Signal feature extraction;
(4) Development of four knowledge bases;
(5) Classification;

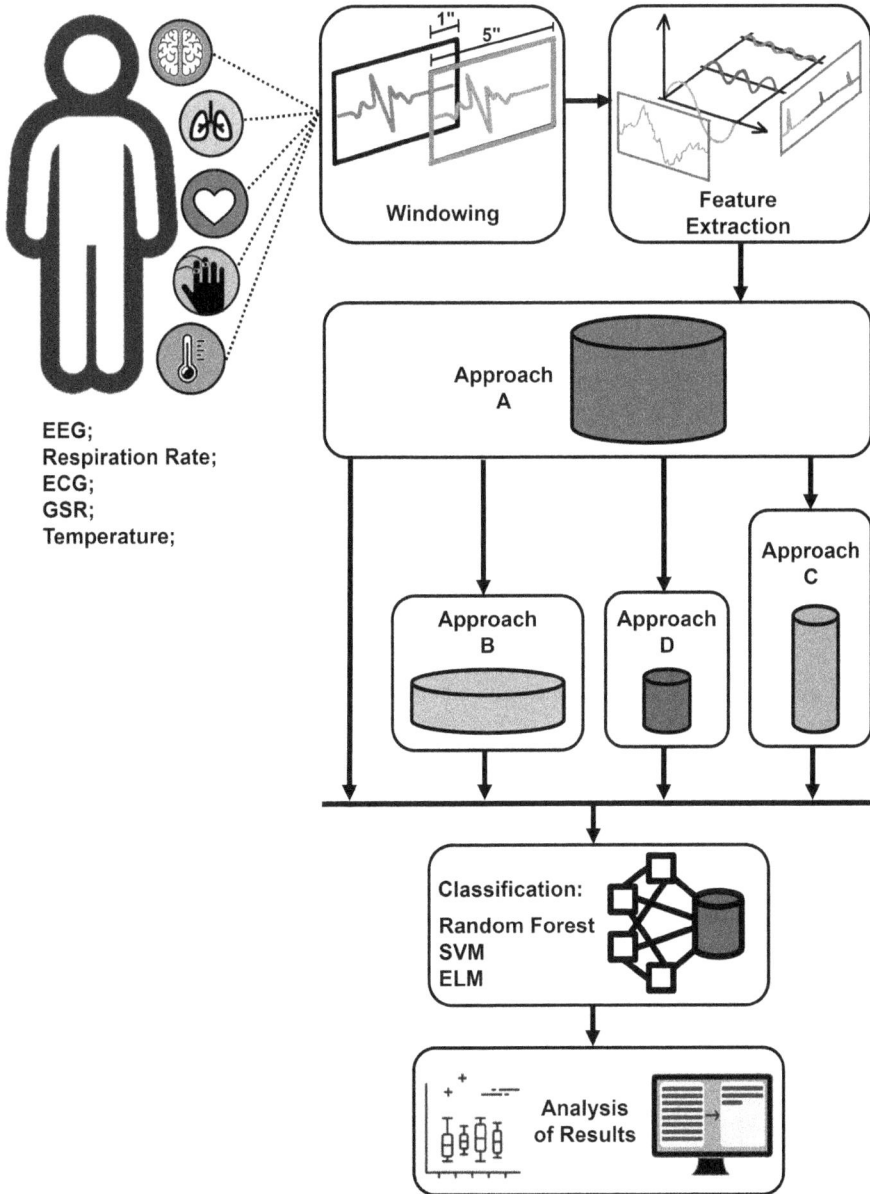

FIGURE 8.1 General proposal of this study. The work was composed of the following steps: database of physiological signals (EEG, ECG, respiration rate, temperature and GSR); signals windowing with overlapping; feature extraction; development of datasets with feature and instances manipulations; training and testing with intelligent models; and results assessment.

experiments were conducted using cross validation (10 folds) and 30 repetitions in order to generate statistically relevant results. Finally, in the last step we display the results in boxplots for the statistical analysis.

8.3.1 SIGNAL DATABASE

In this study, the database used for emotion recognition in physiological signals was the Multimodal Database for Affect Recognition and Implicit Tagging (MAHNOB-HCI), developed by Soleymani et al. (2011). The database has data that were collected using a multimodal configuration, all organized to record data in a synchronized way. The acquired data were: facial videos, audio signals, eye tracking (stare and pupil size), physiological signals from the central nervous system (Electroencephalogram—EEG) and peripheral system (Galvanic Skin Response, Electrocardiogram, respiration rate and skin temperature). MAHNOB-HCI comprises nine classes of emotions, which are: Sadness, Joy, Disgust, Neutral, Fun, Anger, Fear, Surprise and Anxiety.

In instrumental terms, two electrodes in the distal phalanges of the middle and index fingers were used to obtain the electrical activity of the sweat glands (GRS). Moreover, three electrodes were placed in the upper right and left corners of the chest and abdomen, for data on cardiac electrical activity (ECG). Signals of encephalic character (EEG) were recorded from 32 active silver chloride electrodes, including two of reference, positioned according to the international 10–20 system. Data on the participant's skin temperature were also acquired using a sensor. The acquisition of data from the MAHNOB-HCI database was carried out in two stages and involved the participation of 30 volunteers with different cultural backgrounds and of both male and female genders. However, six volunteers did not participate in all acquisition steps.

In the first stage, 20 videos were presented to stimulate emotions in individuals. It is important to note that at the end of each video a neutral clip was played to minimize the emotional bias activated by the previous video and ease participants self-assessed after watching the videos. The evaluation was according a discrete scale with values between 1 to 9 (The dimension of pleasure in the scale from 1 to 9 varies from pleasant (value 1) to unpleasant (value 9). This assessment was based on five different questions: i) What emotion was presented?; ii) What level of pleasure?; iii) What level of activation?; iv) What level of dominance? and v) What is the level of predictability? To classify the videos, the 3D model of inferences of affective states PAD (Pleasure-Activation-Dominance) was used and to answer what emotion the videos presented the participants had to rate the stimuli in a scale from 1 to 9, corresponding to the following emotional states: Neutral, Anxiety, Fun, Sadness, Happiness, Disgust, Anger, Surprise and Fear.

In the second stage, 28 static images and 14 video excerpts were presented, following the steps: 1) an unlabeled stimulus was shown, static images were presented for 5s, so that the participant could become familiarized with the content; 2) then the same stimulus was presented again with a labeling, during this period the individual's reaction is indicated as his reaction to the labeling of the image/video; and finally, 3) a question was asked to the participant if he or she agreed or not with the suggested labeling for the exposed stimulus. It is worth mentioning that at this stage, only the front color camera was used to record the reactions. Finally, from the data collected in the two steps briefly mentioned earlier, Soleymani et al. (2011) built the MAHNOB-HCI database.

8.3.2 SIGNAL SEGMENTATION AND FEATURE EXTRACTION

From the nine emotions in the database, only six had available signals and, therefore, were used in our study. Of these, 55 are related to Happiness emotion, 14 to Sadness, 84 to Neutral, 40 to Disgust, 65 to Amusement and 27 to Anger, thus evidencing the class impairment. This problem was computationally solved later, since it may bias the classification process by providing more samples from a particular class than from another.

Each signal had 47 channels; however, from them nine were empty and 38 had some information. Among these channels, 1 to 32 had EEG signals. On channels 33, 34 and 35 were the ECG. Channel 41 was dedicated to the Galvanic Skin Response, channel 45 to respiration rate and channel 46 to skin temperature.

Then, each complete signal was windowed for 5 seconds, with 1 second overlap between windows. This procedure aims to increase the spectral characteristics of the sample. From this multiplication of signals from the windows, we generated an unbalanced dataset with 1704 instances of Happiness class, 1114 of Neutral, 500 of Sadness, 1222 of Disgust, 2650 of Fun, and 907 of Anger class. Finally, each channel of these instances was subjected to the extraction of 35 features in the time and frequency domain (e.g.: Wavelength, Average Frequency, Entropy and so on). These features are listed in Table 8.2 with their respective mathematical expressions. Both Signal Segmentation and Feature Extraction were performed in the GNU/Octave software, version 4.0.3 (Eaton et al., 2015).

8.3.3 CLASS BALANCING, FEATURE SELECTION AND RESAMPLING

Since the number of instances per class is originally unequal, we carry out a step of class balancing. To balance the number of instances in each of the six classes we use the Synthetic Minority Oversampling TEchnique (SMOTE) (Chawla et al., 2002). SMOTE is a tool to increase the minority group in imbalanced datasets. It creates new artificial instances based on the real ones of a given class. The minority classes are balanced by taking each instance and adding synthetic samples along the line segments joining its k-nearest neighbors. In our approach we set SMOTE with three neighbors. Thus, we end up with a balanced dataset of 15907 instances, being 2649 of Sadness, 2658 of Happiness, 2651 of Disgust, 2651 of Neutral, 2650 of Amusement and 2648 of Anger.

For feature selection and resampling we respectively used the Best First and Gaussian Distribution Based Balance methods. Best First is a heuristic tool for feature selection that searches the space of features subsets to find one with the most relevant features for a given dataset (Xu et al., 1988). Thus, the combination of features that gives the best performance is selected. In this study we set

TABLE 8.2
List of Extracted Features and Their Mathematical Expressions.

Parameter	Equation	Parameter	Equation
Mean (μ)	$\mu = \dfrac{1}{N}\displaystyle\sum_{n=1}^{N} x_n$	Waveform length	$WL = \displaystyle\sum_{n=1}^{N-1} |x_{n+1} - x_n|$
Variance	$var = \dfrac{1}{N-1}\displaystyle\sum_{n=1}^{N}(x_n - \mu)^2$	Zero crossing	$ZC = \displaystyle\sum_{n=1}^{N-1}[sgn(x_n \times x_{n+1}) \cap |x_n - x_{n+1}| \geq threshold]$ $sgn(x) = \begin{cases} 1, if\ x \geq threshold \\ 0, \quad otherwise \end{cases}$
Standard deviation (σ)	$\sigma = \sqrt{\dfrac{1}{N-1}\displaystyle\sum_{n=1}^{N}|x_n - \mu|^2}$	Slope Sign Changes	$SSC = \displaystyle\sum_{n=1}^{N-1}[f(x_n - x_{n-1}) \times (x_n - x_{n+1})]]$ $f(x) = \begin{cases} 1, if\ x \geq threshold \\ 0, \quad otherwise \end{cases}$
Root mean square	$RMS = \sqrt{\dfrac{\sum_{n=1}^{N}(x_n)^2}{N}}$	Hjorth parameter activity	$Hjorth_{activity} = \dfrac{1}{N-1}\displaystyle\sum_{n=1}^{N}(x_n - \mu)^2$
Average Amplitude Changes	$AAC = \dfrac{1}{N}\left(\displaystyle\sum_{n=1}^{N}\left|\dfrac{d\,x(t)}{dt}\right|\right)$	Hjorth parameter mobility	$Hjorth_{mobility} = \sqrt{\dfrac{var\left(\frac{d\,x(t)}{dt}\right)}{var(x(t))}}$
Difference Absolute Deviation	$DASDV = \sqrt{\dfrac{1}{N}\displaystyle\sum_{n=1}^{N}\left(\dfrac{d\,x(t)}{dt}\right)^2}$	Hjorth parameter complexity	$Hjorth_{complexity} = \dfrac{Hjorth_{mobility}\left(\frac{d\,x(t)}{dt}\right)}{Hjorth_{mobility}(x(t))}$
Integrated Absolute Value	$IAV = \displaystyle\sum_{n=1}^{N} x_n$	Mean frequency	$MNF = \dfrac{\sum_{j=1}^{M} f_j P_j}{\sum_{j=1}^{M} P_j}$ Where f_j, P_j are the frequencies and power of the spectrum, respectively, and M is the length of the frequencies
Logarithm Detector	$LOGD = e^{(\frac{1}{N}\sum_{n=1}^{N}\log(|x_n|))}$	Median frequency	$MDF = \dfrac{1}{2}\displaystyle\sum_{j=1}^{M} P_j$
Simple Square Integral	$SSI = \displaystyle\sum_{n=1}^{N} x_n^2$	Mean power	$MNP = \displaystyle\sum_{j=1}^{M}\dfrac{P_j}{M}$
Mean Absolute Value	$MAV = \dfrac{1}{N}\displaystyle\sum_{n=1}^{N}|x_n|$	Peak frequency	$PKF = \max(P_j)$
Mean Logarithm Kernel	$MLOGK = \dfrac{1}{N}\displaystyle\sum_{n=1}^{N}|x_n|$	Power Spectrum ratio	$PSR = \dfrac{PKF}{\sum_{j=1}^{M} P_j}$
Skewness (s)	$s = \dfrac{\frac{1}{N}\sum_{n=1}^{N}(x_n - \mu)^3}{\sigma^3}$	Total Power	$TP = \displaystyle\sum_{j=1}^{M} P_j$
Kurtosis	$kurt = \dfrac{\frac{1}{N}\sum_{n=1}^{N}(x_n - \mu)^4}{\sigma^4}$	First Spectral Moment	$SM1 = \displaystyle\sum_{j=1}^{M} f_j P_j$
Maximum Amplitude	$MAX = \max(x_n)$	Second Spectral Moment	$SM2 = \displaystyle\sum_{j=1}^{M} f_j^2 P_j$
Third Moment	$M3 = \left|\dfrac{1}{N}\displaystyle\sum_{n=1}^{N}(x_n)^3\right|$	Third Spectral Moment	$SM3 = \displaystyle\sum_{j=1}^{M} f_j^3 P_j$
Fourth Moment	$M4 = \left|\dfrac{1}{N}\displaystyle\sum_{n=1}^{N}(x_n)^4\right|$	Variance of Central Frequency	$VCF = \dfrac{SM2}{TP} - \left(\dfrac{SM1}{TP}\right)^2$
Fifth Moment	$M5 = \left|\dfrac{1}{N}\displaystyle\sum_{n=1}^{N}(x_n)^5\right|$	Shannon's entropy	$E = -\displaystyle\sum_{t} S_t^2 \log(S_t^2),\ where\ S\ is\ the\ signal$

the algorithm to start with an empty set of features and search forward. Each different subset is evaluated by considering the individual predictive ability of each feature along with the degree of redundancy between them. Subsets that are highly correlated with the class while having low intercorrelation are preferred. This search process continues until there are five consecutive non-improving subsets.

The Gaussian Distribution Based Balance is a resampling method that was used here to reduce the amount of instances of the original dataset (Bermejo et al., 2011). In this method sampling of instances is performed following a Gaussian distribution learned for each pair (*feature, class label*). Therefore, a subsample is created preserving the statistical behavior of the original instances. Our approach created a subset with 10% of the original amount of instances. All these steps were performed using the Waikato Environment for Knowledge Analysis (WEKA), version 3.8 (Witten and Frank, 2005).

8.3.4 CLASSIFICATION

For the classification step we assess the performances of different configurations of Support Vector Machine (SVM), Random Forest (RF) and Extreme Learning Machine (ELM). SVM is a statistical and supervised paradigm method proposed by Cortes and Vapnik (1995). SVM stands out for its good generalization performance (Wang et al., 2021). Its main goal is to find an ideal/optimal hyperplane capable of separating distinct classes (Pal, 2005; Pham et al., 2019; Skariah et al., 2021). Generally, the hyperplane can be modified by tweaking the kernel functions (Zeng et al., 2021). Some common kernel types are shown in Table 8.3, along with their respective functions. For SVM settings, we used the Linear Kernel, Polynomial Kernel and Radial Basis Function Kernel. The configurations used to find an optimal hyperplane capable of classifying our emotions classes can be seen in Table 8.4.

Random Forest is a supervised machine learning algorithm applied to solve classification and regression problems. RF can be understood as a hierarchical

TABLE 8.3
Kernel Functions Used on SVM and ELM Models.

Kernel type	Kernel function
Linear	$K(x, y) = x \bullet y$
Polynomial	$K(x, y) = (1 + x \bullet y)^d$
Radial Basis Function (RBF)	$K(x, y) = \exp(-\gamma(x - y) \bullet (x - y))$
Sigmoid	$K(x, y) = \tanh(b(x, y) + c)$

TABLE 8.4
Experimental Settings for the Classifiers.

Classifiers	Settings
Random Forest	Trees: 10, 20, 50, 100, 150, 200, 250, 300, and 350 Batch size: 100
SVM	Kernel functions: linear, polynomial (d = 2, d = 3, and d = 4), RBF ($\gamma = 0.50$)
ELM	Kernel functions: linear, polynomial (d = 2, d = 3, and d = 4), RBF and sigmoid. Neurons in the hidden layer: 500.

collection of decision trees (Jackins et al., 2021; Pal, 2005). Briefly, forests are created composed of decision trees responsible for analyzing the input data and assigning the output values to the classes. In order to find the amount of trees that best improve classification rate, we use the settings in Table 8.4, varying the number of trees among the values: 10, 20, 50, 100, 150, 200, 250, 300 and 350. It is worth clarifying that the Random Forest classifier not only handles missing data well, but also reduces the possibility of dataset overfitting. In addition, it usually has better overall performance when compared to Deep Learning methods, due to its simplicity in formulation, ease of implementation and lower computational cost (Pavani et al., 2021).

The Extreme Learning Machines are neural networks proposed by Huang et al. (2004) that stand out for their reduced training time and great generalization power (da Silva and Krohling, 2016). As explained by Arruda and Krohling (2016), the ELM processing starts after assigning random values for the weights of the input layer neurons and biases. Furthermore, the weights of the output layer neurons do not use iterative processes (such as backpropagation), but are analytically calculated. In terms of parameters, for the experiments using ELM we tested the linear, polynomial, RBF and sigmoid kernels. In addition to 500 neurons in the hidden layer, as shown in Table 8.4.

To assess the performance of the three models mentioned earlier, experiments were carried out using the settings shown in Table 8.4. For the experiments with Random Forest and SVM we used the Weka software (Version 3.8) (Witten and Frank, 2005). The experiments with ELM were conducted from the mathematical computing environment GNU/Octave (version 4.0.3) (Eaton et al., 2015).

It is worth mentioning that the k-fold cross-validation method with 10 folds was used during the experiments to avoid overfitting (Siriyasatien et al., 2018). In order to obtain statistical assessment regarding the performance of the models, each algorithm was tested 30 times.

8.3.4.1 Metrics

We chose five main metrics to assess the classifiers' performance: accuracy, kappa statistic, sensitivity, specificity and area under the ROC curve (Table 8.5). Accuracy is a metric that indicates how efficient the classifier is at correctly predicting the class of each instance. It is an index directly proportional to the true positives (TP) and true negatives (TN) rates. The kappa statistic is a metric similar to accuracy. However, kappa takes into account the random hit chance (Artstein and Poesio, 2008). When predictions are purely random, the kappa index has a value of 0 or it can assume negative values. Sensitivity is the metric used to assess the classifier's performance in identifying the true positives. Sensitivity is commonly called the true positive rate (TPR), but it is also known as recall. Specificity, as opposed to sensitivity, is used to assess performance in identifying the true negatives. Thus, it is known as the true negative rate (TNR). The area under the ROC (Receiver Operating Characteristics) curve is also a metric used to assess how well the model performs in the prediction. The ROC curve is a probabilistic curve and the area under it represents the chance that the model correctly predicts the data (Hanley and McNeil, 1982). The curve is

TABLE 8.5

Mathematical Expressions for the Metrics Used to Evaluate the Method. TP, TN, FP and FN are the Quantity of True Positives, True Negatives, False Positives and False Negatives, Respectively. TPR and FPR are the True Positive Rate and False Positive Rate, Respectively.

Metric	Mathematical Expression
Accuracy	$\frac{TP+TN}{TP+TN+FP+FN}$
Sensitivity	$\frac{TP}{TP+FN}$
Specificity	$\frac{TN}{TN+FP}$
Kappa	$\frac{(\rho_o - \rho_e)}{(1 - \rho_e)}$
Area Under the ROC Curve	$AUC = \int TPR\, d(FPR)$

built from the False Positive Rate (FPR) on the x-axis, and the sensitivity (TPR) on the y-axis. In the case of multiclass problems, such as in this work, the AUC can be evaluated in two ways: one vs one; one vs all. In the first one, all the curves of the combination of all classes with each other are plotted in pairs. In the second, the curves of the combination of one class vs all others are plotted.

In Table 8.5, ρ_o is the observed agreement rate, also called accuracy. And ρ_e is the expected agreement rate, defined in the Equation 8.1.

$$\rho_e = \frac{(TP+FP)(TP+FN)+(FN+TN)(FP+TN)}{(TP+FP+FN+TN)^2} \qquad (8.1)$$

8.4   RESULTS AND DISCUSSION

In this study we evaluated four distinct approaches to represent the central and peripheral physiological signals. In approach A, the signals are represented by the entire set of instances and extracted features. In approach B we submit this complete set to a feature selection step. Feature selection reduced the number of attributes from 1330 to 25. Approach C kept the original number of features, but we applied a method of resampling in the set. In this approach, the number of instances became approximately 10% of the total instances in the original dataset. Finally, in approach D, both manipulations were applied. That is, the set of approach D has a reduced amount of both features and instances, when compared to the original set (approach A). Figure 8.2 summarizes these approaches.

In this section we will present the results obtained by each of the four approaches. Tables 8.6–8.9 show the mean values and their respective standard deviations for the metrics of accuracy, kappa statistic, sensitivity, specificity and area under the ROC curve. In these tables, we highlight the best performing configurations for each classifier family (RF, SVM or ELM). Then, we organized the accuracy and kappa results

Approach A

15907 instances

1330 features

A

Approach B

15907 instances

25 features

B

Approach C

1596 instances

1330 features

C

Approach D

1596 instances

25 features

D

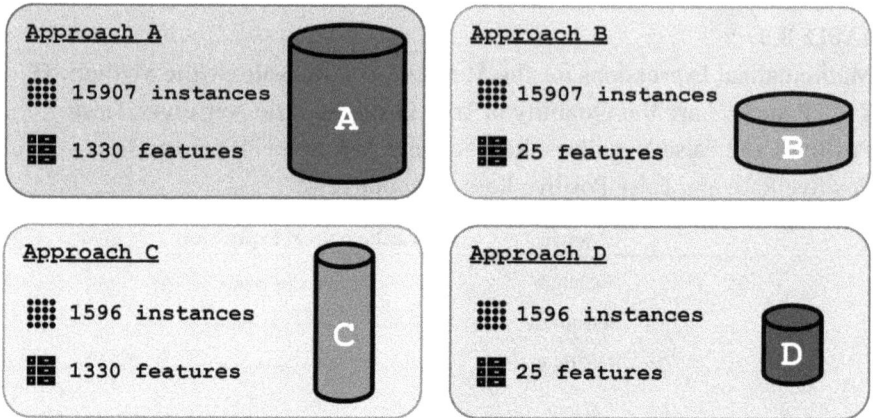

FIGURE 8.2 Proposed approaches. Approach A has the complete set of features and instances; approach B is composed of selected features and all instances; C has the original set of features but has the reduction of the amount of instances; and D has the reduction on both instances and features.

TABLE 8.6
Classification Performance for the Dataset with the Complete Amount of Features and Instances (Approach A).

Classifier		Accuracy (%)	Kappa statistic	Recall/Sensitivity	Specificity	AUC
Random Forest	10 trees	98.08 ± 0.35	0.9770 ± 0.0042	0.9959 ± 0.0037	0.9966 ± 0.0016	0.9996 ± 0.0007
	20 trees	98.28 ± 0.33	0.9793 ± 0.0040	0.9956 ± 0.0040	0.9972 ± 0.0015	0.9998 ± 0.0005
	50 trees	98.40 ± 0.31	0.9808 ± 0.0037	0.9955 ± 0.0039	0.9975 ± 0.0014	0.9999 ± 0.0001
	100 trees	98.44 ± 0.29	0.9813 ± 0.0035	0.9956 ± 0.0039	0.9976 ± 0.0013	0.9999 ± 0.0001
	150 trees	98.46 ± 0.29	0.9815 ± 0.0034	0.9956 ± 0.0039	0.9977 ± 0.0013	0.9999 ± 0.0001
	200 trees	98.47 ± 0.29	0.9816 ± 0.0035	0.9957 ± 0.0039	0.9977 ± 0.0013	0.9999 ± 0.0001
	250 trees	98.47 ± 0.29	0.9816 ± 0.0034	0.9958 ± 0.0039	0.9977 ± 0.0013	0.9999 ± 0.0001
	300 trees	98.48 ± 0.29	0.9817 ± 0.0034	0.9957 ± 0.0039	0.9977 ± 0.0013	0.9999 ± 0.0001
	350 trees	98.47 ± 0.28	0.9817 ± 0.0034	0.9958 ± 0.0039	0.9976 ± 0.0013	0.9999 ± 0.0001
SVM	linear	86.79 ± 0.76	0.8414 ± 0.0092	0.9859 ± 0.0075	0.9751 ± 0.0043	0.9886 ± 0.0022
	poly 2	95.29 ± 0.52	0.9435 ± 0.0062	0.9887 ± 0.0063	0.9942 ± 0.0021	0.9968 ± 0.0014
	poly 3	93.09 ± 1.76	0.9171 ± 0.0212	0.9795 ± 0.0274	0.9906 ± 0.0044	0.9956 ± 0.0029
	poly 4	80.81 ± 2.66	0.7697 ± 0.0319	0.9491 ± 0.0416	0.9727 ± 0.0093	0.9892 ± 0.0048
	RBF	95.11 ± 0.52	0.9414 ± 0.0063	0.9826 ± 0.0074	0.9919 ± 0.0025	0.9868 ± 0.0046
ELM	linear	29.14 ± 7.46	0.1497 ± 0.0896	0.2914 ± 0.2365	0.8583 ± 0.0892	0.5748 ± 0.1134
	poly 2	40.32 ± 5.34	0.2838 ± 0.0640	0.4032 ± 0.2365	0.8806 ± 0.0611	0.6419 ± 0.1144
	poly 3	41.86 ± 7.15	0.3023 ± 0.0858	0.4186 ± 0.2453	0.8837 ± 0.0600	0.6512 ± 0.1218
	poly 4	44.28 ± 5.57	0.3314 ± 0.0669	0.4429 ± 0.2392	0.8886 ± 0.0593	0.6657 ± 0.1165
	RBF	16.67 ± 0.01	0.0000 ± 0.0000	0.1667 ± 0.3728	0.8333 ± 0.3728	0.5000 ± 0.0000
	sigmoid	38.87 ± 6.94	0.2664 ± 0.0832	0.3887 ± 0.2373	0.8777 ± 0.0676	0.6332 ± 0.1171

obtained by these best settings in the graphs of Figures 8.3–8.6 to better assess their statistical behavior.

Overall, the results point to better performances for classifying emotions in those signals using Random Forest classifiers. This classifier was superior to SVM and ELM in approaches A, B and D. In approach C, Random Forest and SVM performed similarly and both were superior to ELM. The good performance of the Random

TABLE 8.7
Classification Performance for the Dataset with Feature Selection (Approach B).

Classifier		Accuracy (%)	Kappa statistic	Recall/Sensitivity	Specificity	AUC
Random Forest	10 trees	97.50 ± 0.37	0.9700 ± 0.0044	0.9934 ± 0.0051	0.9961 ± 0.0017	0.9992 ± 0.0010
	20 trees	97.68 ± 0.35	0.9721 ± 0.0042	0.9940 ± 0.0047	0.9966 ± 0.0016	0.9996 ± 0.0006
	50 trees	97.82 ± 0.33	0.9738 ± 0.0040	0.9946 ± 0.0042	0.9968 ± 0.0015	0.9998 ± 0.0002
	100 trees	97.86 ± 0.33	0.9743 ± 0.0040	0.9948 ± 0.0043	0.9969 ± 0.0015	0.9998 ± 0.0001
	150 trees	97.87 ± 0.33	0.9744 ± 0.0040	0.9948 ± 0.0043	0.9969 ± 0.0015	0.9999 ± 0.0001
	200 trees	97.88 ± 0.33	0.9746 ± 0.0040	0.9949 ± 0.0043	0.9969 ± 0.0015	0.9999 ± 0.0001
	250 trees	97.89 ± 0.33	0.9747 ± 0.0040	0.9949 ± 0.0043	0.9969 ± 0.0015	0.9999 ± 0.0001
	300 trees	97.90 ± 0.33	0.9748 ± 0.0039	0.9949 ± 0.0043	0.9969 ± 0.0015	0.9999 ± 0.0001
	350 trees	97.90 ± 0.33	0.9748 ± 0.0040	0.9950 ± 0.0043	0.9969 ± 0.0015	0.9999 ± 0.0001
SVM	linear	33.15 ± 1.27	0.1978 ± 0.0153	0.5252 ± 0.0604	0.7862 ± 0.0229	0.7553 ± 0.0148
	poly 2	45.03 ± 1.18	0.3404 ± 0.0141	0.6283 ± 0.0453	0.8540 ± 0.0136	0.8623 ± 0.0104
	poly 3	57.98 ± 1.11	0.4958 ± 0.0133	0.8976 ± 0.0186	0.8883 ± 0.0080	0.9378 ± 0.0052
	poly 4	74.90 ± 1.19	0.6988 ± 0.0142	0.9735 ± 0.0102	0.9440 ± 0.0082	0.9748 ± 0.0040
	RBF	43.40 ± 1.12	0.3209 ± 0.0134	0.6170 ± 0.0472	0.8364 ± 0.0195	0.8444 ± 0.0111
ELM	linear	16.34 ± 5.34	-0.0034 ± 0.0641	0.1634 ± 0.1937	0.8327 ± 0.1222	0.4981 ± 0.0933
	poly 2	21.42 ± 7.54	0.0571 ± 0.0904	0.2142 ± 0.2243	0.8428 ± 0.1139	0.5286 ± 0.1091
	poly 3	26.06 ± 6.23	0.1126 ± 0.0747	0.2605 ± 0.2265	0.8521 ± 0.1078	0.5563 ± 0.1001
	poly 4	25.43 ± 6.61	0.1052 ± 0.0793	0.2543 ± 0.2287	0.8509 ± 0.1096	0.5526 ± 0.1047
	RBF	20.11 ± 5.93	0.0413 ± 0.0711	0.2011 ± 0.2283	0.8402 ± 0.1352	0.5207 ± 0.1095
	sigmoid	16.42 ± 5.83	-0.0029 ± 0.0700	0.1642 ± 0.2080	0.8328 ± 0.1370	0.4985 ± 0.1090

TABLE 8.8
Classification Performance for the Dataset with Reduced Number of Instances (Approach C).

Classifier		Accuracy (%)	Kappa statistic	Recall/Sensitivity	Specificity	AUC
Random Forest	10 trees	98.82 ± 0.84	0.9859 ± 0.0101	1.0000 ± 0.0000	0.9999 ± 0.0010	1.0000 ± 0.0000
	20 trees	99.90 ± 0.25	0.9988 ± 0.0030	1.0000 ± 0.0000	1.0000 ± 0.0000	1.0000 ± 0.0000
	50 trees	100.00 ± 0.00	1.0000 ± 0.0000	1.0000 ± 0.0000	1.0000 ± 0.0000	1.0000 ± 0.0000
	100 trees	100.00 ± 0.00	1.0000 ± 0.0000	1.0000 ± 0.0000	1.0000 ± 0.0000	1.0000 ± 0.0000
	150 trees	100.00 ± 0.00	1.0000 ± 0.0000	1.0000 ± 0.0000	1.0000 ± 0.0000	1.0000 ± 0.0000
	200 trees	100.00 ± 0.00	1.0000 ± 0.0000	1.0000 ± 0.0000	1.0000 ± 0.0000	1.0000 ± 0.0000
	250 trees	100.00 ± 0.00	1.0000 ± 0.0000	1.0000 ± 0.0000	1.0000 ± 0.0000	1.0000 ± 0.0000
	300 trees	100.00 ± 0.00	1.0000 ± 0.0000	1.0000 ± 0.0000	1.0000 ± 0.0000	1.0000 ± 0.0000
	350 trees	100.00 ± 0.00	1.0000 ± 0.0000	1.0000 ± 0.0000	1.0000 ± 0.0000	1.0000 ± 0.0000
SVM	linear	100.00 ± 0.00	1.0000 ± 0.0000	1.0000 ± 0.0000	1.0000 ± 0.0000	1.0000 ± 0.0000
	poly 2	100.00 ± 0.00	1.0000 ± 0.0000	1.0000 ± 0.0000	1.0000 ± 0.0000	1.0000 ± 0.0000
	poly 3	100.00 ± 0.00	1.0000 ± 0.0000	1.0000 ± 0.0000	1.0000 ± 0.0000	1.0000 ± 0.0000
	poly 4	100.00 ± 0.00	1.0000 ± 0.0000	1.0000 ± 0.0000	1.0000 ± 0.0000	1.0000 ± 0.0000
	RBF	18.35 ± 2.83	0.0246 ± 0.0338	0.3233 ± 0.4535	0.7177 ± 0.4292	0.5856 ± 0.0936
ELM	linear	99.26 ± 0.66	0.9911 ± 0.0080	0.9926 ± 0.0163	0.9985 ± 0.0030	0.9956 ± 0.0082
	poly 2	97.66 ± 0.9784	0.9719 ± 0.0117	0.9766 ± 0.0316	0.9953 ± 0.0064	0.9860 ± 0.0158
	poly 3	96.93 ± 1.24	0.9632 ± 0.0148	0.9693 ± 0.0387	0.9939 ± 0.0081	0.9816 ± 0.0202
	poly 4	95.30 ± 1.18	0.9435 ± 0.0142	0.9530 ± 0.0442	0.9906 ± 0.0096	0.9718 ± 0.0224
	RBF	16.67 ± 0.00	0.0000 ± 0.0000	0.1667 ± 0.3728	0.8333 ± 0.3728	0.5000 ± 0.0000
	sigmoid	90.76 ± 2.10	0.8892 ± 0.0252	0.9076 ± 0.0667	0.9815 ± 0.0121	0.9446 ± 0.0346

Forest-based classifiers indicates that the problem has a degree of complexity that can make it difficult to generalize.

Nevertheless, all classifiers settings presented low data dispersion. Low dispersion is a positive factor, as it indicates that the results have high reliability. Furthermore, the results of the metrics were consistent with each other. In other words, we didn't get any discrepant results between the metrics.

TABLE 8.9
Classification Performance for the Dataset with Feature Selection and Reduced Number of Instances (Approach D).

Classifier		Accuracy (%)	Kappa statistic	Recall/Sensitivity	Specificity	AUC
Random Forest	10 trees	67.06 ± 3.61	0.6047 ± 0.0433	0.7408 ± 0.0852	0.8851 ± 0.0265	0.9056 ± 0.0275
	20 trees	71.07 ± 3.65	0.6528 ± 0.0438	0.7698 ± 0.0868	0.8994 ± 0.0251	0.9297 ± 0.0218
	50 trees	73.88 ± 3.36	0.6866 ± 0.0403	0.7925 ± 0.0780	0.9068 ± 0.0251	0.9441 ± 0.0180
	100 trees	75.04 ± 3.16	0.7005 ± 0.0380	0.8030 ± 0.0782	0.9111 ± 0.0241	0.9483 ± 0.0168
	150 trees	75.37 ± 3.14	0.7044 ± 0.0377	0.8072 ± 0.0816	0.9114 ± 0.0238	0.9496 ± 0.0163
	200 trees	75.41 ± 3.22	0.7049 ± 0.0386	0.8080 ± 0.0812	0.9136 ± 0.0242	0.9504 ± 0.0161
	250 trees	75.52 ± 3.14	0.7062 ± 0.0376	0.8096 ± 0.0797	0.9136 ± 0.0247	0.9507 ± 0.0160
	300 trees	75.61 ± 3.14	0.7073 ± 0.0376	0.8094 ± 0.0810	0.9143 ± 0.0241	0.9510 ± 0.0162
	350 trees	75.78 ± 3.17	0.7094 ± 0.0381	0.8104 ± 0.0825	0.9141 ± 0.0244	0.9512 ± 0.0160
SVM	linear	42.92 ± 3.92	0.3150 ± 0.0467	0.5088 ± 0.1029	0.8431 ± 0.0315	0.7772 ± 0.0438
	poly 2	43.63 ± 3.75	0.3235 ± 0.0450	0.5056 ± 0.1013	0.8584 ± 0.0302	0.7789 ± 0.0415
	poly 3	42.23 ± 3.51	0.3068 ± 0.0421	0.4226 ± 0.0972	0.8815 ± 0.0290	0.7687 ± 0.0439
	poly 4	39.12 ± 3.42	0.2694 ± 0.0410	0.3809 ± 0.0963	0.9031 ± 0.0273	0.7520 ± 0.0456
	RBF	43.94 ± 3.76	0.3273 ± 0.0451	0.4832 ± 0.0954	0.8578 ± 0.0307	0.7810 ± 0.0432
ELM	linear	42.26 ± 2.46	0.3071 ± 0.0295	0.4226 ± 0.1364	0.8845 ± 0.0387	0.6535 ± 0.0690
	poly 2	42.37 ± 3.97	0.3084 ± 0.0476	0.4237 ± 0.1220	0.8847 ± 0.0420	0.6542 ± 0.0668
	poly 3	42.05 ± 3.80	0.3046 ± 0.0456	0.4205 ± 0.1224	0.8841 ± 0.0354	0.6523 ± 0.0674
	poly 4	41.52 ± 3.04	0.2982 ± 0.0365	0.4152 ± 0.1072	0.8830 ± 0.0382	0.6491 ± 0.0569
	RBF	23.22 ± 2.32	0.0787 ± 0.0279	0.2322 ± 0.2830	0.8464 ± 0.2469	0.5393 ± 0.0394
	sigmoid	42.53 ± 2.60	0.3103 ± 0.0312	0.4253 ± 0.1269	0.8850 ± 0.0407	0.6552 ± 0.0714

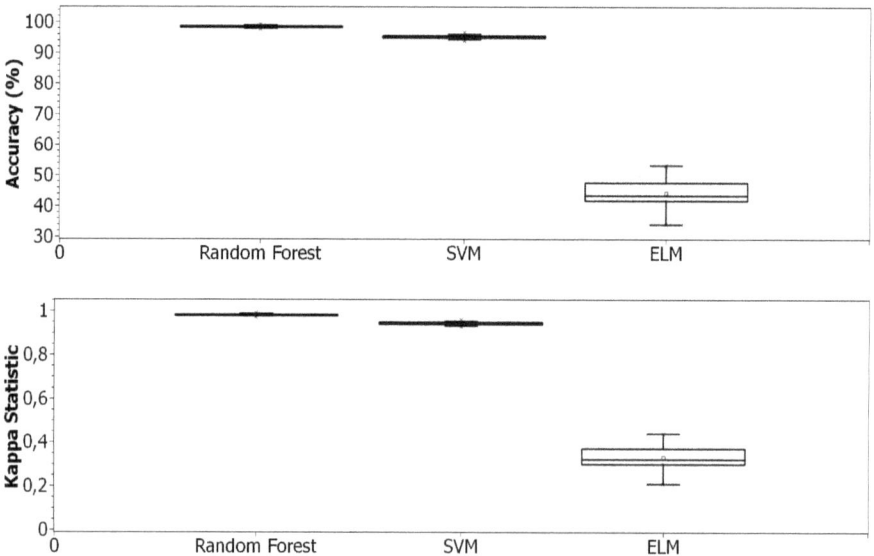

FIGURE 8.3   Classification performance for the dataset with the complete set of features and instances (approach A). Accuracy results are in (a); in (b) are the kappa statistic results.

Regarding the steps of reducing the number of features and instances, we observed that they were positive when applied in isolation. However, there was a significant decrease in the performance of the methods for approach D, where we applied both steps to reduce the dataset. It is likely that the reduction of both factors

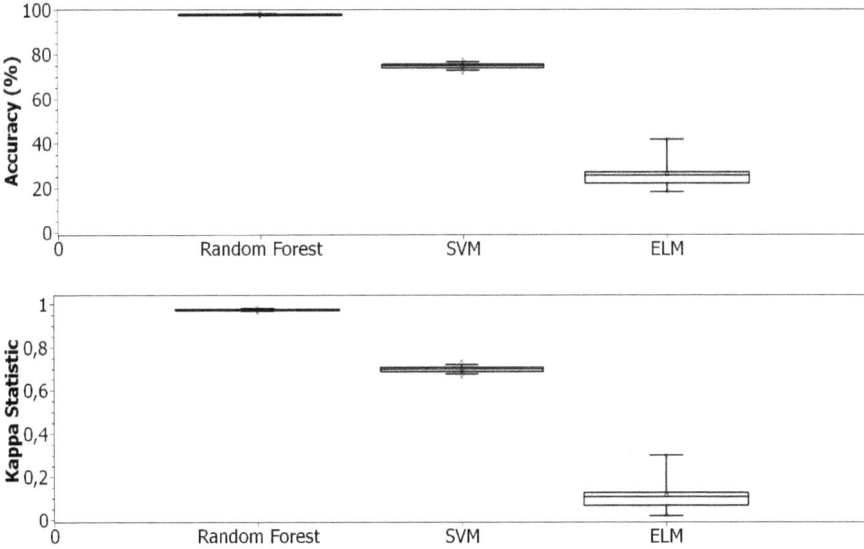

FIGURE 8.4 Classification performance for the dataset with the selected features (approach B). In (a) are the results of accuracy while (b) shows the kappa results.

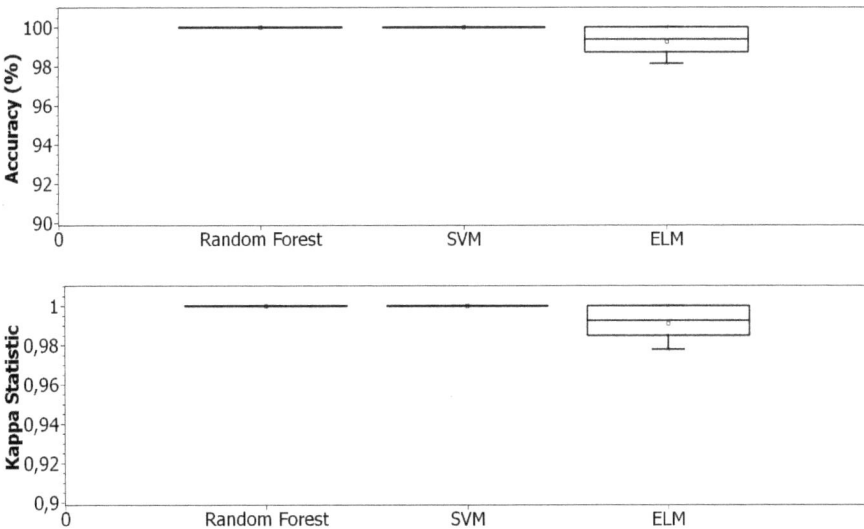

FIGURE 8.5 Classification performance for the dataset with reduced number of instances (approach C). In (a) are the results of accuracy while (b) shows the kappa results.

mischaracterized the problem, making it difficult to differentiate between classes. When we applied only the feature selection in approach B, the performance of Random Forest remained high, but there was a deterioration for the SVM and ELM methods, when compared to the respective results in approach A. This finding points

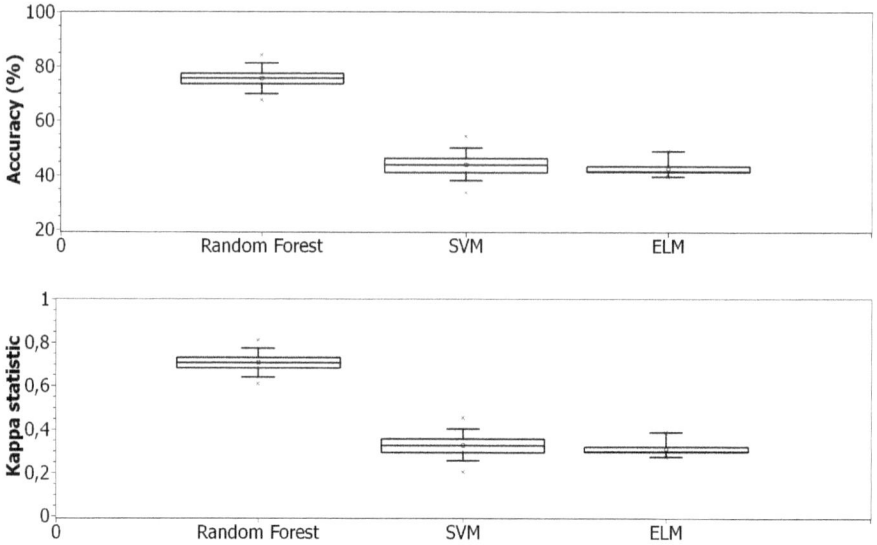

FIGURE 8.6 Classification performance for the dataset with feature selection and reduced number of instances (approach D). In (a) are the results of accuracy while (b) shows the kappa results.

to a relatively low redundancy and a certain degree of interdependence between the features. In approach C, when applying only the resampling, there was a significant increase in the performances of SVM and ELM. This phenomenon may indicate that the resampling process adds another learning step to the method. Possibly this step contributes to the understanding of the patterns distribution of the classes in this problem.

8.4.1 APPROACH A

Table 8.6 presents the performances of different configurations and classifiers for the complete set of instances and features. From these results we found good performance of Random Forest and SVM classifiers. However, ELM showed unsatisfactory results. The best result among the evaluated configurations was using Random Forest with 300 decision trees. This configuration achieved an accuracy of 98.48 ± 0.29, with a kappa of 0.9817 ± 0.0034, sensitivity of 0.9957 ± 0.0039, specificity of 0.9977 ± 0.0013 and AUC of 0.9999 ± 0.0001. The configurations with the SVM classifier also achieved performance close to the maximum values of the metrics, especially with a 2nd-degree polynomial kernel. The worst results were achieved by the ELM method. Among these configurations, ELM with 4th-degree polynomial kernel obtained better performances. The best results with ELM for this dataset were 44.28 ± 5.57 of accuracy, 0.3314 ± 0.0669 of kappa index, 0.4429 ± 0.2392 of sensitivity, specificity of 0.8886 ± 0.0593 and AUC of 0.6657 ± 0.1165.

Furthermore, we plot the best results for each classifier family on the graphs in Figure 8.3. From these graphs, it is possible to see that there is a significant difference between the performance of the tested families of classifiers. The behavior pattern of each method is well defined, with Random Forest presenting better performance than the other methods, SVM with intermediate performance and ELM with the worst result. We also observed that all methods presented low data dispersion, being slightly greater for the ELM method. The low dispersion proves the high reliability of the results, as there was no great variability. In addition, there is consistency between the accuracy and kappa statistic values.

8.4.2 APPROACH B

For approach B we selected the features from the original dataset. This step reduced by 98% the number of features to represent the signals. However, this drastic reduction did not significantly affect the performance of the Random Forest-based classifiers. This indicates that this method can be robust to differentiate classes even when they are described by few attributes. That is, there is a small set of features that provide relevant information for classification using Random Forest. Compared to the other methods, Random Forest achieved significant prominence (see Figure 8.4 for further details). We obtained the best results when using 300 decision trees, with an accuracy of 97.90 ± 0.33, kappa of 0.9748 ± 0.0039, sensitivity of 0.9949 ± 0.0043, 0.9969 ± 0.0015 of specificity and AUC of 0.9999 ± 0.0001. Similar results were found even with only 10 trees, as shown in Table 8.7. This shows that the problem was relatively easy to be solved by the Random Forest method. However, for SVMs and ELMs the classification challenge was greater. The ELM method presented the worst results, with maximum performance obtained when using a 3rd-degree polynomial kernel, with accuracy of 26.06 ± 6.23, kappa of 0.1126 ± 0.0747, 0.2605 ± 0.2265 of sensitivity, 0.8521 ± 0.1078 of specificity and AUC of 0.5563 ± 0.1001.

8.4.3 APPROACH C

For the dataset submitted to the resampling step alone, we reached the results shown in Table 8.8. Note that using 10% of the instances, the classification methods were, in general, very successful. The Random Forest classifier showed impeccable performance from the 50-tree configuration. For the SVM method, metrics also reached their maximum values for almost all configurations, with the only exception being with RBF kernel. The ELM classifier performed a little worse than the other methods. However, the linear kernel ELM performed very close to excellent, with an accuracy of 99.26 ± 0.66, kappa of 0.9911 ± 0.0080, 0.9926 ± 0.0163 of sensitivity, 0.9985 ± 0.0030 of specificity and AUC of 0.9956 ± 0.0082.

Figure 8.5 shows the statistical behaviors of the configurations with the best performance of each classifier family. From these graphs, it is evident that the behavior of the accuracy and kappa statistic metrics are consistent. There was low dispersion of results, especially for the Random Forest and SVM methods. For these methods, the performance remained maximum for the classification of emotions in this context.

These extremely positive results indicate that reducing the number of instances in the base may have added another learning step to the process. It may also indicate a certain level of redundancy in the original instances of the problem. Moreover, we realized that the behavior of the dataset fits well with the Gaussian distribution, since this model was successful in understanding the patterns associated with each class of the problem.

8.4.4 APPROACH D

We obtained the worst ranking performance when manipulating both the number of instances and features of the original dataset. These results are detailed in Table 8.9 and Figure 8.6. From these results we can see that the best performance for this approach was using the Random Forest classifier with 350 trees. This configuration achieved 75.78 ± 3.17 of accuracy, kappa of 0.7094 ± 0.0381, sensitivity of 0.8104 ± 0.0825, specificity of 0.9141 ± 0.0244 and AUC of 0.9512 ± 0.0160. The SVM and ELM methods, in turn, showed an accuracy of less than 50%, with low kappa and sensitivity values as well. Both achieved relatively high specificities and median AUC results. However, the dispersion of results for all methods remained low, despite being a little higher than that obtained from the other approaches. This demonstrates that even with the drop in performance, the methods remain reliable and their performance suffers little variability.

The notable decrease in performance obtained in this approach points to a possible mischaracterization of the problem by reducing the number of features and instances at the same time. Therefore, with the reductions, the set probably became insufficient to characterize the six classes of emotions in the problem.

8.5 CONCLUSION

In our daily lives, emotions are essential for the regulation of social interaction. Ideally, each individual is born with the ability to perceive, understand and recognize different emotions. However, some pathologies may affect this human ability, thus reducing people's social skills and quality of life. In this sense, artificial intelligence applied to emotion recognition becomes an alternative tool to overcome these disabilities. This tool is also fundamental to improve human-machine interactions, which is increasing day by day.

Therefore, in this chapter we propose an approach based on classic intelligent methods (i.e. Random Forest, SVM and ELM) to classify emotions in EEG and peripheral physiological signals. All signals are represented by a set of explicit features based on statistical, time and spectral parameters. This set of features showed to be suitable and relevant to differentiate the six existing emotions with Random Forest being the best architecture to identify these emotions.

During the study we submitted the dataset to feature selection and instance reduction processes. Feature selection slightly reduced classification performance. On the other hand, performances were greater after instances manipulation. These findings may suggest that the process of decreasing the number of instances provides a better understanding of the classes patterns. Moreover, it may also indicate that there is some kind of redundancy in the instances.

The overall classification results with Random Forest were outstanding, overcoming the other models. As for the other classifiers, SVM had moderate performance while ELM showed the worst results. However, these excellent results with Random Forest may indicate a low generalization ability against data from other subjects. Thus, future studies must be conducted with new data to further assess its performance.

8.6 ACKNOWLEDGEMENTS

This study was financed in part by the Coordenação de Aperfeiçoamento de Pessoal de Nível Superior—Brazil (CAPES)—Finance Code 001, and by the Fundação de Amparo à Ciência e Tecnologia do Estado de Pernambuco (FACEPE), Brazil, under the code IBPG-0013–1.03/20.

8.7 DISCLOSURE STATEMENT

No potential conflict of interest was reported by the authors.

REFERENCES

Abdallah, M., M. A. Talib, S. Feroz, Q. Nasir, H. Abdalla, and B. Mahfood. Artificial intelligence applications in solid waste management: a systematic research review. *Waste Management*, 109:231–246, 2020.

Alarcao, S. M. and M. J. Fonseca. Emotions recognition using eeg signals: a survey. *IEEE Transactions on Affective Computing*, 10(3):374–393, 2017.

Ali, M., F. Al Machot, A. Haj Mosa, M. Jdeed, E. Al Machot, and K. Kyamakya. A globally generalized emotion recognition system involving different physiological signals. *Sensors*, 18(6):1905, 2018.

Andrade, M. K., M. A. Santana, G. Moreno, I. Oliveira, J. Santos, M. C. A. Rodrigues, and W. P. dos Santos. An eeg brain-computer interface to classify motor imagery signals. In *Biomedical Signal Processing*, pages 83–98. Berlin: Springer, 2020.

Arruda, V. F. and R. A. Krohling. Otimizaçao via enxame de partículas para treinamento de redes neurais artificiais do tipo elm: Um estudo de caso para prediç ao de séries temporais. In *Simpósio Brasileiro de Pesquisa Operacional—SBPO*, Federal University of Espírito Santo, 2016.

Artstein, R. and M. Poesio. Inter-coder agreement for computational linguistics. *Computational Linguistics*, 34(4):555596, 2008. ISSN 0891–2017.

Atkinson, J. and D. Campos. Improving bci-based emotion recognition by combining eeg feature selection and kernel classifiers. *Expert Systems with Applications*, 47:35–41, 2016.

Bermejo, P., J. A. Gámez, and J. M. Puerta. Improving the performance of naive bayes multinomial in e-mail foldering by introducing distribution-based balance of datasets. *Expert Systems with Applications*, 38(3):2072–2080, 2011.

Bomfim, A. J. d. L., R. A. d. S. Ribeiro, and M. H. N. Chagas. Recognition of dynamic and static facial expressions of emotion among older adults with major depression. *Trends in Psychiatry and Psychotherapy*, 41:159–166, 2019.

Chaturvedi, V., A. B. Kaur, V. Varshney, A. Garg, G. S. Chhabra, and M. Kumar. Music mood and human emotion recognition based on physiological signals: a systematic review. *Multimedia Systems*, 1–24, 2021.

Chawla, N. V., K. W. Bowyer, L. O. Hall, and W. P. Kegelmeyer. Smote: synthetic minority oversampling technique. *Journal of Artificial Intelligence Research*, 16:321–357, 2002.

Chen, J., D. Remulla, J. H. Nguyen, Y. Liu, P. Dasgupta, and A. J. Hung. Current status of artificial intelligence applications in urology and their potential to influence clinical practice. *BJU International*, 124(4):567–577, 2019.

Commowick, O., A. Istace, M. Kain, B. Laurent, F. Leray, M. Simon, S. C. Pop, P. Girard, R. Ameli, J.-C. Ferré, et al. Objective evaluation of multiple sclerosis lesion segmentation using a data management and processing infrastructure. *Scientific Reports*, 8(1):1–17, 2018.

Cortes, C. and V. Vapnik. Support-vector networks. *Machine learning*, 20(3):273–297, 1995.

Cruz, T., T. Cruz, and W. Santos. Detection and classification of lesions in mammographies using neural networks and morphological wavelets. *IEEE Latin America Transactions*, 16(3):926–932, 2018.

da Silva, C. A. S. and R. A. Krohling. Classificaç ao de grandes bases de dados utilizando algoritmo de máquina de aprendizado extremo. In *Simpósio Brasileiro de Pesquisa Operacional—SBPO*, Federal University of Espírito Santo, 2016.

de Freitas Barbosa, V. A., J. C. Gomes, M. A. de Santana, E. d. A. Jeniffer, R. G. de Souza, R. E. de Souza, and W. P. dos Santos. Heg.IA: an intelligent system to support diagnosis of Covid-19 based on blood tests. *Research on Biomedical Engineering*, 1–18, 2021.

de Oliveira, A. P. S., M. A. Santana, M. K. S. Andrade, J. C. Gomes, M. C. Rodrigues, and W. P. dos Santos. Early diagnosis of parkinsons disease using eeg, machine learning and partial directed coherence. *Research on Biomedical Engineering*, 36(3):311–331, 2020.

de Oliveira, E. and P. A. Jaques. Classificação de emoções básicas através de imagens capturadas por webcam. *Revista Brasileira de Computação Aplicada*, 5(2):40–54, 2013.

Doma, V. and M. Pirouz. A comparative analysis of machine learning methods for emotion recognition using eeg and peripheral physiological signals. *Journal of Big Data*, 7(1):1–21, 2020.

Dorneles, S. O. D. S. O., D. N. F. Barbosa, and J. L. V. Barbosa. Sensibilidade ao contexto na identificação de estados afetivos aplicados à educação: um mapeamento sistemático. *RENOTE*, 18(1), 2020.

dos Santos, W. P., F. M. de Assis, R. E. de Souza, and P. B. dos Santos Filho. Evaluation of alzheimer's disease by analysis of mr images using objective dialectical classifiers as an alternative to ADC maps. In *2008 30th Annual International Conference of the IEEE Engineering in Medicine and Biology Society*, pages 5506–5509. IEEE, 2008.

Eaton, J. W., D. Bateman, S. Hauberg, and R. Wehbring. GNU octave version 4.0. 0 manual: a high-level interactive language for numerical computations, 2015. www.gnu.org/software/octave/doc/interpreter, 8:13, 2015.

Ferreira, C. D. and N. Torro-Alves. Reconhecimento de emoções faciais no envelhecimento: uma revisão sistemática. *Universitas Psychologica*, 15(5), 2016.

Gomes, J. C., V. A. d. F. Barbosa, M. A. Santana, J. Bandeira, M. J. S. Valença, R. E. de Souza, A. M. Ismael, and W. P. dos Santos. Ikonos: an intelligent tool to support diagnosis of Covid-19 by texture analysis of x-ray images. *Research on Biomedical Engineering*, 1–14, 2020.

Gupta, V., M. D. Chopda, and R. B. Pachori. Cross-subject emotion recognition using flexible analytic wavelet transform from eeg signals. *IEEE Sensors Journal*, 19(6):2266–2274, 2018.

Hanley, J. A. and B. J. McNeil. The meaning and use of the area under a receiver operating characteristic (roc) curve. *Radiology*, 143(1):29–36, 1982.

Hasanzadeh, F., M. Annabestani, and S. Moghimi. Continuous emotion recognition during music listening using eeg signals: a fuzzy parallel cascades model. *Applied Soft Computing*, 101:107028, 2021.

Huang, G.-B., Q.-Y. Zhu, and C.-K. Siew. Extreme learning machine: a new learning scheme of feedforward neural networks. In *2004 IEEE International Joint Conference on Neural Networks (IEEE Cat. No.04CH37541)*, volume 2, pages 985–990. IEEE, 2004.

Islam, M. R., M. M. Islam, M. M. Rahman, C. Mondal, S. K. Singha, M. Ahmad, A. Awal, M. S. Islam, and M. A. Moni. Eeg channel correlation based model for emotion recognition. *Computers in Biology and Medicine*, 136:104757, 2021a.

Islam, M. R., M. A. Moni, M. M. Islam, M. Rashed-Al-Mahfuz, M. S. Islam, M. K. Hasan, M. S. Hossain, M. Ahmad, S. Uddin, A. Azad, et al. Emotion recognition from eeg signal focusing on deep learning and shallow learning techniques. *IEEE Access*, 9:94601–94624, 2021b.

Izard, C. E. *Human Emotions*. New York: Springer, 1977.

Izard, C. E. *The Psychology of Emotions*. New York: Springer, 1991.

Jackins, V., S. Vimal, M. Kaliappan, and M. Y. Lee. Ai-based smart prediction of clinical disease using random forest classifier and naive bayes. *The Journal of Supercomputing*, 77(5):5198–5219, 2021.

Kang, Q., Q. Gao, Y. Song, Z. Tian, Y. Yang, Z. Mao, and E. Dong. Emotion recognition from deaf eeg signals using stacking ensemble learning frame-work based on a novel brain network. *IEEE Sensors Journal*, 21(20):23245–23255, 2021. doi: 10.1109/JSEN.2021.3108471.

Liu, Y., T. Gedeon, S. Caldwell, S. Lin, and Z. Jin. Emotion recognition through observer's physiological signals. *arXiv, 02*, 2020.

Miguel, F. K., H. A. Ogaki, C. M. Inaba, and D. d. O. Ribeiro. Percepção emocional e inteligência: Contribuições para o modelo chc. *Revista Sul-Americana de Psicologia*, 1(1):36–47, 2013.

Mohammadi, Z., J. Frounchi, and M. Amiri. Wavelet-based emotion recognition system using eeg signal. *Neural Computing and Applications*, 28(8):1985–1990, 2017.

Ong, D. C. An ethical framework for guiding the development of affectively-aware artificial intelligence. *arXiv preprint arXiv:2107.13734*, 2021.

Pal, M. Random forest classifier for remote sensing classification. *International Journal of Remote Sensing*, 26(1):217–222, 2005.

Pavani, G., B. Biswal, and P. Biswal. Robust classification of neovascularization using random forest classifier via convoluted vascular network. *Biomedical Signal Processing and Control*, 66:102420, 2021. ISSN 1746–8094. www.sciencedirect.com/science/article/pii/S1746809421000173.

Pereira, J. M. S., M. A. Santana, R. C. F. Lima, S. M. L. Lima, and W. P. Santos. Method for classification of breast lesions in thermographic images using elm classifiers. In W. P. dos Santos, M. A. de Santana, and W. W. A. da Silva, editors, *Understanding a Cancer Diagnosis*, pages 117–132. New York: Nova Science, 1 edition, 2020a.

Pereira, J. M. S., M. A. Santana, R. C. F. Lima, and W. P. Santos. Lesion detection in breast thermography using machine learning algorithms without previous segmentation. In W. P. dos Santos, M. A. de Santana, and W. W. A. da Silva, editors, *Understanding a Cancer Diagnosis*, pages 81–94. New York: Nova Science, 1 edition, 2020b.

Pereira, J. M. S., M. A. Santana, W. W. A. Silva, R. C. F. Lima, S. M. L. Lima, and W. P. Santos. Dialectical optimization method as a feature selection tool for breast cancer diagnosis using thermographic images. In W. P. dos Santos, M. A. de Santana, and W. W. A. da Silva, editors, *Understanding a Cancer Diagnosis*, pages 95–118. New York: Nova Science, 1 edition, 2020c.

Pham, B. T., I. Prakash, K. Khosravi, K. Chapi, P. T. Trinh, T. Q. Ngo, S. V. Hosseini, and D. T. Bui. A comparison of support vector machines and bayesian algorithms for landslide susceptibility modelling. *Geocarto International*, 34(13):1385–1407, 2019.

Portela, N. M., C. A. B. Mello, W. P. dos Santos, A. L. B. Candeias, C. M. G. de Gusmão, S. C. S. Machado, and M. A. B. Rodrigues. A new algorithm for segmenting and counting aedes aegypti eggs in ovitraps. In *2009 Annual International Conference of the IEEE Engineering in Medicine and Biology Society*, pages 6714–6717. IEEE, 2009.

Rodrigues, A. L., M. A. de Santana, W. W. Azevedo, R. S. Bezerra, V. A. Barbosa, R. C. de Lima, and W. P. dos Santos. Identification of mammary lesions in thermographic images: feature selection study using genetic algorithms and particle swarm optimization. *Research on Biomedical Engineering*, 35(3):213–222, 2019.

Santana, M. A., J. C. Gomes, G. M. de Souza, A. Suarez, A. S. Torcate, F. S. Fonseca, G. M. M. Moreno, and W. P. dos Santos. Reconhecimento automático de emoçoes a partir de sinais multimodais e inteligência artificial. *Anais do IV Simpósio de Inovação em Engenharia Biomédica-SABIO 2020*, page 43. Editora Universitária da UFPE, 2020a.

Santana, M. A., J. M. S. Pereira, R. C. F. Lima, and W. P. Santos. Breast lesions classification in frontal thermographic images using intelligent systems and moments of haralick and zernike. In W. P. dos Santos, M. A. de Santana, and W. W. A. da Silva, editors, *Understanding a Cancer Diagnosis*, pages 65–80. New York: Nova Science, 1 edition, 2020b.

Santana, M. A., J. M. S. Pereira, F. L. d. Silva, N. M. d. Lima, F. N. d. Sousa, G. M. S. d. Arruda, R. d. C. F. d. Lima, W. W. A. d. Silva, and W. P. d. Santos. Breast cancer diagnosis based on mammary thermography and extreme learning machines. *Research on Biomedical Engineering*, 34:45–53, 2018. ISSN 2446–4740.

Sepúlveda, A., F. Castillo, C. Palma, and M. Rodriguez-Fernandez. Emotion recognition from ecg signals using wavelet scatte. machine learning. *Applied Sciences*, 11, 2021. www.mdpi.com/2076-3417/11/11/4945.

Siriyasatien, P., S. Chadsuthi, K. Jampachaisri, and K. Kesorn. Dengue epidemics prediction: a survey of the state-of-the-art based on data science processes. *IEEE Access*, 6:53757–53795, 2018.

Skariah, A., R. Pradeep, R. Rejith, and C. Bijudas. Health monitoring of rolling element bearings using improved wavelet cross spectrum technique and support vector machines. *Tribology International*, 154:106650, 2021.

Soleymani, M., J. Lichtenauer, T. Pun, and M. Pantic. A multimodal database for affect recognition and implicit tagging. *IEEE Transactions on Affective Computing*, 3(1):42–55, 2011.

Song, T., G. Lu, and J. Yan. Emotion recognition based on physiological signals using convolution neural networks. In *Proceedings of the 2020 12th International Conference on Machine Learning and Computing*, page 161165, New York. Association for Computing Machinery, 2020. https://doi.org/10.1145/3383972.3384003.

Sorinasa, J., J. C. Fernandez-Troyano, M. Val-Calvo, J. M. Ferrández, and E. Fernandez. A new model for the implementation of positive and negative emotion recognition, 2020. https://arxiv.org/abs/1905.00230

Torcate, A. S., M. A. Santana, G. M. M. Moreno, A. Suarez, J. C. Gomes, W. P. dos Santos, F. S. Fonseca, and G. M. de Souza. Intervenções e impactos da musicoterapia no contexto da doença de alzheimer: Uma revisão de literatura sob a perspectiva da computação afetiva. *Anais do IV Simpósio de Inovação em Engenharia Biomédica-SABIO 2020*, page 31. Editora Universitária da UFPE, 2020.

Wang, X., S. Wang, Z. Huang, and Y. Du. Condensing the solution of support vector machines via radius-margin bound. *Applied Soft Computing*, 101:107071, 2021.

Witten, I. H. and E. Frank. *Data Mining: Pratical Machine Learning Tools and Technique.* San Francisco, CA: Morgan Kaufmann Publishers, 2005.

Woyciekoski, C. and C. S. Hutz. Inteligência emocional: teoria, pesquisa, medida, aplicações e controvérsias. *Psicologia: Reflexão e Crítica*, 22:1–11, 2009.

Xu, L., P. Yan, and T. Chang. Best first strategy for feature selection. In *9th International Conference on Pattern Recognition*, pages 706–707. IEEE Computer Society, 1988.

Zawacki-Richter, O., V. I. Marín, M. Bond, and F. Gouverneur. Systematic review of research on artificial intelligence applications in higher education—where are the educators? *International Journal of Educational Technology in Higher Education*, 16(1):1–27, 2019.

Zeng, J., P. C. Roussis, A. S. Mohammed, C. Maraveas, S. A. Fatemi, D. J. Armaghani, and P. G. Asteris. Prediction of peak particle velocity caused by blasting through the combinations of boosted-chaid and svm models with various kernels. *Applied Sciences*, 11(8), 2021. ISSN 2076–3417. www.mdpi.com/2076-3417/11/8/3705.

Zhao, M., F. Adib, and D. Katabi. Emotion recognition using wireless signals. *Communications of the ACM*, 61(9):91100, 2018. ISSN 0001–0782. https://doi.org/10.1145/3236621.

Zhao, S., F. Blaabjerg, and H. Wang. An overview of artificial intelligence applications for power electronics. *IEEE Transactions on Power Electronics*, 36(4):4633–4658, 2020. doi: 10.1109/TPEL.2020.3024914.

Zheng, W.-L., J.-Y. Zhu, and B.-L. Lu. Identifying stable patterns over time for emotion recognition from EEG. *IEEE Transactions on Affective Computing*, 10(3):417–429, 2017.

9 Identification of Emotion Parameters in Music to Modulate Human Affective States
Towards Emotional Biofeedback as a Therapy Support

Maíra A. Santana, Ingrid B. Nunes,
Flávio S. Fonseca, Arianne S. Torcate,
Amanda Suarez, Vanessa Marques,
Nathália Córdula, Juliana C. Gomes,
Giselle M. M. Moreno and
Wellington Pinheiro dos Santos

9.1 INTRODUCTION

In recent years, the study of affectivity has been growing exponentially, especially in the computational field. This is mostly due to the fact that emotions play the role of directing human behaviors and interactions (Suhaimi, Mountstephens, and Teo, 2020). Therefore, if emotional states play a central role in our social life, guiding motivations, communication, memory and learning, it is natural and important that there is also progress in the study of affect through artificial intelligence (Suhaimi et al., 2020). In this sense, emotions are understood as subjective, physiological and cognitive reactions to environmental, internal or external events, which define the behavior of an individual when faced with a stimulus (Fonseca, 2016). In general, the literature considers six emotions as basic and universal: joy, sadness, anger, disgust, fear and surprise (Meska, Mano, Silva, Pereira, and Mazzo, 2020). Furthermore, the neutral state is regularly used in field as an intermediate condition that precedes the other emotional states (Meska et al., 2020).

It is important to highlight that there are several forms of emotional expression, such as changes in the individual's vocal features (rhythm and frequency) (Miguel, 2015). Studies even prove a strong relationship between these features and the

DOI: 10.1201/9781003201137-11

music purpose of communicating these emotional states (Juslin and Laukka, 2003). However, in addition to expressing affectivity, the human ability to recognize these states through bodily, sound, physiological and other signals is an essential aspect of a healthy social dynamic.

Physiologically, humans are able to demonstrate and recognize emotional behaviors due to some neural structures and connections. These structures make up the brain's limbic system, whose main structures are: the cingulate gyrus, parahippocampal gyrus, hypothalamus, thalamus, hippocampus, amygdala, septum, prefrontal area and cerebellum (Rolls, 2015). The activation of these structures is also part of the brain response to musical stimuli, since music is a way of expressing emotions (Hsieh, Hornberger, Piguet, and Hodges, 2012).

Furthermore, damage to these brain structures is linked to some dysfunctions and diseases, such as Alzheimer's disease, Parkinson's disease, Autistic Spectrum Disorder and depression. In the case of Alzheimer's, Hsieh et al. (2012) shows that with the progression of the disease, the individual usually presents a certain deficiency in the recognition of emotions from facial expressions (Hsieh et al., 2012). A similar behavior was also found in the study by Li, Wang, Tian, Zhou, Li, Wang, and Yu (2016), which shows that there are dysfunctions in the connection networks on the limbic system of patients with Alzheimer's, hindering their ability to perceive and express emotions (Li et al., 2016). On the other hand, these pathologies rarely affect brain regions associated with auditory processing. Therefore, auditory stimuli, particularly music, can be used to support perception, control, induction and expression of emotions in audiences like these.

As one of the oldest ways of human expression and communication, music has the potential to universally and cross-culturally affect human beings (Cowen, Fang, Sauter, and Keltner, 2020; Juslin, 2013; Song, Dixon, Pearce, and Halpern, 2016). The effects of music on the human body are diverse. Such effects start in structures of the Central Nervous System and require complex cognitive functions. Music is capable of activating areas all over the brain, especially those related to the reward system, and to auditory, motor and visual processing (Arjmand, Hohagen, Paton, and Rickard, 2017; Koelsch, 2018). In addition, musical interaction triggers physiological, hormonal and mood-modulating responses (Arjmand et al., 2017; Schaefer, 2017). Furthermore, during musical composition, melodic parameters are able to incorporate emotional aspects in the music, thus conveying emotions (Juslin, 2013; Song et al., 2016).

The modulation of affective state by music is a tool that has been explored in the context of music therapy for various cognitive and behavioral disorders. Music therapy is a therapeutic approach in which musical stimuli are used to achieve non-musical goals (de l'Etoile, 2008). In recent years it has been better studied and explored for various applications. Several studies have sought to understand the effects of musical stimuli in the context of therapy for people with Autistic Spectrum Disorder and learning difficulties (Safonicheva and Ovchinnikova, 2021). In these groups, aspects such as attention, socialization and trust are strongly developed. Other studies have identified a number of benefits of these musical stimuli in elderly people with Parkinson's disease, Alzheimer's disease and other neurodegenerative disorders (de l'Etoile, 2008). In these groups, music plays an important role in slowing down the progression of the disease. Studies also showed that adequate musical stimuli may have positive effects in reducing depression and anxiety (de l'Etoile, 2008). One of the main musical parameters that allow this modulation of human affective state is the emotions conveyed by the music itself. For example, if music expresses happiness, resonance effects induced by it on brain waves is likely to improve the mood of this individual (Schaefer, 2017). Similar processes may occur for other affective states.

Nevertheless, perceiving, controlling, inducing and expressing emotions consist in some of the most complex and fundamental aspects of every human being development. These complex abilities also guide our social skills and our interaction with the world elements around us. One of the reasons for this complexity is that an emotion manifests itself differently in each individual. Therefore, studies developed in the last century came up with two main ways to represent affective states: the categorical and the dimensional approaches (Yu, Wang, Lai, and Zhang, 2015). The dimensional approach has particularly drawn attention since it can provide a more fine-grained emotional assessment. In this approach, emotions are represented as continuous numerical values such as valence-arousal two-dimensional space (Russell, 1980). Valence values are related to the degree of positive and negative emotions, while arousal represents the activation degree (excitement and calmness) (Russell, 1980). Based on this emotion representation proposed by Russell (1980), many studies describe emotions as function of these parameters (Cheuk, Luo, Balamurali, Roig, and Herremans, 2020; Grekow, 2018; Vatolkin and Nagathil, 2019; Yu et al., 2015). However, they usually acquire valence and arousal rating from human assessment, which is highly subjective, nonspecific and subject-dependent. Thus, it is important to work on ways to automatically define valence and arousal ratings. Artificial Intelligence (AI) techniques may be of great help in this context. From this tool, we can train an algorithm to learn how to associate some stimuli to valence-arousal values, thus acquiring emotional information from these stimuli.

Therefore, this chapter proposes an approach to the automatic identification of emotions expressed by music of different genres based on the prediction of valence and arousal parameters. We explore AI regression tools in an attempt to find an optimal configuration to associate musical stimuli to valence and arousal values. This method may be of great help to support human emotional perception of music or other kinds of content. It may also be useful to improve the affective component of human-computer interaction in several applications, such as recommendation systems. Moreover, it may address some music therapy issues related to creating and recommending personalized content to better achieve patient needs and, consequently, improve the therapeutic intervention.

The chapter is organized as follows: Section 2 will present some related studies that also seek to assess musical content by its emotional effect in human beings. Then, in Section 3, we present the experimental materials and methods proposed in this chapter. This section is followed by the results and discussions, and at the last section we provide some conclusions, highlighting limitations and future perspectives.

9.2 RELATED STUDIES

The research carried out by Sharma, Gupta, Sharma, and Purwar (2020) aimed to classify emotions in songs through the prediction of valence and arousal, based on the Russell model. The study was divided into two parts. In the first part, they performed a comparative study between classical algorithms such as SVM, Logistic Regression, KNN, Random Forest, Naive Bayes and decision trees on audio resources. In the second part, a hybrid model based on MLP was developed to increase the accuracy of the forecasts. The database used in their experiments was the PMEmo 2019. Their results demonstrate that, both for prediction of arousal and valence, the SVM algorithm with RBF kernel stood out in terms of accuracy, obtaining, respectively, 68.75% and 61.97%. The proposed model managed to increase the overall prediction accuracy of some algorithms, such as the Linear SVM, which was equivalent to 50.33%, and with the application of the model, the accuracy managed to reach 63%.

In order to classify emotions in music, Chen and Li (2020) proposes a stack-based multifunctional combined network, applying CNN-LSTM (Convolutional neural networks-long term long memory) with a combination of 2D features input and 1D features input through DNN (deep neural networks). The authors used the Last.fm database, composed of 1 million songs. Following Thayer's emotion model, four emotional tags were extracted. Emotional classes were angry, happy, relaxed and sad, and 500 songs were collected from each emotion category, resulting in 2,000 songs. Their experiments show that the classification of music audios into emotions using the proposed model reached a 68% precision rate. On the other hand, song lyrics rating achieved a precision of 74%. Overall, the average precision obtained by their model was 78%.

The study by Sobeeh, Öztürk, and Hamed (2019) investigates the effect of listening to two pieces of high-excitement music. The snippets have the same square root of the mean amplitude and different valences (positive happy valence and negative anxious valence), in reaction time and Simon effect. The high arousal anxious (AHA) group showed faster reaction time and greater Simon effect than the other groups. Thus, this larger Simon effect was interpreted by poor interference control through an incongruous conflict situation. The high arousal group showed faster reaction time and greater Simon effect compared to the silent control group. Although the study results are consistent with previous studies of poor interference control with positive mood, it provides a new finding that negative anxious mood tracks faster reaction time, while interference control with positive mood was limited and with greatest Simon effect.

Hizlisoy, Yildirim, and Tufekci (2021) proposed an approach to the recognition of musical emotions based on the deep convolutional neural network architecture of long short-term memory. The authors used features extracted by feeding convolutional neural network layers with log-mel filterbank energies and mel-frequency cepstrum coefficients, in addition to standard acoustic resources. The results show that adding new features to the default audio attributes improves classification performance. Improved performance was achieved after selecting features by applying the correlation-based feature selection method.

In the study of Vatolkin and Nagathil (2019) they propose a computational method to associate emotions with music. They tested different linear regression methods to predict arousal and valence values. The features extracted from the audio signals were combinations of energy, harmony, rhythm and timbre parameters. The greatest contribution of this work was the inclusion of a feature selection step. After extracting the features, they applied a Minimum Redundancy-Maximum Relevance (MRMR) algorithm for feature selection. From this selection they could assess different sets of features to represent audio data. Therefore, the authors found that timbre features are the most relevant in arousal prediction, while valence prediction is mostly associated to rhythm features. In this study RMSE values were close to 1 in most of the tested feature combinations. Other sets of audio features and regression algorithms may be included to improve results.

Cheuk et al. (2020) studied a novel approach of Triplet Neural Netwoks (TNNs) in the context of emotion prediction from music. They assess these networks performance in representing musical signals in a regression task by using this representation as input for regression models based on Support Vector Machines and Gradient Boosting Machines. The authors compared the representation provided by TNNs to others feature selection methods such as Principal Component Analysis (PCA) and Autoencoders (AE). Their approach overcame the other methods with higher prediction rates for both valence and arousal, possibly providing new music categorization and recommendation approaches.

The work from Grekow (2018) also investigates different feature combinations in an attempt to find an optimal set of features to represent audio signals in order to improve music emotion detection. In their study they applied regression algorithms to estimate valence and arousal values. The authors used 31 features to represent the signals, such as Zero Crossings, Spectral Centroid, Spectral Flux, Spectral Rolloff, Mel-Frequency Cepstral Coefficients (MFCC), chroma, mean, geometric mean, power mean, median, moments up to the 5th order, energy, root mean square (RMS), spectrum, flatness, crest, variance, skewness, kurtosis of probability distribution and a single Gaussian estimate. They tested different combinations of these features, which were used as input of SMOreg, REPTree, M5P-SMOreg regression algorithms. The authors found that feature selection process improved regression performance in almost 50%. For arousal prediction, their approach achieved Pearson's correlation coefficient of 0.79, and mean absolute error of 0.09. Regarding valence prediction, this best model showed moderate performance, with correlation coefficient 0.58, and 0.10 of error.

The article by Yang and Hirschberg (2018) aimed to perform a study on the continuous tracking of emotions, following Russell's circumplex model of affect, in order to predict arousal and valence. The combination of raw wave form signals inputs and log-mel filterbank features was performed to assess their joint effects. The neural network architecture used contains a set of convolutional neural network layers (CNN) and bidirectional long-term memory layers (BLSTM) to take into account the temporal and spectral variation and the textual content model. In order to evaluate the performance of the model, the SEMAINE database and the RECOLA database were chosen. For all of these experiments, the correlation of agreement coefficient of tension (CCC) was used as the objective function to train the models. The results demonstrate that all models perform significantly better than the baseline model, which indicates that the models can learn important features for the arousal and valence of the data of both databases. Furthermore, the 'W Only' model outperforms the 'S Only' model in predicting arousal, while the 'S Only' model outperforms the 'W Only' model in predicting valence. Finally, by combining the waveform and spectrogram inputs, the 'W + S' model provides further improvement in predicting both arousal (0.680 for SEMAINE database and 0.692 for RECOLA) and valence (0.506 for SEMAINE and 0.423 for RECOLA data), which shows that waveforms and spectrograms contain complementary emotion information. Comparing results, the CCC for predicting valence in SEMAINE is systematically higher than in RECOLA.

In the study of Tan, Villarino, and Maderazo (2019), the objective was to create a system capable of detecting the mood of a song based on the four quadrants of Russell's model. To achieve this, the researchers used two classification algorithms (SVM and Naïve Bayes) to train separate classifier models for valence and arousal using selected audio resources for SVM and lyrical resources for Naïve Bayes. This process returns four trained valence and arousal models for each algorithm. The dataset used for this study consisted of 180 songs that had lyrics annotated while 162 songs had audio annotations. The number of songs that contained both lyrics and audio annotations was 133. Arousal detection is highly accurate when used with audio features, while valence detection is highly accurate when using letters. The arousal is easily distinguishable when heard, as its range would be from top to bottom. This study focused more on the use of time to detect arousal using audio resources. Valence detection using lyrics with Naïve Bayes resulted in greater accuracy than using audio because it is difficult to distinguish and analyze melody and the positivity or negativity of a word as they cannot be correctly distinguished.

In Table 9.1 we present the main information from these related studies, such as their main goal, the computation techniques used and their main findings. At the last line of this table is our method, proving that our proposal is well contextualized in the state-of-the-art.

9.3 MATERIALS AND METHODS

Our proposed method of recognizing emotions in music consists of five main steps (Figure 9.1): (1) acquisition and organization of the music database; (2) preprocessing; (3) feature extraction; (4) development of a knowledge dataset; and (5) performing regression experiments using the Waikato Environment for Knowledge Analysis (WEKA), version 3.8 (Witten and Frank, 2005). Initially, we requested access to the database "Emotions in Music" (Soleymani, Caro, Schmidt, Sha, and Yang, 2013), which consists of 45-second excerpts of 744 songs available in the "1 to 1000" package from the Free Music Archive (FMA). All excerpts were pre-processed, then, valence and arousal ratings were assigned to each by the participants.

During the following phases, we performed a 5-second windowing, with a 1-second overlap in the available exertions, in addition to extracting 35 features useful to the regression problem. Subsequently, we created two knowledge bases, one with the values of valence and the other for arousal values. In the fifth step, we assess

TABLE 9.1
Summary of Related Works.

Work	Main goal	Method	Main results
Sharma et al. (2020)	Emotion classification in musics through the prediction of valence and arousal	Using the PMEmo 2019 database, the authors compared the performance of SVM, Logistic Regression, KNN, Random Forest, Naive Bayes and decision tree architectures.	They found accuracies up to 68.75% to predict arousal, and 61.97% for valence prediction.
Chen and Li (2020)	Classification of music into 4 emotions categories using 2,000 songs from the Last.fm database.	Their classifier was a hybrid approach combining LSTM to features extracted by deep neural networks.	Emotion classification was better when using the song lyrics (precision of 74%) than when using only the melody (68%).
Sobeeh et al. (2019)	Measure the effect of listening two pieces of high-arousal songs, one with positive valence (happy), and the other with negative valence (anxious).	They compared the results in terms of the reaction time and Simon effect.	The main finding of this study was that negative anxious mood tracks faster reaction time than positive happy music stimuli.
Hizlisoy et al. (2021)	Recognition of musical emotions based on arousal and valence parameters.	The authors built their own database composed of 124 Turkish traditional excerpts of 30s. They represented the songs using a combination of CNN and standard acoustic features. It was then used as input to a LSTM classifier	They achieved an overall accuracy of 99.19%. However, the amount of songs in the database is small. This is especially problematic when using deep learning methods and can easily lead to overfitting.
Vatolkin and Nagathil (2019)	Associate emotions with music based on the prediction of arousal and valence values.	The main proposal of this study was the inclusion of a feature selection step using a Minimum Redundancy-Maximum Relevance algorithm. Than, the authors tested different linear regression methods to assess arousal and valence prediction.	They achieved RMSE values close to 1 after adding the feature selection approach, also finding that that timbre features are the most relevant in arousal prediction, while valence prediction is mostly associated to rhythm features.
Cheuk et al. (2020)	Emotion prediction from music in terms of valence and arousal	The authors proposed a novel feature selection approach based on Triplet Neural Networks (TNN). They use 744 music signals from the DEAM database.	Their music representation using TNN overcame other widely used methods such as PCA and autoencoder. The proposed method achieved higher Pearson's correlation coefficient rates for both valence (0.367 ± 0.113) and arousal (0.662 ± 0.065) when compared to these other methods
Grekow (2018)	Improve emotion detection from music by assessing different features combination applied to regression algorithms to estimate valence and arousal values.	The authors used different combination of 31 features to represent the signals. As regression algorithms they performed experiments with SMOreg, REPTree, and M5P-SMOreg.	They found that feature selection process improved regression performance in almost 50%. Their approach achieved Pearsons correlation coefficient of 0.79 for arousal and of 0.58 for valence prediction.
Yang and Hirschberg (2018)	Perform a study on the continuous tracking of emotions in terms of arousal and valence.	The authors used convolutional neural network (CNN) and bidirectional long-term memory layers (BLSTM) for feature extraction and classification using two databases: SEMAINE and RECOLA.	Their approach achieved concordance coefficient up to 0.680 for arousal prediction and 0.506 in predicting valence using SEMAINE data. The results were slightly lesse satisfying for RECOLA data for both arousal (0.692) and valence (0.423)
Tan et al. (2019)	Develop a system capable of detecting emotions in songs based on the four quadrants of Russell's model.	The authors used two classification algorithms (SVM and Naïve Bayes) to train separate classifier models for valence and arousal. The dataset consisted of 180 songs with lyrics, 162 songs with only audio annotations, and 133 songs with both audio and lyrics.	They found that SVM was better to predict arousal while Random Forest showed better performance when predicting valence.
Our approach	Emotion prediction in terms of valence and arousal from music signals.	We applied SVM, ELM, MLP, Random Forest, and Linear regression algorithms to predict both arousal and valence parameters	Random Forest outperformed the other models with Pearson's correlation coefficient of 0.76 ± 0.02 for valence prediction, and 0.85 ± 0.01 for predicting arousal values.

regression performances from the following algorithms: Random Forest, SVM, MLP, Linear and ELM. All experiments were conducted using k-fold cross-validation with 10 folds and 30 repetitions in order to generate statistically relevant results. Later, at the end of the experiments, to assess the statistical behavior, we plotted the results in boxplot and tables with average and standard deviation values. All these steps are illustrated in Figure 9.1.

Our approach of predicting emotional parameters may be applied to situations such as the one represented in Figure 9.2, where the trained regression model is able to effectively predict arousal and valence values or recommend musical content from these emotional parameters. Therefore, one may use the system to support human emotional perception of music or other kind of content. It is particularly useful to guide and optimize personalized therapeutic approaches such as music therapy by supporting the perception, control, induction and expression of emotional states. This way of automatically rating content based on valence and arousal parameters also may be applied to improve human-machine affective interactions, thus refining custom content recommendation systems. The system may also act as the core of

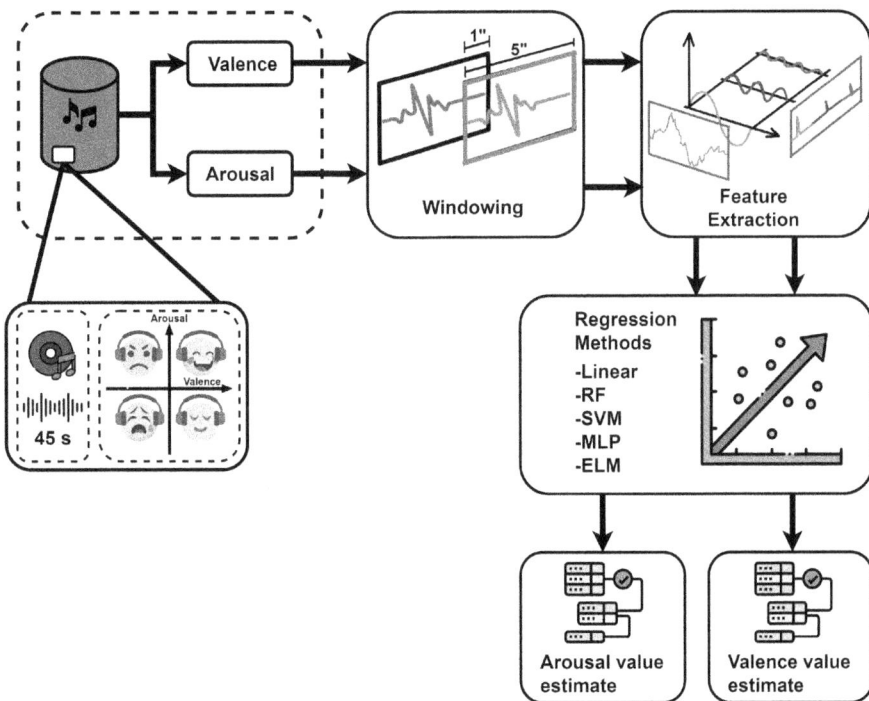

FIGURE 9.1 Diagram of the proposed method. First of all, we have a database containing 744 music excerpts of 45s with arousal and valence annotations. Then, all songs with their respective valence and arousal data were submitted to a windowing process with windows of 5s and overlapping of 1s. After segmentation, we extracted explicit features from these signals. Finally, both arousal and valence knowledge datasets were used as input to assess the performance of different regression models in estimating their values.

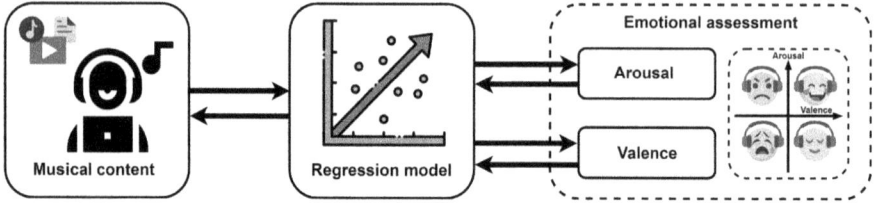

FIGURE 9.2 System usability. The optimal trained regression model is able to predict arousal and valence values from musical content, providing emotional information. The system may also recommend musical content from a given emotional state given in terms of valence and arousal.

assistive solutions so that people with social, learning and communication disabilities would be able to benefit from it.

9.3.1 MUSIC DATABASE

The database was created by Soleymani et al. (2013) from the music database of the Free Music Archive (FMA) website, which contains several songs of different genres. In all, 1,000 songs were selected. For the selection, the 300 songs classified as best are downloaded in MP3 format, and correspond to the genres: Blues, Electronic, Rock, Classical, Folk, Jazz, Country and Pop. Tracks with more than 10 minutes or less than 1 minute were excluded, resulting in 125 songs of each genre. Due to some redundancies, the dataset has shrunk to 744 songs. After selecting the songs, the authors extracted 45s excerpts from each of them. Snippets are pulled from a random (evenly distributed) starting point for each song. Then, the excerpts were all converted to the same sampling frequency (44100Hz) (Soleymani et al., 2013).

During the assembly of the database, the authors also collected annotations regarding parameters associated with emotions. For these notes, the participants received the definitions of arousal (emotional intensity) and valence (positive emotion versus negative emotion). They were also asked to provide personal information, including gender, age and nationality. These participants were subjected to two short audio clips of music that contained highly dynamic emotion shifts; they then indicated whether arousal and valence were increasing or decreasing. After annotating songs continuously, the participants were further asked to rate the level of arousal or valence for the entire clip on a 9-point scale.

9.3.2 MUSIC SEGMENTATION AND FEATURE EXTRACTION

All instances, in both valence and arousal datasets were submitted to a 5-second windowing process. In order to not lose potentially useful data when transitioning between windows, an overlap of 1 second was included between windows. After that, we extracted 35 time, frequency and statistical features from all instances. The list of features can be seen in Figure 9.3. It is important to mention that all procedures reported in this subsection were performed in the GNU/Octave mathematical computing environment, version 4.0.3 (Eaton, Bateman, Hauberg, and Wehbring,

Parameter	Equation	Parameter	Equation
Mean (μ)	$\mu = \dfrac{1}{N}\sum_{n=1}^{N} x_n$	Waveform length	$WL = \sum_{n=1}^{N-1}\lvert x_{n+1} - x_n\rvert$
Variance	$var = \dfrac{1}{N-1}\sum_{n=1}^{N}(x_n - \mu)^2$	Zero crossing	$ZC = \sum_{n=1}^{N-1}[sgn(x_n \times x_{n+1}) \cap \lvert x_n - x_{n+1}\rvert \geq threshold]$ $sgn(x) = \begin{cases}1, & if\ x \geq threshold\\ 0, & otherwise\end{cases}$
Standard deviation (σ)	$\sigma = \sqrt{\dfrac{1}{N-1}\sum_{n=1}^{N}\lvert x_n - \mu\rvert^2}$	Slope Sign Changes	$SSC = \sum_{n=1}^{N-1}[f(x_n - x_{n-1}) \times (x_n - x_{n+1})]]$ $f(x) = \begin{cases}1, & if\ x \geq threshold\\ 0, & otherwise\end{cases}$
Root mean square	$RMS = \sqrt{\dfrac{\sum_{n=1}^{N}(x_n)^2}{N}}$	Hjorth parameter activity	$Hjorth_{activity} = \dfrac{1}{N-1}\sum_{n=1}^{N}(x_n - \mu)^2$
Average Amplitude Changes	$AAC = \dfrac{1}{N}\left(\sum_{n=1}^{N}\left\lvert\dfrac{d\,x(t)}{dt}\right\rvert\right)$	Hjorth parameter mobility	$Hjorth_{mobility} = \sqrt{\dfrac{var\left(\frac{d\,x(t)}{dt}\right)}{var(x(t))}}$
Difference Absolute Deviation	$DASDV = \sqrt{\dfrac{1}{N}\sum_{n=1}^{N}\left(\dfrac{d\,x(t)}{dt}\right)^2}$	Hjorth parameter complexity	$Hjorth_{complexity} = \dfrac{Hjorth_{mobility}\left(\frac{d\,x(t)}{dt}\right)}{Hjorth_{mobility}(x(t))}$
Integrated Absolute Value	$IAV = \sum_{n=1}^{N} x_n$	Mean frequency	$MNF = \dfrac{\sum_{j=1}^{M} f_j P_j}{\sum_{j=1}^{M} P_j}$ Where f_j, P_j are the frequencies and power of the spectrum, respectively, and M is the length of the frequencies
Logarithm Detector	$LOGD = e^{\left(\frac{1}{N}\sum_{n=1}^{N}\log(\lvert x_n\rvert)\right)}$	Median frequency	$MDF = \dfrac{1}{2}\sum_{j=1}^{M} P_j$
Simple Square Integral	$SSI = \sum_{n=1}^{M} x_n^{2}$	Mean power	$MNP = \sum_{j=1}^{M}\dfrac{P_j}{M}$
Mean Absolute Value	$MAV = \dfrac{1}{N}\sum_{n=1}^{N}\lvert x_n\rvert$	Peak frequency	$PKF = \max(P_j)$
Mean Logarithm Kernel	$MLOGK = \dfrac{1}{N}\left\lvert\sum_{n=1}^{N} x_n\right\rvert$	Power Spectrum ratio	$PSR = \dfrac{PKF}{\sum_{j=1}^{M} P_j}$
Skewness (s)	$s = \dfrac{\frac{1}{N}\sum_{n=1}^{N}(x_n - \mu)^3}{\sigma^3}$	Total Power	$TP = \sum_{j=1}^{M} P_j$
Kurtosis	$kurt = \dfrac{\frac{1}{N}\sum_{n=1}^{N}(x_n - \mu)^4}{\sigma^4}$	First Spectral Moment	$SM1 = \sum_{j=1}^{M} f_j P_j$
Maximum Amplitude	$MAX = \max(x_n)$	Second Spectral Moment	$SM2 = \sum_{j=1}^{M} f_j^{2} P_j$
Third Moment	$M3 = \left\lvert\dfrac{1}{N}\sum_{n=1}^{N}(x_n)^3\right\rvert$	Third Spectral Moment	$SM3 = \sum_{j=1}^{M} f_j^{3} P_j$
Fourth Moment	$M4 = \left\lvert\dfrac{1}{N}\sum_{n=1}^{N}(x_n)^4\right\rvert$	Variance of Central Frequency	$VCF = \dfrac{SM2}{TP} - \left(\dfrac{SM1}{TP}\right)^2$
Fifth Moment	$M5 = \left\lvert\dfrac{1}{N}\sum_{n=1}^{N}(x_n)^5\right\rvert$	Shannon's entropy	$E = -\sum_l S_l^2 \log(S_l^2)$, where S is the signal

FIGURE 9.3 List of the 34 extracted features with their mathematical representations.

2015). This step resulted in a dataset with 8,184 instances and 35 attributes for both valence and excitability data.

9.3.3 REGRESSION MODELS

To carry out the experiments we used the following algorithms: Random Forest (RF), Support Vector Machine (SVM), Linear Regression, Multilayer Perceptron (MLP) and Extreme Learning Machine (ELM). As shown in Table 9.2, each of the regressors we used had different settings.

Random Forest is an algorithm that works by combining decision trees. Each of these trees is composed by vectors of clustered samples, randomly obtained. The

TABLE 9.2

Configuration of the Regression Models.

Regressor	Settings
Random Forest	Trees: 10, 20, 50, 100, 150, 200, 250, 300 and 350
	Batch size: 100
SVM	Kernel functions: linear, polynomial (d = 2, d = 3, and d = 4), and RBF (γ = 0.50)
ELM	Kernel functions: linear, polynomial (d = 2, d = 3, and d = 4), RBF and sigmoid
	Neurons in the hidden layer: 500.
Linear Regression	-
MLP	1 hidden layer: 10, 20, 50, and 100 neurons
	2 hidden layers with 20 neurons each (20, 20)
	2 hidden layers with 100 neurons each (100, 100)

vectors have a similar distribution for all trees in the forest and select subsets of random samples from the initial data (Biau and Scornet, 2016; Breiman, 2001). This algorithm uses the supervised learning paradigm and are widely applied to solve classification and regression problems. However, for classification problems, the chosen prediction is given by the class most voted by the trees. In regression problems, the prediction is given by the average of the values obtained by each tree (Biau and Scornet, 2016). Due to the operating mode of this algorithm, RF can be applied in databases with large data volumes. In addition, it also adapts well to databases with missing data and does not tend to overfit the model, even with the increase in the number of trees (Biau and Scornet, 2016; Breiman, 2001; da Silva, de Lima, da Silva, Silva, Marques, de Araújo, Júnior, de Souza, de Santana, Gomes et al., 2021; de Lima, da Silva, da Silva, Luiz Silva, Marques, de Araújo, Albuquerque Júnior, de Souza, de Santana, Gomes et al., 2020).

The Support Vector Machine is also a supervised machine learning algorithm. SVM algorithm was initially developed by Vapnik and has been applied to classification and regression problems. This algorithm can be used in databases with more than two classes, through the one against all technique (Ahmad, Basheri, Iqbal, and Rahim, 2018; Patle and Chouhan, 2013). SVM works by creating a hyperplane capable of separating the different classes of input data. The optimal hyperplane will be the one that separates the classes in the best way and that is farthest from the support vectors, that is, hyperplanes that have larger margins (Ahmad et al., 2018). These hyperplanes are defined by kernel functions. Kernel functions allow mapping of linearly non-separable data into linearly separable spaces (Patle and Chouhan, 2013). In addition, they help to find the optimal hyperplane by maximizing the hyperplane margins (Ahmad et al., 2018). For the experiments of this research, we tested the SVM with linear kernel, polynomial kernel and Radial Basis Function (RBF) kernel. Such kernel types are mathematically described in Table 9.3.

Linear Regression is another machine learning technique used to create a prediction model that already is well studied in the field of statistics. This model is given

TABLE 9.3

Kernel Functions Used in SVM and ELM Architectures.

Kernel type	Kernel function
Linear	$K(x, y) = x \bullet y$
Polynomial	$K(x, y) = (1 + x \bullet y)^d$
Radial Basis Function (RBF)	$K(x, y) = \exp(-\gamma (x - y) \bullet (x - y))$
Sigmoid	$K(x, y) = \tanh(b(x, y) + c)$

by the correlation obtained between dependent and independent variables, through a linear adjustment of the data (Kavitha, Varuna, and Ramya, 2016). The model obtained by the Linear Regression seeks to reduce the error associated with the least square of the residuals, given by the original data and the data obtained by the model prediction (Schuld, Sinayskiy, and Petruccione, 2016).

Multilayer Perceptron is one of the neural network algorithms that operate following the feedforward model. Therefore, it has input, hidden and output layers. All these layers are made up of several neurons. These neurons can be present in varying amounts in each of the layers. When MLP has more hidden layers in its structure, it is possible to apply it to solve non-linear problems (Desai and Shah, 2020; Driss, Soua, Kachouri, and Akil, 2017). In addition, each of these neurons have weights associated with each other, which are adjusted during the training phase usually using a backpropagation procedure. This training phase observed in MLP is of the supervised type. It allows the weights associated with neurons to be adjusted and updated by subtracting the output response (Desai and Shah, 2020; Driss et al., 2017). However, MLP has certain drawbacks in addressing problems. One of them is the fact that it is not possible to have an approximation of the function format when modeling a problem, the commonly called 'black box'. Furthermore, it is important to point out that a greater number of layers and neurons does not guarantee better performance and tends to require more processing time and computational power (Singh and Husain, 2014).

The Extreme Learning Machine, like the MLP, is a type of neural network that operates through the feedforward process. In addition, ELM is also built by input, hidden and output layers. Unlike MLP, the ELM model does not use the backpropagation technique during the training stage. Moreover, the input weights are randomly initialized which leads to a decrease in processing time when compared to MLP (Ahmad et al., 2018). In ELM, the output weights determination is analytically performed. This process does not occur iteratively, which results in faster processing and good applicability to more complex problems (de Moraes Ramos, Santana, Sousa, Assunção Ferreira, dos Santos, and Lira Soares, 2021). Furthermore, ELM presents a good performance generalization, due to the tendency to have a lower norm of its weights (Deng, Zheng, and Chen, 2009; Huang, Zhu, and Siew, 2006).

All these models were trained using the k-fold cross-validation method with 10 folds to avoid overfitting (Jung and Hu, 2015; Siriyasatien, Chadsuthi, Jampachaisri, and Kesorn, 2018). In addition, we trained the architectures 30 times to assess their statistical behavior and thus measure the reliability of the results. At this stage we used the Weka software, version 3.8, for experiments with Linear Regression, SVM, Random Forest and MLP (Witten and Frank, 2005). For the ELM experiments, we implemented the models in the GNU/Octave programming environment (version 4.0.3) (Eaton et al., 2015).

9.3.4 Metrics

To evaluate the performance obtained by the algorithms, we used seven metrics, which are: Pearson's correlation coefficient (PCC), Kendall's coefficient of concordance (KCC), Spearman's correlation (SCC), mean absolute error (MAE), root mean squared error (RMSE), relative absolute error (RAE) and root relative squared error (RRSE). Table 9.4 presents each metric along with their respective mathematical expressions.

According to Zhou, Deng, Xia, and Fu (2016), the Pearson's correlation coefficient is historically the first formal measure of correlation, and it is one of the most used relationship measures in regression problems. PCC is also known as the product-momentum correlation coefficient. The main objective is to measure the strength of the linear association/relation between two variables (Adler and Parmryd, 2010; Sedgwick, 2012). Briefly, Zhou et al. (2016) explains that PCC highlights the strength of the linear relationship between the two random variables x and y. Where, the value of the correlation coefficient is positive if the variables are directly related ($r_{xy} = 1$), and negative if the variables are considered uncorrelated ($r_{xy} = 0$).

Kendall's coefficient of agreement can be understood as a measure of agreement among m judges who rank a set of n entities (Field, 2005). This metric quantitatively assesses the coherence of the collective classification (S) provided by the (Franceschini and Maisano, 2021) models. The main intention is that Kendall's W presents results ranging from 0 (when there is no agreement) to 1 (when there is complete agreement).

Spearman's correlation coefficient is a distribution-free rank statistic, being a measure of the strength of an association between two variables (Hauke and Kossowski,

---

TABLE 9.4
Metrics and Their Mathematical Expressions.

Metric	Mathematical expression
Pearson's Correlation Coefficient	$r_{xy} = \frac{\sqrt{\sum(x_i - \bar{x})}\sqrt{\sum(y_i - \bar{y})}}{\sum(x_i - \bar{x})^2 \ \sum(y_i - \bar{y})^2}$
Kendalls Coefficient of Concordance	$W = \frac{12S}{m^2(n^3 - n)}$
Spearman's Correlation	$\rho = 1 - \frac{6\sum d_i^2}{n(n^2 - 1)}$
Mean absolute error	$MAE = \frac{1}{n}\sum_{i=1}^{n} |x_i - x|$
Root mean squared error	$RMSE = \sqrt{\frac{1}{n}\sum_{i=1}^{n}(S_i - O_i)^2}$
Relative absolute error	$RAE = \frac{\sum_{i=1}^{n}|p_i - a_i|}{\sum_{i=1}^{n}|\bar{a} - a_i|}$
Root relative squared error	$RRSE = \sqrt{\frac{\sum_{i=1}^{n}(p_i - a_i)^2}{\sum_{i=1}^{n}(a_i\bar{a})^2}}$

2011). Xiao, Ye, Esteves, and Rong (2016) explain that, basically, Spearman's correlation assesses the strength of the relationship between two variables. In other words, it evaluates monotonous relationships, which can be linear or not. Spearman's correlation (ρ) between two variables (d and n) will have a similar (or identical) rank when the correlation is equal to 1, and will have a dissimilar rank when the observations are completely opposite, with correlation value equal to -1.

The mean absolute error is constantly used to evaluate regression models (Chai and Draxler, 2014; Qi, Du, Siniscalchi, Ma, and Lee, 2020) and is derived from a measure of the mean error. In other words, MAE is the mean of all absolute errors. Sammut and Webb (2011) explains that the MAE for a given testing set is the mean of the absolute values of the individual prediction errors over all n instances of that set. That is, each forecast error is the difference between the expected value (x_i) and the predicted value (x) for the instance.

The root mean square error is a widely used statistical metric to measure the performance of models (Chai and Draxler, 2014). The RMSE is the square root of the mean square of all errors. This metric stands out due to its general purpose for numerical predictions (Neill and Hashemi, 2018). In other words, RMSE is a measure that calculates the root mean square of the n errors between real values (S_i) and predictions (O_i). Chai and Draxler (2014) point out that this metric is useful and provides an overview of the error distribution.

Relative absolute error is the metric that returns the absolute error between actual and expected values between variables (Dineva and Atanasova, 2019; Mishra, Paygude, Chaudhary, and Idate, 2018). RAE is often used to assess the performance of predictive models, particularly in areas such as machine learning, data mining and Artificial Intelligence (Subasi, El-Amin, Darwich, and Dossary, 2020). The main goal of this metric is to return an error measure regarding the variability of the n results detected ($p_i - a_i$) in relation to the real values ($\bar{a} - a_i$) (Jeyasingh and Veluchamy, 2017).

Subasi et al. (2020) explains that the root relative squared error refers to the result of a simple predictor if it had been used. A simple predictor can be thought as the real values average. Thus, the RRSE takes the total squared error ($(p_i - a_i)^2$) and normalizes it, dividing it by the total squared error of the simple predictor ($(a_i\bar{a})^2$). Later, by taking the square root of the relative squared error, the error is reduced to the same dimensions as the quantity being predicted.

9.4   RESULTS AND DISCUSSION

In this section we first present the results associated to valence prediction in Table 9.5 and Figure 9.4. Then, Table 9.6 and Figure 9.5 show the regression performance in predicting arousal values from the music.

Overall, the models achieved reasonable results for the metrics that perform global assessments, such as Pearson's correlation coefficient, Spearman's coefficient, mean absolute error and RMSE. However, Kendall's coefficient, RAE and RRSE usually performed worse, since they are more sensitive to local variations.

Moreover, we observed that there is an agreement between the valence and arousal results. There is even consistency between the best settings on most models.

TABLE 9.5

Performance of Different Models in the Prediction of Valence Values.

Metric	Regression Model				
	Linear Regression	SVM	ELM	Random Forest	MLP
	-	poly 4	linear	350 trees	(20, 20) neurons
PCC	0.4644 ± 0.0240	0.4603 ± 0.1202	0.2592 ± 0.0091	0.7569 ± 0.0153	0.5293 ± 0.0271
KCC	0.3223 ± 0.0180	0.3761 ± 0.0173	0.1768 ± 0.0053	0.5648 ± 0.0148	0.3709 ± 0.0192
SCC	0.4725 ± 0.0250	0.5398 ± 0.0235	0.2582 ± 0.0070	0.7517 ± 0.0166	0.5352 ± 0.0258
MAE	0.1721 ± 0.0042	0.1738 ± 0.0902	0.3022 ± 0.0033	0.1228 ± 0.0033	0.1836 ± 0.0266
RMSE	0.2138 ± 0.0047	0.2379 ± 0.0832	0.1343 ± 0.0008	0.1614 ± 0.0043	0.2278 ± 0.0291
RAE (%)	86.00 ± 1.38	76.49 ± 5.33	55.58 ± 0.26	61.35 ± 1.37	91.73 ± 13.17
RRSE (%)	88.57 ± 1.27	81.01 ± 3.71	25.04 ± 0.45	66.88 ± 1.46	94.39 ± 11.96

In addition, for both predictions of valence and arousal, the model that best fits the data was Random Forest with 350 trees. Furthermore, we noticed that arousal prediction was slightly better than valence prediction, with higher correlation coefficients and lower errors.

This finding agrees to the ones in Yang and Hirschberg (2018), Grekow (2018) and Cheuk et al. (2020). However, our approach performed better than these other regression models. Random Forest achieved Pearson's correlation up to 0.76 for valence prediction, with Kendall's of 0.56, and 0.75 for Spearman's correlation. In the arousal context, results were even better for the coefficients of Pearson (0.85), Kendall (0.66) and Spearman (0.84). These promising performances demonstrate that the adopted method is suitable for the representation of this kind of data and for the prediction of the parameters of interest. Furthermore, 350-trees Random Forest's good performance confirms the complexity of the emotion recognition problem at hand.

9.4.1 VALENCE PREDICTION

Table 9.5 presents the general performance of the best configurations from the five different families of regression algorithms. These performances are associated to valence prediction. Detailed results are shown in Table 9.A1. For this problem of predicting valence values, SVM with 4th-degree polynomial kernel overcame the other SVM architectures. Performance was also better for ELM of linear kernel, Random Forest with 350 trees and MLP with two layers of 20 neurons.

In this context, Random Forest model fits valence data better than the others, with the highest correlations. This model reached an average Pearson's correlation of 0.7569, Kendall's correlation of 0.5648 and 0.7517 for Spearman's coefficient. Moreover, data dispersion was minimal for this model, such as may be seen in Figure 9.4. Random Forest also achieved the lowest MAE and RMSE, with average values of 0.1228 and 0.1614, respectively. However, the percentage errors associated with this model had high values: 61.35% for relative absolute error and 66.88% for root relative squared error.

On the opposite side, ELM showed the worse performance regarding correlation coefficients, with low correlation and more data dispersion. Its MAE and RMSE results were reasonable; relative absolute error was statistically similar to the result of Random Forest; and root relative squared error was the best overall, with average value of 25.04%.

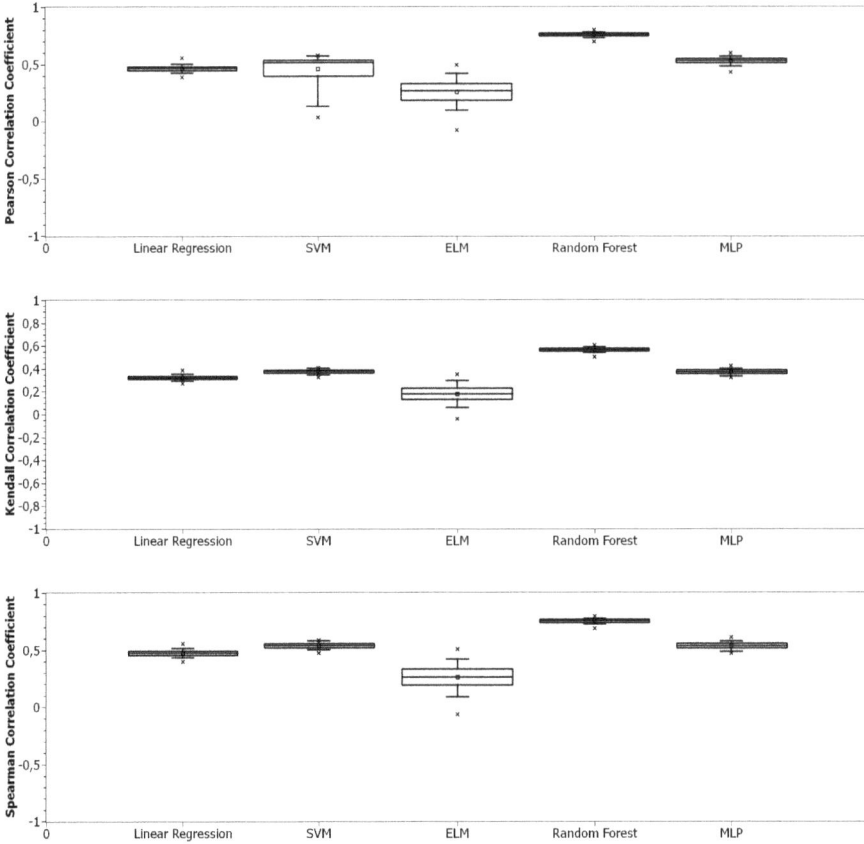

FIGURE 9.4 Correlation coefficients from different models to predict valence in songs. Pearson's coefficient is in (a), (b) shows Kendall's coefficient, and Spearman's coefficient is in (c).

9.4.2 Arousal Prediction

In predicting arousal values, the SVM configuration with the best performance was with the RBF kernel, ELM with the linear kernel, Random Forest with 350 trees and MLP with two layers of 20 neurons each. These results are shown in Table 9.6 (see Table 9.A2 for other architectures). In this context Random Forest also outperformed the other regression models, with the highest correlations coefficients and lowest errors, except for the root relative squared error, which was better for ELM. Linear Regression, SVM and MLP achieved similar results with low data dispersion, as showed in Figure 9.5. Data dispersion for ELM was also low, but greater than the other models.

Using Random Forest architecture with 350 trees our approach achieved an average Pearson's correlation of 0.8470, an average of 0.6600 for Kendall's correlation and 0.8428 for average Spearman's correlation. MAE reached an average value of 0.1180, and RMSE was around 0.1571. The relative absolute error associated with this model was 47.82%, and root relative squared error of 53.63%.

TABLE 9.6

Performance of Different Models in the Prediction of Arousal Values.

Metric	Linear Regression	SVM	ELM	Random Forest	MLP
	-	RBF	linear	350 trees	(20, 20) neurons
PCC	0.7017 ± 0.0159	0.6379 ± 0.1807	0.2988 ± 0.0112	0.8470 ± 0.0107	0.7443 ± 0.0165
KCC	0.5144 ± 0.0143	0.5503 ± 0.0152	0.1928 ± 0.0057	0.6600 ± 0.0124	0.5467 ± 0.0152
SCC	0.7118 ± 0.0166	0.7461 ± 0.0171	0.2838 ± 0.0076	0.8428 ± 0.0116	0.7440 ± 0.0167
MAE	0.1674 ± 0.0042	0.1585 ± 0.0109	0.2626 ± 0.0028	0.1180 ± 0.0031	0.1727 ± 0.0220
RMSE	0.2088 ± 0.0049	0.2810 ± 0.1676	0.1561 ± 0.0012	0.1571 ± 0.0044	0.2152 ± 0.0249
RAE (%)	67.84 ± 1.70	95.74 ± 56.64	51.16 ± 0.20	47.82 ± 1.42	69.98 ± 8.90
RRSE (%)	71.27 ± 1.59	75.23 ± 6.07	25.01 ± 0.34	53.63 ± 1.55	73.48 ± 8.47

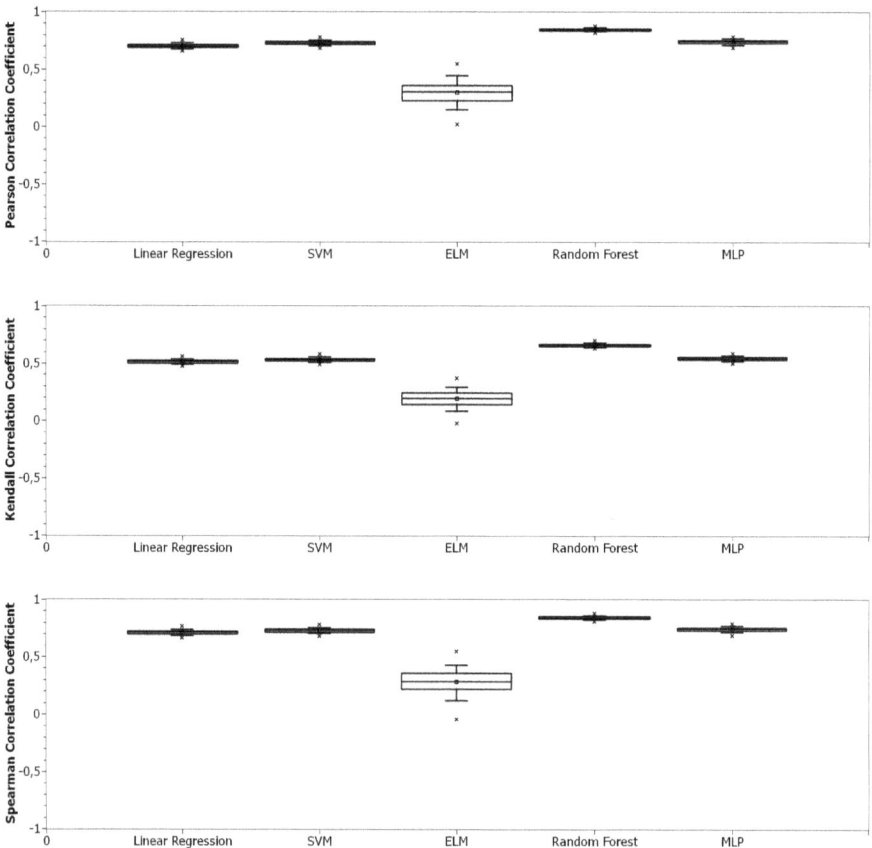

FIGURE 9.5 Correlation coefficients from different models to predict arousal in songs. Pearson's coefficient is in (a), (b) shows Kendall's coefficient, and Spearman's coefficient is in (c).

9.5 CONCLUSION

Emotions are a complex feature of every human being. They mainly affect social skills due to their direct link with the spontaneity and reliability of interpersonal relationships. One of the most popular ways of representing emotions is as a function of valence-arousal values. However, defining degrees of valence and arousal is usually an extremely subjective task since it is highly human-dependent. Thus, in this chapter we proposed an AI-based approach to the automatic identification of emotions in music of different genres. This identification uses regression algorithms to predict valence and arousal parameters. Therefore, we assessed the performance of different models based on linear regression, SVM, ELM, Random Forest and MLP.

In this context, Random Forest model performed better for both arousal and valence predictions. This model reached a Pearson's correlation of 0.7569 for valence and of 0.8470 for arousal. In Random Forest model for regression, each decision tree that builds the forest is responsible for estimating an output value. Thus, the actual output is the average of these estimated values for each tree. The fact that the output is an average makes the Random Forest a smooth model. In other words, it is a model that tends to present a good global estimation but tends to fail locally, justifying the high RMSE values found with this model.

The fact that this model was the best one also points to a possible low generalization ability of the problem of predicting valence and arousal values from music. This ability needs to be further investigated in future works. Low generalization may also be associated with the features used for representing the musical signals. From these results we cannot be sure whether valence and arousal are directly functions of these extracted features. They may be functions of other variables and therefore may require more complex representation models.

In this way, other studies with a greater variety of data and models should be conducted, in addition to more in-depth investigations into the relationship of the features with valence and arousal parameters. Even so, the obtained results were promising to enable the prediction of valence and arousal from music. This automatic prediction is certainly useful to provide less subjective assessment of human emotional perception of music. Therefore, this type of prediction can have important contributions in the fields of emotion recognition and personalized content recommendation.

9.6 ACKNOWLEDGEMENTS

This study was financed in part by the Coordenação de Aperfeiçoamento de Pessoal de Nível Superior—Brazil (CAPES)—Finance Code 001, and by the Fundação de Amparo à Ciência e Tecnologia do Estado de Pernambuco (FACEPE), Brazil, under the code IBPG-0013–1.03/20.

9.7 DISCLOSURE STATEMENT

No potential conflict of interest was reported by the authors.

APPENDIX A

REGRESSION RESULTS FOR ALL TESTED CONFIGURATIONS

TABLE 9.A1

Performance for All Tested Configurations of Different Models in the Prediction of Valence Values

Table A1. Performance for all tested configurations of different models in the prediction of valence values.

Regression Model		PCC	KCC	SCC	MAE	RMSE	RAE (%)	RRSE (%)
Linear Regression		0.4644 ± 0.0240	0.3223 ± 0.0180	0.4725 ± 0.0250	0.1721 ± 0.0042	0.2138 ± 0.0047	86.00 ± 1.38	88.57 ± 1.27
SVM	linear	0.4512 ± 0.0260	0.3156 ± 0.0189	0.4608 ± 0.0265	0.1721 ± 0.0043	0.2161 ± 0.0051	86.00 ± 1.54	89.55 ± 1.50
	poly 2	0.4889 ± 0.0433	0.3485 ± 0.0185	0.5044 ± 0.0254	0.1665 ± 0.0045	0.2130 ± 0.0237	83.18 ± 1.89	88.26 ± 10.05
	poly 3	0.5079 ± 0.0647	0.3641 ± 0.0186	0.5243 ± 0.0251	0.1642 ± 0.0099	0.2257 ± 0.1631	82.06 ± 5.05	93.61 ± 68.75
	poly 4	0.4603 ± 0.1202	0.3761 ± 0.0173	0.5398 ± 0.0235	0.1738 ± 0.0902	0.2379 ± 0.0832	76.49 ± 5.33	81.01 ± 3.71
	RBF	0.5039 ± 0.0255	0.3519 ± 0.0189	0.5076 ± 0.0260	0.1646 ± 0.0042	0.2088 ± 0.0051	82.23 ± 1.62	86.52 ± 1.60
ELM	linear	0.2592 ± 0.0091	0.1768 ± 0.0053	0.2582 ± 0.0070	0.3022 ± 0.0033	0.1343 ± 0.0008	55.58 ± 0.26	25.04 ± 0.45
	poly 2	0.1614 ± 0.0320	0.1760 ± 0.0223	0.2573 ± 0.0317	0.4107 ± 0.1001	0.9047 ± 0.3076	109.68 ± 18.43	168.93 ± 57.26
	poly 3	0.0419 ± 0.0259	0.0993 ± 0.0301	0.1467 ± 0.0440	1.9223 ± 1.4020	5.1993 ± 4.0065	443.25 ± 252.01	990.93 ± 754.48
	poly 4	0.0162 ± 0.0278	0.0565 ± 0.0332	0.0836 ± 0.0494	1.36E+3 ± 7.43E+3	7.32E+3 ± 39.36E+3	1.55E+5 ± 8.42E+5	1.42E+5 ± 7.67E+5
	RBF	0.1496 ± 0.0178	0.0267 ± 0.0153	0.0288 ± 0.0188	0.9624 ± 0.0032	0.2322 ± 0.0006	96.36 ± 0.24	43.43 ± 0.62
	sigmoid	0.1372 ± 0.0201	0.0931 ± 0.0200	0.1345 ± 0.0280	1.83E+5 ± 8.61E+5	1.45E+6 ± 7.53E+6	1.14E+8 ± 6.00E+8	2.94E+8 ± 1.53E+9
Random Forest	10 trees	0.7003 ± 0.0179	0.5118 ± 0.0164	0.6935 ± 0.0193	0.1304 ± 0.0036	0.1725 ± 0.0045	65.14 ± 1.65	71.47 ± 1.67
	20 trees	0.7283 ± 0.0165	0.5373 ± 0.0157	0.7222 ± 0.0182	0.1265 ± 0.0034	0.1668 ± 0.0043	63.21 ± 1.50	69.10 ± 1.55
	50 trees	0.7459 ± 0.0155	0.5541 ± 0.0148	0.7403 ± 0.0169	0.1242 ± 0.0032	0.1634 ± 0.0042	62.04 ± 1.38	67.71 ± 1.47
	100 trees	0.7523 ± 0.0153	0.5603 ± 0.0147	0.7469 ± 0.0166	0.1234 ± 0.0033	0.1623 ± 0.0043	61.64 ± 1.36	67.23 ± 1.45
	150 trees	0.7545 ± 0.0153	0.5625 ± 0.0147	0.7492 ± 0.0166	0.1231 ± 0.0034	0.1618 ± 0.0043	61.50 ± 1.38	67.06 ± 1.46
	200 trees	0.7555 ± 0.0154	0.5636 ± 0.0148	0.7503 ± 0.0167	0.1229 ± 0.0034	0.1616 ± 0.0043	61.42 ± 1.39	66.97 ± 1.47
	250 trees	0.7561 ± 0.0154	0.5642 ± 0.0149	0.7509 ± 0.0167	0.1229 ± 0.0034	0.1615 ± 0.0043	61.39 ± 1.39	66.93 ± 1.48
	300 trees	0.7566 ± 0.0154	0.5646 ± 0.0149	0.7514 ± 0.0167	0.1228 ± 0.0034	0.1615 ± 0.0043	61.37 ± 1.38	66.90 ± 1.47
	350 trees	0.7569 ± 0.0153	0.5648 ± 0.0148	0.7517 ± 0.0166	0.1228 ± 0.0033	0.1614 ± 0.0043	61.35 ± 1.37	66.88 ± 1.46
MLP	10 neurons	0.4870 ± 0.0355	0.3494 ± 0.0222	0.5064 ± 0.0303	0.1920 ± 0.0292	0.2366 ± 0.0325	95.91 ± 14.40	98.01 ± 13.34
	20 neurons	0.4875 ± 0.0446	0.3506 ± 0.0307	0.5075 ± 0.0434	0.1904 ± 0.0279	0.2359 ± 0.0314	95.1288 ± 13.8311	97.74 ± 12.96
	50 neurons	0.4418 ± 0.1326	0.3230 ± 0.1019	0.4682 ± 0.1477	0.1962 ± 0.0494	0.2474 ± 0.0735	98.0037 ± 24.6803	102.53 ± 30.51
	100 neurons	0.4354 ± 0.1653	0.3066 ± 0.1371	0.4442 ± 0.1997	3.05E+74 ± 5.28E-75	8.72E+75 ± 1.51E+77	1.52E+77 ± 2.63E+78	3.62E+78 ± 6.27E+79
	(20, 20) neurons	0.5293 ± 0.0271	0.3709 ± 0.0192	0.5352 ± 0.0258	0.1836 ± 0.0266	0.2278 ± 0.0291	91.73 ± 13.17	94.39 ± 11.96
	(100, 100) neurons	0.1565 ± 0.2435	0.1334 ± 0.2111	0.1916 ± 0.3055	0.2092 ± 0.0323	0.2569 ± 0.0413	104.48 ± 15.87	106.45 ± 17.02

TABLE 9.A2

Performance for All Tested Configurations of Different Models in the Prediction of Arousal Values

Table A2. Performance for all tested configurations of different models in the prediction of arousal values.

Regression Model		PCC	KCC	SCC	MAE	RMSE	RAE (%)	RRSE (%)
Linear Regression	-	0.7017 ± 0.0159	0.5144 ± 0.0143	0.7118 ± 0.0166	0.1674 ± 0.0042	0.2088 ± 0.0049	67.84 ± 1.70	71.27 ± 1.59
SVM	linear	0.6976 ± 0.0166	0.5090 ± 0.0148	0.7055 ± 0.0172	0.1669 ± 0.0044	0.2103 ± 0.0052	67.66 ± 1.84	71.79 ± 1.74
	poly 2	0.7199 ± 0.0322	0.5311 ± 0.0142	0.7281 ± 0.0160	0.1597 ± 0.0044	0.2041 ± 0.0144	64.73 ± 1.85	69.69 ± 4.78
	poly 3	0.7190 ± 0.0729	0.5399 ± 0.0144	0.7364 ± 0.0161	0.1588 ± 0.0144	0.2329 ± 0.2734	64.36 ± 5.78	79.42 ± 92.59
	poly 4	0.6379 ± 0.1807	0.5503 ± 0.0152	0.7461 ± 0.0171	0.1585 ± 0.0109	0.2810 ± 0.1676	95.74 ± 56.64	75.23 ± 6.07
	RBF	0.7309 ± 0.0153	0.5338 ± 0.0145	0.7302 ± 0.0163	0.1579 ± 0.0041	0.2001 ± 0.0050	63.99 ± 1.81	68.30 ± 1.71
ELM	linear	0.2988 ± 0.0112	0.1928 ± 0.0057	0.2838 ± 0.0076	0.2626 ± 0.0028	0.1561 ± 0.0012	51.16 ± 0.20	25.01 ± 0.34
	poly 2	0.1886 ± 0.0358	0.1659 ± 0.0234	0.2422 ± 0.0342	0.6160 ± 0.2175	1.1730 ± 0.4356	114.92 ± 23.16	187.81 ± 69.27
	poly 3	0.0402 ± 0.0319	0.0786 ± 0.0198	0.1167 ± 0.0294	2.3917 ± 1.2484	5.7969 ± 2.9025	400.69 ± 131.24	930.57 ± 466.48
	poly 4	0.0195 ± 0.0327	0.0554 ± 0.0244	0.0820 ± 0.0352	6.8994 ± 4.9853	16.4990 ± 11.3780	1004.70 ± 565.66	2640.30 ± 1817.80
	RBF	NaN ± NaN	NaN ± NaN	NaN ± NaN	0.9655 ± 0.0039	0.2856 ± 0.0009	96.59 ± 0.30	45.80 ± 0.49
	sigmoid	0.1282 ± 0.0242	0.0988 ± 0.0159	0.1430 ± 0.0230	1.70E+4 ± 4.59E+4	6.78E+5 ± 1.88E+5	3.28E+6 ± 9.00E+6	1.11E+7 ± 3.05E+7
Random Forest	10 trees	0.8210 ± 0.0124	0.6297 ± 0.0137	0.8160 ± 0.0134	0.1251 ± 0.0035	0.1673 ± 0.0048	50.70 ± 1.64	57.13 ± 1.74
	20 trees	0.8341 ± 0.0117	0.6446 ± 0.0132	0.8295 ± 0.0127	0.1215 ± 0.0032	0.1621 ± 0.0045	49.24 ± 1.53	55.35 ± 1.66
	50 trees	0.8422 ± 0.0111	0.6542 ± 0.0127	0.8379 ± 0.0120	0.1193 ± 0.0031	0.1589 ± 0.0045	48.35 ± 1.46	54.25 ± 1.60
	100 trees	0.8449 ± 0.0108	0.6575 ± 0.0124	0.8407 ± 0.0117	0.1185 ± 0.0031	0.1579 ± 0.0044	48.05 ± 1.44	53.90 ± 1.56
	150 trees	0.8460 ± 0.0107	0.6588 ± 0.0123	0.8417 ± 0.0116	0.1182 ± 0.0031	0.1575 ± 0.0044	47.93 ± 1.42	53.76 ± 1.55
	200 trees	0.8464 ± 0.0107	0.6593 ± 0.0123	0.8422 ± 0.0116	0.1181 ± 0.0031	0.1573 ± 0.0044	47.88 ± 1.41	53.70 ± 1.55
	250 trees	0.8466 ± 0.0107	0.6596 ± 0.0124	0.8424 ± 0.0117	0.1181 ± 0.0031	0.1572 ± 0.0044	47.86 ± 1.42	53.67 ± 1.55
	300 trees	0.8468 ± 0.0107	0.6599 ± 0.0123	0.8427 ± 0.0116	0.1180 ± 0.0031	0.1571 ± 0.0044	47.83 ± 1.41	53.64 ± 1.55
	350 trees	0.8470 ± 0.0107	0.6600 ± 0.0124	0.8428 ± 0.0116	0.1180 ± 0.0031	0.1571 ± 0.0044	47.82 ± 1.42	53.63 ± 1.55
MLP	10 neurons	0.7157 ± 0.0230	0.5303 ± 0.0167	0.7270 ± 0.0184	0.1871 ± 0.0301	0.2320 ± 0.0336	75.8235 ± 12.2587	79.1906 ± 11.4993
	20 neurons	0.6970 ± 0.1017	0.5154 ± 0.0972	0.7066 ± 0.1333	0.1964 ± 0.0884	0.2429 ± 0.0926	79.5956 ± 35.9354	82.9243 ± 31.6468
	50 neurons	0.6823 ± 0.1529	0.5136 ± 0.1019	0.7037 ± 0.1392	6.29E+134 ± 1.09E+136	1.80E+136 ± 3.12E+137	2.53E+137 ± 4.39E+138	6.11E+138 ± 1.06E+140
	100 neurons	0.6790 ± 0.1711	0.4995 ± 0.1486	0.6836 ± 0.2069	9.02E+6 ± 1.56E+8	5.63E+7 ± 9.74E+8	3.67E+9 ± 6.35E+10	1.92E+10 ± 3.34E+11
	(20, 20) neurons	0.7443 ± 0.0165	0.5467 ± 0.0152	0.7440 ± 0.0167	0.1727 ± 0.0220	0.22 ± 0.02	69.98 ± 8.90	73.4750 ± 8.4705
	(100, 100) neurons	0.2162 ± 0.3487	0.1995 ± 0.3523	0.2694 ± 0.4913	0.2389 ± 0.0486	0.2931 ± 0.0663	96.80 ± 19.48	100.01 ± 22.31

"NaN" results correspond to outputs that tend to ±∞, suffering underflow or overflow. In these cases, we used the errors as comparison parameters.

REFERENCES

Adler, J. and I. Parmryd. Quantifying colocalization by correlation: the pearson correlation coefficient is superior to the mander's overlap coefficient. *Cytometry Part A*, 77(8):733–742, 2010.

Ahmad, I., M. Basheri, M. J. Iqbal, and A. Rahim. Performance comparison of support vector machine, random forest, and extreme learning machine for intrusion detection. *IEEE Access*, 6:33789–33795, 2018.

Arjmand, H.-A., J. Hohagen, B. Paton, and N. S. Rickard. Emotional responses to music: shifts in frontal brain asymmetry mark periods of musical change. *Frontiers in Psychology*, 8:2044, 2017.

Biau, G. and E. Scornet. A random forest guided tour. *Test*, 25(2):197–227, 2016.

Breiman, L. Random forests. *Machine Learning*, 45(1):5–32, 2001.

Chai, T. and R. R. Draxler. Root mean square error (rmse) or mean absolute error (mae)?—arguments against avoiding rmse in the literature. *Geoscientific Model Development*, 7(3):1247–1250, 2014.

Chen, C. and Q. Li. A multimodal music emotion classification method based on multifeature combined network classifier. *Mathematical Problems in Engineering*, 2020, 2020.

Cheuk, K. W., Y.-J. Luo, B. Balamurali, G. Roig, and D. Herremans. Regression-based music emotion prediction using triplet neural networks. In *2020 International Joint Conference on Neural Networks (IJCNN)*, pages 1–7. IEEE, 2020.

Cowen, A. S., X. Fang, D. Sauter, and D. Keltner. What music makes us feel: at least 13 dimensions organize subjective experiences associated with music across different cultures. *Proceedings of the National Academy of Sciences*, 117(4):1924–1934, 2020.

da Silva, C. C., C. L. de Lima, A. C. G. da Silva, E. L. Silva, G. S. Marques, L. J. B. de Araújo, L. A. A. Júnior, S. B. J. de Souza, M. A. de Santana, J. C. Gomes, et al. Covid-19 dynamic monitoring and real-time spatio-temporal forecasting. *Frontiers in Public Health*, 9, 2021.

de l'Etoile, S. Processes of music therapy: clinical and scientific rationales and models. In *The Oxford Handbook of Music Psychology*. Oxford, Oxford University Press, 2008.

de Lima, C. L., C. C. da Silva, A. C. G. da Silva, E. Luiz Silva, G. S. Marques, L. J. B. de Araújo, L. A. Albuquerque Júnior, S. B. J. de Souza, M. A. de Santana, J. C. Gomes, et al. Covid-sgis: a smart tool for dynamic monitoring and temporal forecasting of Covid-19. *Frontiers in Public Health*, 8:761, 2020.

de Moraes Ramos, R. T., M. A. d. Santana, P. A. Sousa, M. R. Assunção Ferreira, W. P. dos Santos, and L. A. Lira Soares. Multivariate regression and artificial neural network to predict phenolic content in schinus terebinthifolius stem bark through tlc images. *Journal of Liquid Chromatography & Related Technologies*, 1–8, 2021.

Deng, W., Q. Zheng, and L. Chen. Regularized extreme learning machine. In *2009 IEEE Symposium on Computational Intelligence and Data Mining*, pages 389–395. IEEE, 2009.

Desai, M. and M. Shah. An anatomization on breast cancer detection and diagnosis employing multilayer perceptron neural network (MLP) and convolutional neural network (CNN). *Clinical eHealth*, 4:1–11, 2020.

Dineva, K. and T. Atanasova. Regression analysis on data received from modular iot system. In *Proceedings of the European Simulation and Modelling Conference ESM*. Spain: Palma de Mallorca, 2019.

Driss, S. B., M. Soua, R. Kachouri, and M. Akil. A comparison study between MLP and convolutional neural network models for character recognition. In *Real-Time Image and Video Processing 2017*, volume 10223, page 1022306. International Society for Optics and Photonics, 2017.

Eaton, J. W., D. Bateman, S. Hauberg, and R. Wehbring. GNU octave version 4.0. 0 manual: a high-level interactive language for numerical computations, 2015. www.gnu.org/software/octave/doc/interpreter, 8:13, 2015.

Field, A. *Kendalls Coefficient of Concordance, Encyclopedia of Statistics in Behavioral Science.* John Wiley, 2005.

Fonseca, V. d. Importância das emoções na aprendizagem: uma abordagem neuropsicopedagógica. *Revista Psicopedagogia*, 33(102):365–384, 2016.

Franceschini, F. and D. Maisano. Aggregating multiple ordinal rankings in engineering design: the best model according to the kendalls coefficient of concordance. *Research in Engineering Design*, 32(1):91–103, 2021.

Grekow, J. Audio features dedicated to the detection and tracking of arousal and valence in musical compositions. *Journal of Information and Telecommunication*, 2(3):322–333, 2018.

Hauke, J. and T. Kossowski. Comparison of values of pearsons and spearmans correlation coefficient on the same sets of data. *Quaestiones Geographicae*, 30(2):87–93, 2011.

Hizlisoy, S., S. Yildirim, and Z. Tufekci. Music emotion recognition using convolutional long short term memory deep neural networks. *Engineering Science and Technology, an International Journal*, 24(3):760–767, 2021.

Hsieh, S., M. Hornberger, O. Piguet, and J. Hodges. Brain correlates of musical and facial emotion recognition: evidence from the dementias. *Neuropsychologia*, 50(8):1814–1822, 2012.

Huang, G.-B., Q.-Y. Zhu, and C.-K. Siew. Extreme learning machine: theory and applications. *Neurocomputing*, 70(1–3):489–501, 2006.

Jeyasingh, S. and M. Veluchamy. Modified bat algorithm for feature selection with the wisconsin diagnosis breast cancer (wdbc) dataset. *Asian Pacific Journal of Cancer Prevention: APJCP*, 18(5):1257, 2017.

Jung, Y. and J. Hu. A K-fold averaging cross-validation procedure. *Journal of Nonparametric Statistics*, 27(2):167–179, 2015. ISSN 10290311.

Juslin, P. N. What does music express? Basic emotions and beyond. *Frontiers in Psychology*, 4:596, 2013.

Juslin, P. N. and P. Laukka. Communication of emotions in vocal expression and music performance: different channels, same code? *Psychological bulletin*, 129(5):770, 2003.

Kavitha, S., S. Varuna, and R. Ramya. A comparative analysis on linear regression and support vector regression. In *2016 Online International Conference on Green Engineering and Technologies (ICGET)*, pages 1–5. IEEE, 2016.

Koelsch, S. Investigating the neural encoding of emotion with music. *Neuron*, 98(6):1075–1079, 2018.

Li, X., H. Wang, Y. Tian, S. Zhou, X. Li, K. Wang, and Y. Yu. Impaired white matter connections of the limbic system networks associated with impaired emotional memory in alzheimer's disease. *Frontiers in aging neuroscience*, 8:250, 2016.

Meska, M. H. G., L. Y. Mano, J. P. Silva, G. A. Pereira, and A. Mazzo. Emotional recognition for simulated clinical environment using unpleasant odors: quasi-experimental study. *Revista latinoamericana de enfermagem*, Miguel, F. K. Psicologia das emoções: uma proposta integrativa para compreender a expressão emocional. *Psico-usf*, 20:153–162, 2015.

Mishra, S., P. Paygude, S. Chaudhary, and S. Idate. Use of data mining in crop yield prediction. In *2018 2nd International Conference on Inventive Systems and Control (ICISC)*, pages 796–802. IEEE, 2018.

Neill, S. P. and M. R. Hashemi. Ocean modelling for resource characterization. In *Fundamentals of Ocean Renewable Energy*, pages 193–235. Philadelphia, USA: Elsevier, 2018.

Patle, A. and D. S. Chouhan. SVM kernel functions for classification. In *2013 International Conference on Advances in Technology and Engineering (ICATE)*, pages 1–9. IEEE, 2013.

Qi, J., J. Du, S. M. Siniscalchi, X. Ma, and C.-H. Lee. On mean absolute error for deep neural network based vector-to-vector regression. *IEEE Signal Processing Letters*, 27:1485–1489, 2020.

Rolls, E. T. Limbic systems for emotion and for memory, but no single limbic system. *Cortex*, 62:119–157, 2015.

Russell, J. A. A circumplex model of affect. *Journal of Personality and Social Psychology*, 39(6):1161, 1980.

Safonicheva, O. G. and M. A. Ovchinnikova. Movements and development. Art-therapy approach in the complex rehabilitation of children with intellectual disorders, including autism spectrum disorder. In *Emerging Programs for Autism Spectrum Disorder*, pages 243–264. Philadelphia, USA: Elsevier, 2021.

Sammut, C. and G. I. Webb. *Encyclopedia of Machine Learning*. Berlin: Springer Science & Business Media, 2011.

Schaefer, H.-E. Music-evoked emotionscurrent studies. *Frontiers in Neuroscience*, 11:600, 2017.

Schuld, M., I. Sinayskiy, and F. Petruccione. Prediction by linear regression on a quantum computer. *Physical Review A*, 94(2):022342, 2016.

Sedgwick, P. Pearsons correlation coefficient. *BMJ*, 345, 2012.

Sharma, H., S. Gupta, Y. Sharma, and A. Purwar. A new model for emotion prediction in music. In *2020 6th International Conference on Signal Processing and Communication (ICSC)*, pages 156–161. IEEE, 2020.

Singh, P. K. and M. S. Husain. Methodological study of opinion mining and sentiment analysis techniques. *International Journal on Soft Computing*, 5(1):11, 2014.

Siriyasatien, P., S. Chadsuthi, K. Jampachaisri, and K. Kesorn. Dengue epidemics prediction: a survey of the state-of-the-art based on data science processes. *IEEE Access*, 6:53757–53795, 2018.

Sobeeh, M., G. Öztürk, and M. Hamed. Effect of listening to high arousal music with different valences on reaction time and interference control: evidence from simon task. *IBRO Reports*, 6:S441, 2019.

Soleymani, M., M. N. Caro, E. M. Schmidt, C.-Y. Sha, and Y.-H. Yang. 1000 songs for emotional analysis of music. In *Proceedings of the 2nd ACM International Workshop on Crowdsourcing for Multimedia*, pages 1–6. ACM, 2013.

Song, Y., S. Dixon, M. T. Pearce, and A. R. Halpern. Perceived and induced emotion responses to popular music: categorical and dimensional models. *Music Perception: An Interdisciplinary Journal*, 33(4):472–492, 2016.

Subasi, A., M. F. El-Amin, T. Darwich, and M. Dossary. Permeability prediction of petroleum reservoirs using stochastic gradient boosting regression. *Journal of Ambient Intelligence and Humanized Computing*, 1–10, 2020.

Suhaimi, N. S., J. Mountstephens, and J. Teo. Eeg-based emotion recognition: a state-of-the-art review of current trends and opportunities. *Computational Intelligence and Neuroscience*, 2020, 2020.

Tan, K., M. Villarino, and C. Maderazo. Automatic music mood recognition using russells twodimensional valence-arousal space from audio and lyrical data as classified using svm and naïve bayes. In *IOP Conference Series: Materials Science and Engineering*, volume 482, page 012019. IOP Publishing, 2019.

Vatolkin, I. and A. Nagathil. Evaluation of audio feature groups for the prediction of arousal and valence in music. In *Applications in Statistical Computing*, pages 305–326. Berlin: Springer, 2019.

Witten, I. H. and E. Frank. *Data Mining: Pratical Machine Learning Tools and Technique.* San Francisco, CA: Morgan Kaufmann Publishers, 2005.

Xiao, C., J. Ye, R. M. Esteves, and C. Rong. Using spearman's correlation coefficients for exploratory data analysis on big dataset. *Concurrency and Computation: Practice and Experience*, 28(14): 3866–3878, 2016.

Yang, Z. and J. Hirschberg. Predicting arousal and valence from waveforms and spectrograms using deep neural networks. In *INTERSPEECH*, pages 3092–3096, 2018. https://www.isca-speech.org/iscaweb/index.php/conferences/interspeech

Yu, L.-C., J. Wang, K. R. Lai, and X.-J. Zhang. Predicting valence-arousal ratings of words using a weighted graph method. In *Proceedings of the 53rd Annual Meeting of the Association for Computational Linguistics and the 7th International Joint Conference on Natural Language Processing (Volume 2: Short Papers)*, pages 788–793, 2015. https://aclanthology.org/P15-2129/

Zhou, H., Z. Deng, Y. Xia, and M. Fu. A new sampling method in particle filter based on pearson correlation coefficient. *Neurocomputing*, 216:208–215, 2016.

Section 3

Gait—Balance Signal Processing

10 Updated ICA Weight Matrix for Lower Limb Myoelectric Classification

Ganesh R. Naik

10.1 INTRODUCTION

Electromyography (EMG) is the process of recording the muscle activity from the skeletal muscles. Surface EMG (sEMG) has been widely used for both upper and lower limb gesture and prosthetic applications [1, 2]. Surface EMG signals related to lower limb movements are more complex in nature, and hence the classification of lower limb movements is found to be more challenging than hand or finger movements. Moreover, the muscles responsible for these movements are in the intermediate and deep layers of the leg/thigh muscles [3–5]. In the recent past, several research studies were conducted to classify both upper and lower limb sEMG data using various signal processing and pattern recognition methods [6–8], where the classification outcome is used to derive input commands for artificial devices [9–11]. Multivariate techniques such as principal component analysis (PCA) and independent component analysis (ICA) are widely used for such tasks [12–14].

ICA is one of the so-called blind source separation (BSS) techniques which utilise both lower and higher order statistics to estimate linearly mixed variables into their independent components (ICs) [15–17]. One of the advantages of using ICA for EMG is that it decomposes the linearly mixed several muscle activities into its constituent ICs or motor units [13, 14]. However, it is anticipated that all the decomposed ICs may not contribute to the extraction of useful EMG features. One of the processes would be extraction of useful ICs using updated unmixing matrix (W) of ICA. Using this method, useful sources (ICs) are estimated by back projecting the unmixing matrix to the original input space.

Nearly all the previous research studies mainly focussed on the classification of lower limb data using healthy subjects only [8, 18, 19]. In this chapter, we propose a classification scheme based on ICA and linear discriminant analysis (LDA) to identify movements associated with the lower limb data using both healthy patients and patients suffering from knee problems. We classify three different classes which include: walking (gait), sitting (leg extension from a sitting position) and standing (flexion of the leg up) by processing sEMG signals.

The rest of the chapter is organised as follows: the ICA and updated ICA matrix principles are briefly summerised in the first part of the chapter. The method, feature

DOI: 10.1201/9781003201137-13

extraction and selection are then explained in detail. The classification and evaluation of the proposed method are explained in the third part. The final part of the chapter discusses the outcome of the proposed research and discusses future directions of the research.

10.2 INDEPENDENT COMPONENT ANALYSIS

ICA is one of the BSS techniques and its goal is to separate/estimate instantaneously mixed N unknown sources s (latent observations) from the channel matrix x (M observations), such that

$$x(t) = As(t) \tag{10.1}$$

where $x(t) = \left[x_1(t), x_2(t), x_3(t), \cdots, x_M(t) \right]^T$, and $s(t) = \left[s_1(t), s_2(t), s_3(t) \cdots, s_N(t) \right]^T$ denote the source and mixing vectors respectively, with an unknown full rank mixing matrix (without any prior information) and T the transpose of the vector. ICA computes the unmixing matrix W and estimates the sources $y(t)$ as:

$$y(t) = Wx(t) \tag{10.2}$$

provided that they are statistically independent from each other [18, 19].

ICA can be solved using both higher order and lower order statistical techniques such as FastICA [15, 16], Infomax [17], JADE, SOBI, etc. This research uses the FastICA algorithm, which adapts a fixed point iteration scheme to maximise the non-Gaussianity of the sources. More details of FastICA can be found in [15, 16].

10.2.1 SOURCE COMPUTATION USING UPDATED UNMIXING MATRIX

Unmixing matrix W is computed using the FastICA algorithm. In order to compute the best estimate of the ICs the rows of the W are re-ordered in descending order (largest to smallest) and the row elements of W which are less than the threshold value ($w < 0.1$) are set to zero (the optimum threshold value was selected based on several empirical analysis). By doing so the unmixing matrix $\hat{W}$ becomes sparser and it is expected to provide the sparse independent sources during the back projection. The new updated sources $\hat{y}$ are estimated using the updated un-mixing matrix $\hat{W}$ as:

$$ICs = \hat{y}(t) = \hat{W}x(t) \tag{10.3}$$

where $\hat{W}$ and $x(t)$ are the updated unmixing matrix and the original recordings respectively.

The correction made in this study is indeed the post-processing after the ICA, and it can also be regarded as further processing on certain element of the unmixing matrix W. Hence, the correction does not change the configuration of real sources (muscles) in the lower limb muscles. However, corrections made indeed help in making data sparse and to extract useful features from the ICA separated sources.

10.3 METHODS

Experiments were conducted to evaluate the performance of the proposed lower limb EMG classification system. Time domain features were extracted from ICA separated EMG sources. The features were then classified using linear discriminant analysis (LDA).

10.3.1 DATA ACQUISITION

For this research, sEMG data was taken from UCI machine learning repository [20]. This database contains samples from 11 healthy subjects and 11 subjects with knee pathology. These data were collected for 22 male subjects with EMG and goniometry equipment MWX8 Datalog Biometrics. The subjects performed three movements, they are: (i) walking (gait), (ii) sitting (leg extension from a sitting position) and (iii) standing (flexion of the leg up). The data acquisition process was conducted using four electrodes connected to the four leg muscles which include: *Vastus Medialis*, *Semitendinosus*, *Biceps Femoris* and *Rectus Femoris*. The goniometer was connected on the knee.

Surface EMG signals were sampled at 1000 Hz and these data were sent directly to the computer and transmitted in Real-time Datalog software through Bluetooth adapter [20]. These (sEMG) signals were band pass filtered using an FIR filter with a frequency range between 20 and 450 Hz.

10.3.2 DATA PROCESSING—FEATURE EXTRACTION AND FEATURE REDUCTION

Prior to feature extraction, the ICA pre-processed (ICA separated data) data was divided into overlapping windows of 256 samples length with a 64-samples increment (25% overlapping) between windows. This segmentation scheme was used for all numerical experiments in this study. To extract the useful information from the segmented sEMG signals, five time domain features were extracted which include root mean square (RMS), mean absolute value (MAV), variance (VAR), waveform length (WL) and zero crossing (ZC).

- *Root mean square* (RMS)

Root mean square (RMS) is one of the widely used features for the analysis of the sEMG signals. For N samples RMS is given as:

$$RMS = \sqrt{\frac{1}{N}\sum_{i=1}^{N} x_i^2} \qquad (10.4)$$

- *Mean absolute value (MAV)*

This feature calculates absolute mean value of lower limb sEMG and is defined as:

$$MAV_i = \frac{1}{N}\sum_{j=1}^{N-1}\left(x_i(j)\right) \tag{10.5}$$

where $x_i(j)$ is the jth signal sample, of the ith window and N represents the number of samples in the window.

- *Variance (VAR)*

This feature calculates the power of the time domain sEMG signal and is represented as:

$$\sigma_i^2 = \frac{1}{N-1}\sum_{j=1}^{N}x_j^2 \tag{10.6}$$

- *Waveform length (WL)*

This feature calculates sEMG waveform amplitude, frequency and duration and is defined as:

$$WL_i = \sum_{j=1}^{N-1}\left(x_i(j)-x_i(j+1)\right) \tag{10.7}$$

- *Zero crossing* (ZC)

The ZC represents the number of points in the window where the sign of a function changes and is defined as:

$$ZC_i = \sum_{j=1}^{N} f(j) \tag{10.8}$$

where

$$f(x) = \begin{cases} 1, & fx_i(j)\times x_i(j+1)\langle 0 \ and \left| x_i(j)\times x_i(j+1)\right| \rangle x_{th} \\ 0, & otherwise \end{cases}$$

These features were chosen based on the fact that time domain features such as RMS and WL can achieve higher performance than that of other feature extraction methods [23].

10.3.3 FEATURE SELECTION AND REDUCTION

Among the five sets of features extracted for healthy and knee pathology data, the more useful features for the classification were selected using Fisher discriminant analysis (FDA). FDA was calculated using the following equation:

$$D_p = \frac{\sum_{i=1}^{N}\left|m_{p,i}-m_p\right|^2}{\sum_i^{Q} s_{p,i}^2} \tag{10.9}$$

TABLE 10.1
FDA Scores for Five Time Domain Features.

Time Domain Features	FDA Scores (Mean ± SD)
RMS	3.4 ± 0.24
ZC	3.6 ± 0.31
WL	3.1 ± 0.25
MAV	2.1 ± 0.51
VAR	1.8 ± 0.34

where D_p is the discriminability score and the subscript p indicates the sEMG electrode. The mean and standard deviation (SD) of each movement is represented as $m_{p,i}$ and $s_{p,i}$ respectively. The m_p is the mean of all movements. The subscript i indicates the movement and N is the number of movements, which is 3. The numerator represents the difference between the means of each movement, whereas the denominator represents the sum of variance of each movement. When D_p is high, the signals from the pth electrode provide effective information that can help in achieving higher classification accuracy. The results of the discriminability values calculated for five time domain features are shown in Table 10.1. From the results (scores), it is evident that three features RMS, ZC and WL have higher discriminability scores than the other features (VAR and MAV). Hence, only three features (RMS, ZC and WL) were chosen for data classification.

10.4 RESULTS AND DISCUSSION

10.4.1 DATA CLASSIFICATION AND CROSS-VALIDATION USING LDA

Classification of lower limb sEMG is performed using LDA classifier. It is a supervised statistical method, which transforms a set of high-dimensional features into a lower-dimensional space while preserving the class structure. It minimises the within-class distance and at the same time maximises the between-class distance, thus achieving maximum discrimination. The advantage of using LDA is that it does not necessitate iterative training, which prevents the potential classification problems such as under or over-training of the data. Furthermore, using LDA, a high dimensionality problem can be well linearized during feature reduction. The odd number trials were used for training set to optimise the LDA model parameters and the even number trials were used for test set. Each time, the order of training and testing trials were completely randomised to utilise all possible combinations.

The performance of the proposed method was evaluated using overall classification accuracy (CA):

$$CA = \frac{No.\ of\ correlctly\ classfied\ testing\ data}{total\ no.\ of\ applied\ testing\ data} \times 100\% \qquad (10.10)$$

In addition, confusion matrix (CM) which calculates the test results between predicted classes and the actual classes was computed.

Table 10.2 presents the classification accuracies with mean and standard deviation (SD) for lower limb sEMG data using healthy subjects. From the results (Table 10.2), it can be seen that the proposed modified ICA-LDA classification model is able to classify three different lower limb movements with > 85% accuracy. Similarly, Table 10.3 presents the average classification accuracies (Mean ± SD) of lower limb sEMG data for the subjects suffering from knee pathology. The classification accuracies in Table 10.3 are encouraging, considering the variations from the knee pathology data. The overall classification results with (Mean ± SD) are shown in Figure 10.1. Similarly, overall classification (average) results using confusion matrices are shown in Figure 10.2 and Figure 10.3 respectively.

Possibly the most interesting result was the similar classification error among three classes (standing, sitting and walking) using EMG combination (Table 10.2) recorded from the healthy subjects. The knee pathology patients' sEMG data patterns did not essentially follow the configurations of healthy subjects (Table 10.3); nevertheless, since the proposed updated ICA weight matrix algorithm adapts to everyone, LDA algorithm was able to provide reasonably high classification accuracy with the knee pathology subjects as well.

TABLE 10.2

Classification Rates of Lower Limb Data for Healthy Subjects Using Source Separation (Updated ICA) Technique: Walking, Sitting and Standing.

| | Lower Limb Gestures (Healthy Participants) | | |
Participants	Walking (gait)	Sitting (leg extension from a sitting position)	Standing (flexion of the leg up)
subject 1	88.5	86.9	85.8
subject 2	87.7	87.8	84.1
subject 3	85.6	89.5	84.6
subject 4	85.2	87.3	85.6
subject 5	87.3	89.2	86.8
subject 6	86.1	87.6	86.7
subject 7	84.6	85.8	84.9
subject 8	88.2	89.4	85.8
subject 9	87.8	85.1	84.2
subject 10	86.3	87.8	84.3
subject 11	88.9	88.3	87.3
MEAN ± SD	86.9 ± 1.44	87.7 ± 1.41	85.5 ± 1.13

TABLE 10.3

Classification Rates of Lower Limb Data for Subjects with Knee Pathology Using Source Separation (Updated ICA) Technique: Walking, Sitting and Standing.

Participants	Lower Limb Gestures (Subjects with Knee Pathology)		
	Walking (gait)	Sitting (leg extension from a sitting position)	Standing (flexion of the leg up)
subject 1	78.2	81.6	75.8
subject 2	78.1	78.1	77.3
subject 3	76.3	80.5	74.6
subject 4	76.2	79.6	73.2
subject 5	79.1	80.2	76.3
subject 6	77.5	78.4	74.1
subject 7	76.7	79.3	73.9
subject 8	78.1	78.7	76.1
subject 9	74.1	80.1	75.1
subject 10	75.7	81.2	76.2
subject 11	76.1	78.3	77.2
MEAN ± SD	76.9 ± 1.43	79.6 ± 1.19	75.4 ± 1.36

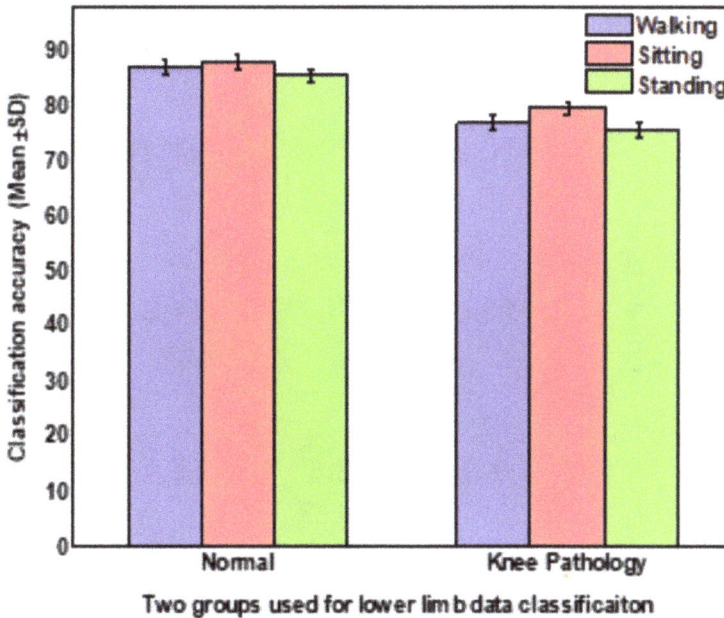

FIGURE 10.1 Overall classification results (Mean ± SD) for healthy and subjects with knee pathology.

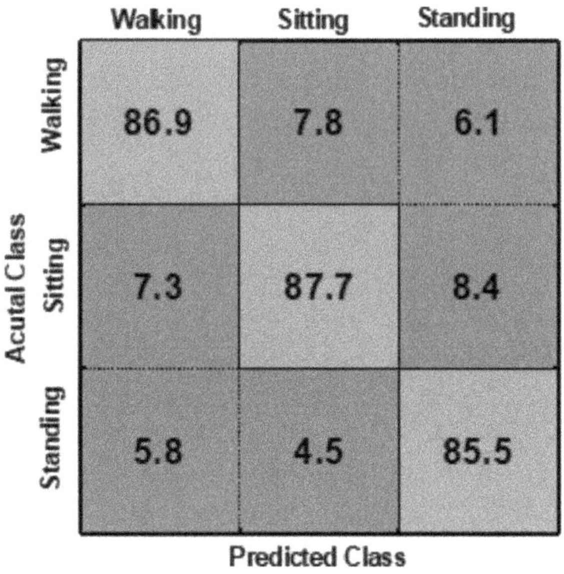

FIGURE 10.2 Confusion matrix showing the error distribution for the lower limb movements' classification for healthy subjects. Here, the results in the diagonal are the correct classification rates while the results outside the diagonal are the errors.

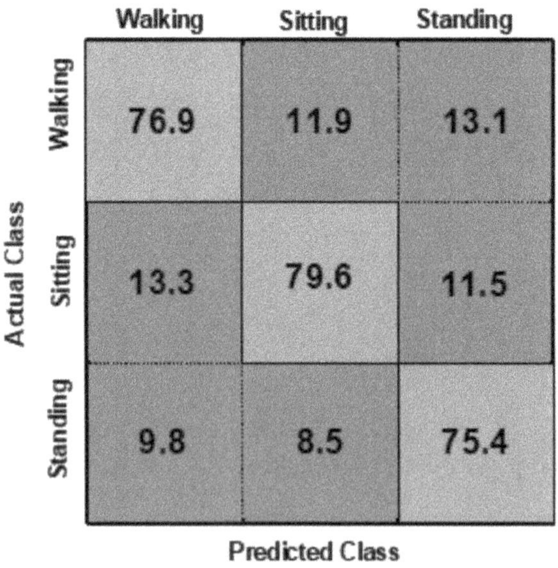

FIGURE 10.3 Confusion matrix showing the error distribution for the lower limb movements' classification for subjects with knee pathology. Here, the results in the diagonal are the correct classification rates while the results outside the diagonal are the errors.

10.5 CONCLUSION

This chapter presented novel modified ICA based classification scheme for lower limb sEMG data using healthy subjects and subjects suffering with knee pathology. The independence and linearity of updated ICA weight matrix were adopted to extract various lower limb sEMG features. FDA was used for feature selection and feature reduction followed by LDA, which was used for classification of different lower limb patterns based on feature values of sEMG signals.

The results of this preliminary investigation suggest that adequate neural control information for better classification of various lower limb tasks can be extracted from sEMG signals recorded using four sensors for both healthy subjects and subjects with knee pathology. While the classification accuracies achieved in this study are promising, they are undoubtedly not adequate for prosthetic limb control. Further investigations are warranted for future clinical applications. Keeping this in mind, our future study will investigate testing the described framework with multiple subjects and possibly on amputee subjects by adding new activity approaches, such as stair ascent and descent.

REFERENCES

[1] J. Rafiee, M. Rafiee, F. Yavari, and M. Schoen, "Feature extraction of forearm EMG signals for prosthetics," *Expert Systems with Applications*, vol. 38, pp. 4058–4067, 2011.

[2] R. N. Khushaba, S. Kodagoda, M. Takruri, and G. Dissanayake, "Toward improved control of prosthetic fingers using surface electromyogram (EMG) signals," *Expert Systems with Applications*, vol. 39, pp. 10731–10738, 2012.

[3] G. S. Murley, H. B. Menz, K. B. Landorf, and A. R. Bird, "Reliability of lower limb electromyography during overground walking: a comparison of maximal-and submaximal normalisation techniques," *Journal of Biomechanics*, vol. 43, pp. 749–756, 2010.

[4] M. O. Ericson, R. Nisell, and J. Ekholm, "Quantified electromyography of lower-limb muscles during level walking," *Scandinavian Journal of Rehabilitation Medicine*, vol. 18, pp. 159–163, 1985.

[5] A. Rainoldi, G. Melchiorri, and I. Caruso, "A method for positioning electrodes during surface EMG recordings in lower limb muscles," *Journal of Neuroscience Methods*, vol. 134, pp. 37–43, 2004.

[6] S. Mulroy, J. Gronley, W. Weiss, C. Newsam, and J. Perry, "Use of cluster analysis for gait pattern classification of patients in the early and late recovery phases following stroke," *Gait & Posture*, vol. 18, pp. 114–125, 2003.

[7] H. Huang, F. Zhang, L. J. Hargrove, Z. Dou, D. R. Rogers, and K. B. Englehart, "Continuous locomotion-mode identification for prosthetic legs based on neuromuscular—mechanical fusion," *IEEE Transactions on Biomedical Engineering*, vol. 58, pp. 2867–2875, 2011.

[8] G. Bovi, M. Rabuffetti, P. Mazzoleni, and M. Ferrarin, "A multiple-task gait analysis approach: kinematic, kinetic and EMG reference data for healthy young and adult subjects," *Gait & Posture*, vol. 33, pp. 6–13, 2011.

[9] R. Merletti and P. A. Parker, *Electromyography: Physiology, Engineering, and Non-Invasive Applications*, vol. 11. New Jersey, USA: John Wiley & Sons, 2004.

[10] D. C. Preston and B. E. Shapiro, *Electromyography and Neuromuscular Disorders: Clinical-Electrophysiologic Correlations (Expert Consult-Online)*. Philadelphia, USA: Elsevier Health Sciences, 2012.

[11] G. S. Murley, K. B. Landorf, H. B. Menz, and A. R. Bird, "Effect of foot posture, foot orthoses and footwear on lower limb muscle activity during walking and running: a systematic review," *Gait & Posture*, vol. 29, pp. 172–187, 2009.

[12] P. Geethanjali, "Comparative study of PCA in classification of multichannel EMG signals," *Australasian Physical & Engineering Sciences in Medicine*, vol. 38, no. 2, pp. 331–343.

[13] G. R. Naik and D. K. Kumar, "Estimation of independent and dependent components of non-invasive EMG using fast ICA: validation in recognising complex gestures," *Computer Methods in Biomechanics and Biomedical Engineering*, vol. 14, pp. 1105–1111, 2011.

[14] G. R. Naik and D. K. Kumar, "Identification of hand and finger movements using multi run ICA of surface electromyogram," *Journal of Medical Systems*, vol. 36, pp. 841–851, 2012.

[15] A. Hyvärinen, J. Karhunen, and E. Oja, *Independent Component Analysis*, vol. 46. New Jersey, USA: John Wiley & Sons, 2004.

[16] A. Hyvarinen, "Fast and robust fixed-point algorithms for independent component analysis," *IEEE Transactions on Neural Networks*, vol. 10, pp. 626–634, 1999.

[17] T.-W. Lee, *Independent Component Analysis*. Boston: Springer, 1998.

[18] C. D. Joshi, U. Lahiri, and N. V. Thakor, "Classification of gait phases from lower limb EMG: application to exoskeleton orthosis," in *Point-of-Care Healthcare Technologies (PHT)*. IEEE, 2013, pp. 228–231.

[19] J. H. Ryu and D. H. Kim, "Multiple gait phase recognition using boosted classifiers based on sEMG signal and classification matrix," *Presented at the Proceedings of the 8th International Conference on Ubiquitous Information Management and Communication*, Siem Reap, Cambodia, 2014.

[20] O. F. A. Sanchez, L. R. Sotelo, M. H. Gonzales, and G. A. M. Hernandez, *"EMG dataset in lower limb data set"*, UCI Machine Learning Repository: 2014-02-05.

11 Cortical Correlates of Unilateral Transfemoral Amputees during a Balance Control Task with Vibrotactile Feedback

Aayushi Khajuria, Upinderpal Singh, and Deepak Joshi

11.1 INTRODUCTION

The loss of sensory input at the amputated limb results in poor gait and balance coordination among lower-limb amputees [1,2]. Therefore, various balance rehabilitation programs were developed to improve their movement coordination. Among these, voluntary postural sway is a well-known protocol for improving balance and reducing the rate of falling in unilateral lower-limb amputees [3]. However, due to the limited availability of physical therapists, access to clinical facilities and cost issues often prohibit their participation in these regimens [4]. Therefore, utilizing biofeedback technology in rehabilitation is the current area of interest. It develops new pathways in the central nervous system, promotes neuroplasticity in basal ganglia, and improves motor skills [5–7]. Previous literature shows the balance improvement in people with impaired balance using visual, auditory, and vibratory feedback [7, 8], though vibrotactile biofeedback has shown effective outcomes [9–11]. It has also been proved recently that the improved limit of stability in transfemoral amputees with vibratory feedback was possibly due to the higher contribution of somatosensory receptors [12]. However, the cortical correlates involved in the processing of vibratory information to improve postural stability remain unclear. This chapter aims to understand the cortical mechanism of balance improvement with vibrotactile feedback in transfemoral amputees.

Somatosensation is generally mediated in part by the somatosensory and posterior parietal cortices of the brain. They underlie the ability to perceive the vibratory stimulus, create meaning about the sensations, and formulate the body actions related to those sensations [13]. The contribution of the primary somatosensory cortex (S1) to the perception of vibration stimulus on the body has been well documented in some former neuroscientific findings [14, 15]. However, a recent study suggests that the S1 is not always involved in vibrotactile detection and depends upon the task demands [16]. Specifically, the S1 is necessary to build up and maintain representations of tactile stimuli (e.g., stimulus frequency or intensity) in both tactile detection and discrimination tasks. On the contrary, there is persuasive evidence that suggests that the secondary somatosensory cortex (S2) is critically involved in the neural processing for high frequency (100–400 Hz) vibrotaction [17]. It has also been proved that the Lateral Sulcus of S2 highlighted more frequency-dependent voxels than S1 when vibrotactile stimulation with different frequencies (20–200 Hz) was applied on the index finger [18]. In addition to this, for the execution of motor response necessary for maintaining the postural stability, the fronto-central region was found to play a crucial role [19–21]. The spectral power and event-related potentials of various EEG rhymes were found to vary over the frontal and midline-central regions to execute the postural stability during the balance control [19, 22–23]. Therefore, the consideration of the S2 for perceiving and processing vibratory information and the fronto-central region for executing the motor response to achieve postural stability can be done in the present case of study. Additionally, functional connectivity between the two regions of interest (secondary somatosensory cortex and fronto-central region) will also be investigated to find if they are functionally connected during the perception of vibrotactile feedback.

11.2 METHODOLOGY

11.2.1 Designing of the Vibrotactile Feedback System

The vibrotactile feedback system is designed to provide proprioception feedback to transfemoral amputees through vibrotactile stimulation on their stump (residual thigh). The stimulation strategy reflects the center of pressure (CoP) displacements under the prosthetic foot. The CoP is determined from the measurements made by the in-house developed pair of insoles with an array of force-sensing resistors (FSRs) [24]. A weighted average formula was utilized to calculate the x and y coordinates of CoP of each foot as given in equation 11.1.

$$COP(x,y) = \left(\frac{\sum F_i x_i}{\sum F_i}, \frac{\sum F_i y_i}{\sum F_i} \right) \tag{11.1}$$

Where F_i is the force at the ith force-sensitive resistor (FSR) and x_i and y_i are the coordinates of the ith FSR. An Arduino microcontroller reads the force data from the insole, uses it to calculate the CoP, and then determines the vibratory feedback to be applied. With this information, the amputee can consciously make adjustments in their balance, allowing them to maintain better stability and prevent falls.

The system consists of three main components i.e., the insoles, the data acquisition unit, and the vibrating tactors as shown in figure 11.1. Two tactors (Vibration Motor 310–113, Precision Microdrives Ltd, London, U.K.) were positioned on the front (towards quadriceps muscles) and backside (towards hamstrings muscles) of

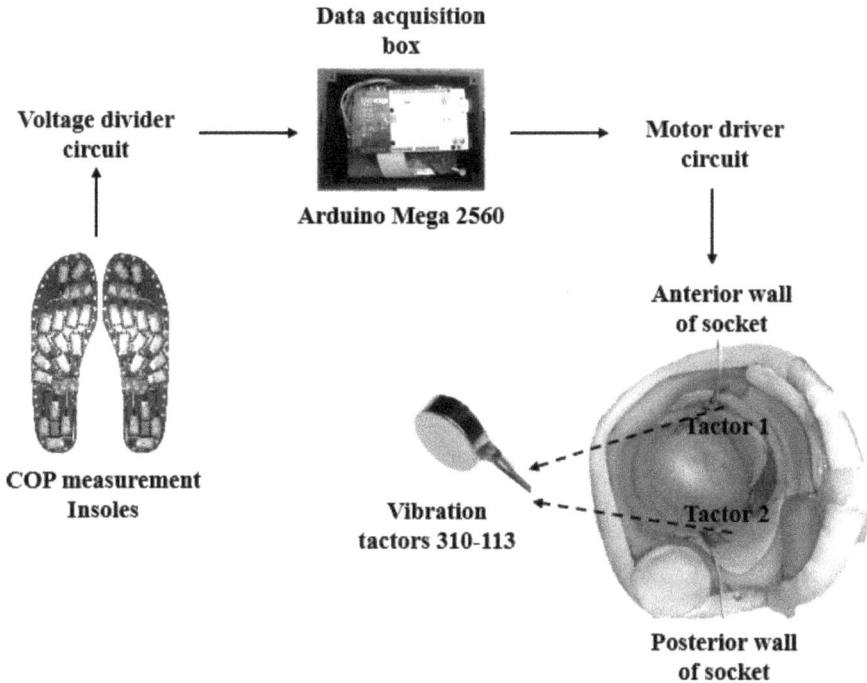

FIGURE 11.1   Flow diagram of the system.

the prosthesis's socket (figure 11.1). The placement of tactors was done in such a way that they should be at 2/3rd length of the stump from the greater trochanter. These tactors were instrumented with the data acquisition system of insoles to determine the feedback.

For feedback, the vertical length of the insole (prosthetic foot) was divided into three zones i.e., posterior zone (20%), mid zone (30%), and anterior zone (50%) (figure 11.2 a). This was following the prior literature where it was shown that the CoP displacement during normal standing occurs at 36.6 ± 7.8% of the foot length in anterior-posterior direction starting with 0% at the heel edge and 100% at the toe edge [25]. This includes 28.8–44.4% of the CoP range which constitutes the midfoot in healthy individuals. To accommodate the amputee population, the midfoot range was extended to 20–50% of foot length for normal standing. These divisions were used to operate the two vibratory tactors based on the displacement of CoP under the prosthetic foot.

If $CoP_Y < 20\%$ of foot length—posterior tactor (2) ON.

$CoP_Y > 20\%$ and $< 50\%$ of foot length—all tactors OFF.

$CoP_Y > 50\%$ of foot length—anterior tactor (1) ON. Therefore, by constantly mapping the COP during the balance tasks vibratory feedback was perceived by the participants.

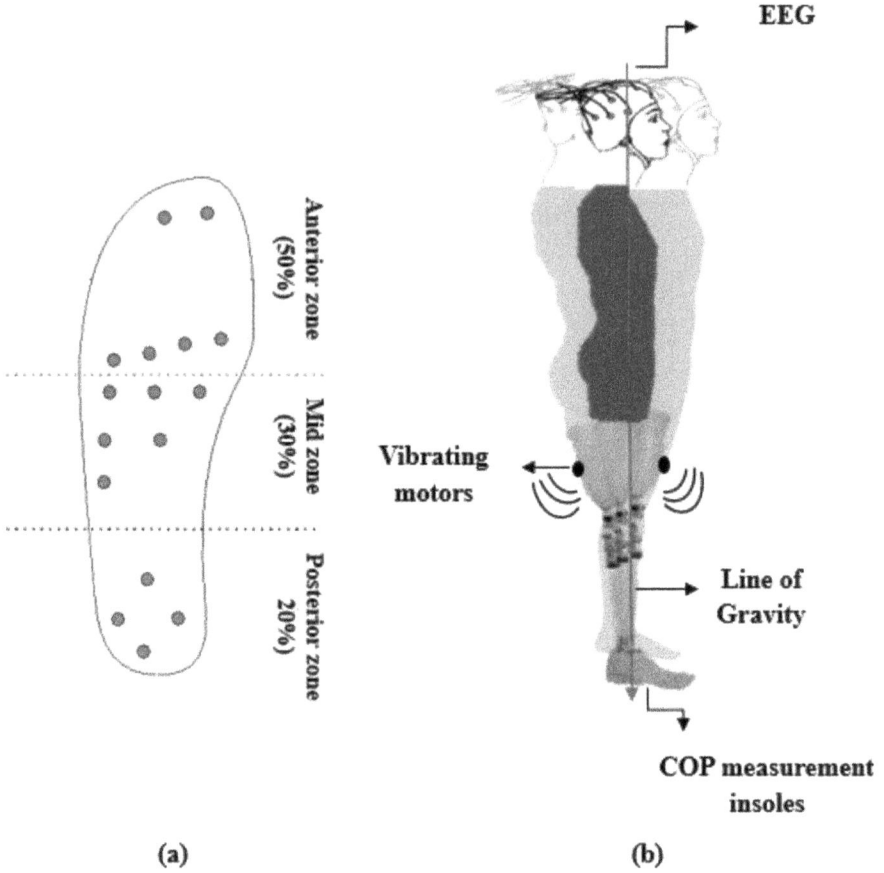

FIGURE 11.2 (a) Division of insole in different zones for vibratory feedback; (b) amputee participant performing the voluntary sway balance task with vibrotactile feedback.

11.2.2 ELECTROENCEPHALOGRAPHY AND ITS SYNCHRONIZATION WITH INSOLES

The EEG data were acquired wirelessly with 64-channel Ag/AgCl active electrode EEG setup (actiCHamp, Brain Products GmbH, Germany). The sampling frequency was kept at 1000 Hz and the electrodes were placed according to the 10% electrode placement system. Channel Fz was kept as an online reference and a conductive EEG gel was applied underneath each electrode to keep the resistance below 20 Kilo ohms to provide a high signal-to-noise ratio. A USB to TTL (PL2303HX) serial converter was used with the instrument control toolbox in MATLAB 2020b for the time synchronization between EEG and CoP data. The USB signal was converted to $0 \rightarrow +5$volt (max 10 mA) TTL signal through which digital markers were placed between the data to synchronize both the recordings.

11.2.3 Experimental Protocols

Six traumatic, right-sided transfemoral amputees were directed to perform a balance control task. The participants were first instructed to stand normally for 5 seconds, then lean forward (anterior) as much as they could for 10 seconds, and then come to a normal upright stance. Similarly, they were directed to lean backward (posterior) for 10 seconds while keeping in mind to not bend their hip joints and to not raise their toes or heels (figure 11.2 b). They were then instrumented with the wireless EEG system and Center of pressure (CoP) measurement insoles along with a vibrotactile feedback system. Data was recorded for two sessions in eyes-closed condition, with and without feedback.

11.2.4 Data Processing

The CoP data collected from insoles at a sampling rate of 100 Hz was filtered with a 4th order low-pass Butterworth filter with a cut-off frequency of 10 Hz to eliminate the measurement noise [26, 27]. Limit of stability (LOS) was derived from the displacement of the center of pressure (CoP) during the voluntary sway in the anterior and posterior direction. LOS was estimated by subtracting the average value of CoP_{NS} (CoP during normal standing) from the average value of CoP_{VS} (CoP during voluntary sway). The data were normalized with the average of sound and prosthetic foot size. Equation 11.2 shows the LOS calculation formula.

$$LOS(\%) = \mid \frac{COP_{VS} - COP_{NS}}{Foot\ size} \mid \times 100 \qquad (11.2)$$

The acquired EEG data were analyzed offline with MATLAB2020b using custom scripts and incorporating functions from EEGLAB software [28]. Firstly, the data were down sampled to 100 Hz and then filtered at 0.1 Hz using 4th order Butterworth filter with an additional notch filter to remove the line noise (50 Hz). Then artifact subspace reconstruction (ASR) was applied to detect and reconstruct the artifactual sections in EEG data [29]. The data were then re-referenced to the common average. Adaptive mixtures independent component analysis (AMICA) was used to estimate the source level activity [30]. A three-shell boundary element head model included in the DIPFIT toolbox within EEGLAB was used and an equivalent current dipole matched to the scalp projections of each independent component (IC) was estimated [30]. Each IC scalp projection, location of its equivalent dipole, and power spectrum was inspected and the ICs related to brain activity (typical EEG power spectra and equivalent current dipole located inside of the head and with low residual variance (< 15%)) were selected [31]. All the ICs were clustered according to the spatial location of their associated equivalent current dipole at the region of interest, their power spectral density (between 1 and 50 Hz), and the topography of their scalp projection. The clustering routine relied on the k-means algorithm. Lastly, data for each trial were segmented into the trials during which the subjects were instructed to sway in the interval −5 to 10 seconds with a baseline latency of −5 seconds. The spectrograms showing the power modulations of multiple cortical rhythms with time and frequency were evaluated finally for both sessions.

For estimating the functional connectivity, phase-locking value (PLV) was extracted from the electrode pairs constituting the fronto-central region (F1, F2, FC1, FCz, FC2) and the secondary somatosensory cortex (CP1, CP3, CP5, P1, P3, P5). All the pairs of combinations were compared in the presence and absence of vibrotactile feedback during the balance task.

11.2.5 STATISTICAL ANALYSIS

The statistical evaluations between the two sessions were carried out by paired T-test and the significance level was set at $\alpha = 0.05$. To avoid the multiple comparison problems, the significance level was corrected by Bonferroni correction. At the fronto-central region, the significance level was set at $\alpha = 0.005$ (0.05/10) and at the somatosensory cortex the significance level was set at $\alpha = 0.0033$ (0.05/15). Similarly, for the inter-regional functional connectivity analysis, the significance level was set at $\alpha = 0.0016$ (0.05/30).

11.3 RESULTS

Results show that vibrotactile sensory feedback effectively helped the participants to improve their stability. With vibrotactile feedback, the LOS of the sound foot was found to be significantly higher (t-test, $p = 0.01$) than in no feedback sessions during forward voluntary sway (figure 11.3). Additionally, two clusters were evaluated containing independent components from all the participants. In Cluster 1 i.e., secondary somatosensory cortex contralateral to amputated side, average spectral power reduces significantly (shown in blue, $p < 0.05$) in the theta band (3–7 Hz) during

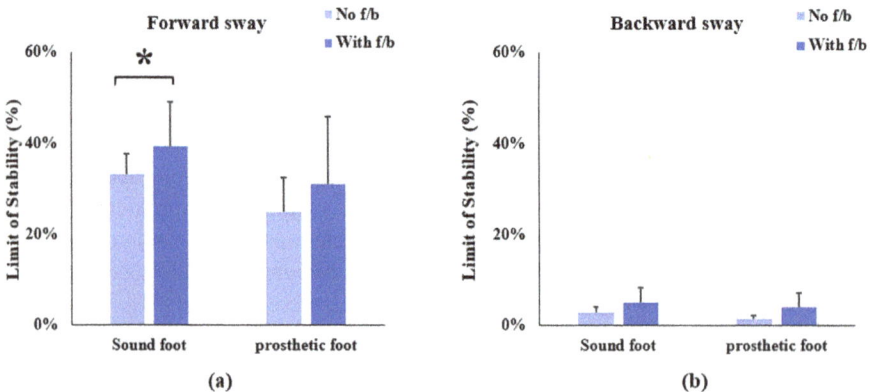

FIGURE 11.3 Limit of stability (LOS) during (a) forward and (b) backward sway between the two sessions averaged across the participants' error bars indicate the standard deviation; *$p = 0.01$.

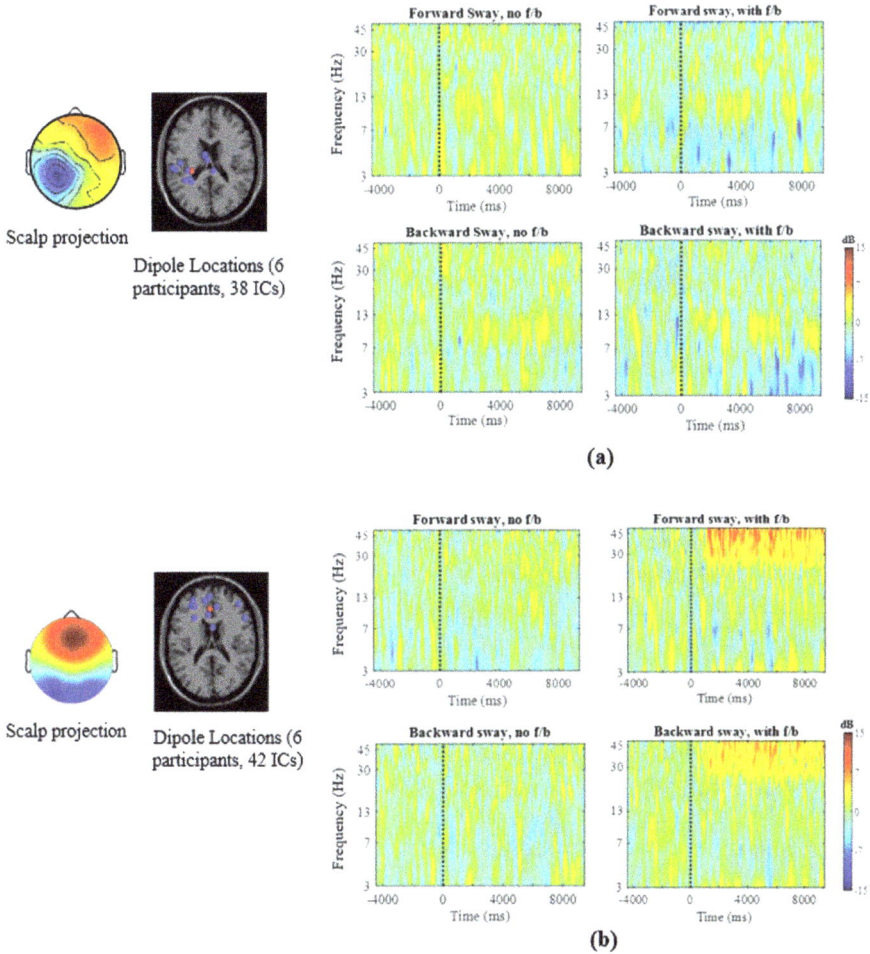

FIGURE 11.4 Time-frequency maps at (a) secondary somatosensory cortex contralateral to amputated side and (b) fronto-central region during the forward and backward sway in both the sessions.

voluntary sway in both forward and backward direction with vibrotactile feedback in comparison to no feedback sessions (figure 11.4 a). Similarly, at Cluster 2 i.e., fronto-central region average spectral power increases significantly (shown in red, $p < 0.05$) in the gamma band (30–50 Hz) during voluntary sway with vibrotactile feedback in both forward and backward directions in comparison to no feedback sessions (figure 11.4 b).

The inter- and intra-regional functional connectivity is shown in figure 11.5. The pairs of gamma connectivity i.e., FCz-FC2, FCz-F2, FC2-F2, and theta connectivity namely: CP3-P3, P5-P3, P1-P3 were found to be significantly stronger in the presence of vibrotactile feedback, compared to the case when no feedback was applied (t-test, $p < 0.003$).

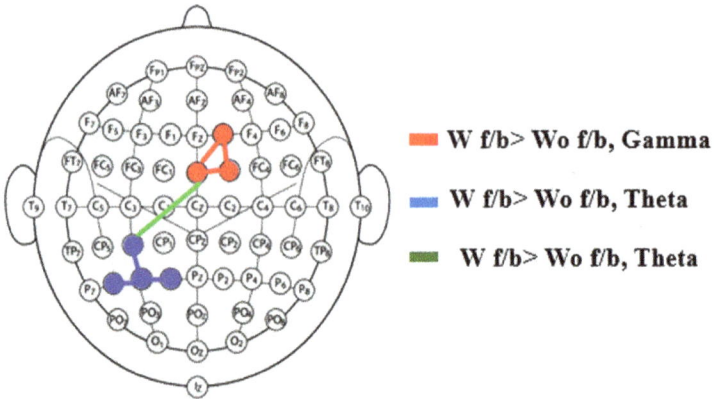

FIGURE 11.5 Functional connectivity at fronto-central region and secondary somatosensory cortex. Red and blue lines indicate the intra-regional functional connectivity in gamma and theta band resp. The inter-regional FC between the two sessions is indicated in the green line. *W f/b > Wo f/b* denote that the connectivity in the presence of vibrotactile feedback is stronger than in the absence of feedback.

11.4 DISCUSSION

The results demonstrated that the improved limit of stability with vibrotactile feedback was in response to theta and gamma oscillations. Gamma oscillations were found in the fronto-central region and theta oscillations were found to be associated with secondary somatosensory cortex contralateral to amputated side. Also, the two regions of interest were strongly connected while perceiving the vibratory feedback. The S2 processes the sensory feedback by significantly reducing the theta spectral power and the fronto-central region executed the motor response to improve the LOS by significantly increasing the gamma power.

The LOS was improved with vibratory feedback at the sound foot of participants during forward voluntary sway. The non-significant improvement at the prosthetic foot was due to the fixed ankle joint of the SACH prosthetic foot. Hence, due to the rigid nature of the prosthetic foot, the participants increase their sound foot LOS. In a recent study [32], it has been proved that the vibrotactile feedback was more effective in transfemoral amputees using a movable ankle joint compared with a fixed ankle joint. Therefore, in the future, the variations in postural performance and cortical activity can also be evaluated for amputees with different prosthetic foot designs (single-axis, multi-axis, energy storage, return type, etc.). In the backward direction, the participants may have more fear of falling due to which they were not able to improve their stability.

The secondary somatosensory cortex was shown to process the different sensory inputs [33, 34]. The reduction in theta power found within the contralateral S2 confirms that the vibratory feedback was well perceived by the participants. The gamma synchronization in the fronto-central region with vibrotactile feedback has been associated with motor execution to improve LOS [35]. It has also been proved that the frontal cortex plays a critical role in tactile decision processing [36]. Therefore, the perceived vibratory information in the secondary somatosensory cortex reported through theta desynchronization might have triggered the fronto-central region to improve the stability by indicating gamma synchronization.

The increased intra-regional functional connectivity at the fronto-central region indicates the activation of brain networks related to movement preparation. This also shows that there is active communication at the fronto-central area and the presence of vibrotactile feedback develops precise voluntary movements which leads to increased LOS. Similarly, the increased theta connectivity at the contralateral S2 was due to the processing of sensory information during the balance task. With vibrotactile feedback, the functional connectivity (CP3-FCz) becomes stronger in theta range during forward voluntary sway suggesting that the relevant sensory information is sent to the fronto-central region from the somatosensory cortex which results in the improvement of LOS.

Overall, through this study, the cortical mechanism of balance improvement in transfemoral amputees has been well understood. As the sensory information is perceived in the contralateral S2, theta spectral power decreases. Gamma power enhancement within the fronto-central region was due to the execution of motor response for improved LOS. Through functional connectivity analysis, stronger theta connectivity was found between the contralateral secondary somatosensory cortex and fronto-central region in presence of vibrotactile feedback. Therefore, by using vibratory feedback in the prosthesis and externally stimulating the highlighted brain regions, neuroplasticity might be promoted [37] which leads to the reduction in the training time for the efficient rehabilitation of amputees.

REFERENCES

[1] Geurts AC, Mulder TW, Nienhuis B, Rijken RA, Postural reorganization following lower limb amputation. Possible motor and sensory determinants of recovery. *Scand J Rehabil Med* 1992; 24: 83–90.

[2] Isakov E, Mizrahi J, Susak Z, Ona I, Hakim N, Influence of prosthesis alignment on the standing balance of below-knee amputees. *Clin Biomech* 1994; 9: 258–262.

[3] Barnett CT, Vanicek N, Polman RCJ, Postural responses during volitional and perturbed dynamic balance tasks in new lower-limb amputees: a longitudinal study. *Gait Posture* 2013; 37: 319–325.

[4] Dulmen SV, Sluijs E, Dijk LV, Ridder D, Heerdink R, Bensing J, Patient adherence to medical treatment: a meta-review. *BMC Health Serv Res* 2007; 7.

[5] Petzinger GM, Fisher BE, Van Leeuwen JE, Vukovic M, Akopian G, Meshul CK, Holschneider DP, Nacca A, Walsh JP, Jakowec MW, Enhancing neuroplasticity in the basal ganglia: the role of exercise in Parkinson's disease. *Mov Disord* 2010; 25: S141–S145. https://doi.org/10.1002/mds.22782

[6] Sigrist R, Rauter G, Riener R, Wolf P, Augmented visual, auditory, haptic, and multi-modal feedback in motor learning: a review. *Psychon Bull Rev* 2013; 20: 21–53.

[7] Giggins OM, Persson UM, Caulfield B, Biofeedback in rehabilitation. *J Neuroeng Rehabilitation* 2013; 10: 60.

[8] Huang H, Wolf SL, He J, Recent developments in biofeedback for neuromotor rehabilitation. *J Neuroeng Rehabilitation* 2006; 3. https://doi.org/10.1186/1743-0003-3-11

[9] Crea S, Cipriani C, Donati M, Carrozza MC, Vitiello N, Providing time-discrete gait information by wearable feedback apparatus for lower-limb amputees: usability and functional validation. *IEEE Trans Neural Syst Rehabil Eng* 2015; 23 (2): 250–257.

[10] Vimal AK, Godiyal AK, Singh U, Bhasin S, Joshi D, Transfemoral amputee's limit of stability and sway analysis during weight shifting exercise with a vibrotactile feedback system. *Somatosens Mot Res* 2019; 36: 31–41.

[11] Rusaw D, Hagberg K, Nolan L, Ramstrand N. Can vibratory feedback be used to improve postural stability in persons with transtibial limb loss? *J Rehabil Res Dev* 2012; 49: 1239–1254.

[12] Khajuria A, Joshi D, Effects of vibrotactile feedback on postural sway in transfemoral amputees: a wavelet analysis. *J Biomech* 2021; 115: 110145. https://doi.org/10.1016/j.jbiomech.2020.110145.

[13] Romo R, Hernández A, Salinas E, et al. From sensation to action. *Behav Brain Res* 2002; 135: 105–118.

[14] Morley JW, Vickery RM, Stuart M, Turman AB, Suppression of vibrotactile discrimination by transcranial magnetic stimulation of the primary somatosensory cortex. *Eur J Neurosci* 2007; 26: 1007–1010.

[15] Romo R, Hernández A, Zainos A, Brody C, Salinas E, Exploring the cortical evidence of a sensory-discrimination process. *Philos Trans R Soc Lond Ser B Biol Sci* 2002; 357: 1039–1051.

[16] Luigi T, Holmes NP, Involvement of human primary somatosensory cortex in vibrotactile detection depends on task demand. *NeuroImage* 2016; 138: 184–196. https://doi.org/10.1016/j.neuroimage.2016.05.056.

[17] Ryun S, Kim JS, Lee H, Chung CK, Tactile frequency-specific high-gamma activities in human primary and secondary somatosensory cortices. *Sci Rep* 2017; 7: 15442. https://doi.org/10.1038/s41598-017-15767-x

[18] Chung YG, Kim J, Han SW, Kim HS, Hyun Choi M, Chung SC, Park JY, Kim SP, Frequency-dependent patterns of somatosensory cortical responses to vibrotactile stimulation in humans: a fMRI study. *Brain Res* 2013; 1504: 47–57. https://doi.org/10.1016/j.brainres.2013.02.003.

[19] Semyon Slobounov, Mark Hallett, Cheng Cao and Karl Newell, Modulation of cortical activity as a result of voluntary postural sway direction: an EEG study. *Neurosci Lett* 2008; 443: 309–313.

[20] Varghese JP, Marlin A, Beyer KB, Staines WR, Mochizuki G, McIlroy WE, Frequency characteristics of cortical activity associated with perturbations to upright stability. *Neurosci Lett* 2014; 578: 33–38.

[21] Varghese JP, Beyer KB, Williams L, Miyasike-daSilva V, and McIlroy WE, Standing still: Is there a role for the cortex? *Neurosci Lett* 2015; 590: 18–23.

[22] Goel R, Ozdemir RA, Nakagome S, Contreras-Vidal JL, Paloski WH, Parikh PJ, Effects of speed and direction of perturbation on electroencephalographic and balance responses. *Exp Brain Res* 2018; 236: 2073–2083.

[23] Hulsdunker T, Mierau A, Neeb C, Kleinoder H, Struder HK, Cortical processes associated with continuous balance control as revealed by EEG spectral power. *Neurosci Lett* 2015; 592: 1–5.

[24] Tiwari A, Joshi D, Template-based insoles for the center of pressure estimation in different foot sizes. *IEEE Sens Lett* 2020; 4: 1–4. https://doi.org/10.1109/LSENS.2020.3010373.

[25] Jonsson E, Henriksson M, Hirschfeld H, Does the functional reach test reflect stability limits in elderly people? *J Rehabil Med* 2003; 35: 26–30.

[26] Ruhe A, Fejer R, Walker B, The test-retest reliability of center of pressure measures in bipedal static task conditions—a systematic review of the literature. *Gait Posture* 2010; 32: 436–445.

[27] Kirchner M, Schubert P, Schmidtbleicher D, Haas CT, Evaluation of the temporal structure of postural sway fluctuations based on a comprehensive set of analysis tools. *Physica A* 2012; 391: 4692–4703.

[28] Delorme A, Makeig S, EEGLAB: an open-source toolbox for analysis of single-trial EEG dynamics including independent component analysis. *J Neurosci Methods* 2004; 134: 9–21.

[29] Chang C, Hsu S, Pion-Tonachini L, Jung T, Evaluation of artifact subspace reconstruction for automatic artifact components removal in multi-channel EEG recordings. *IEEE Trans Biomed Eng* 2020; 67: 1114–1121. https://doi.org/10.1109/TBME.2019.2930186.

[30] Oostenveld R, Oostendorp TF, Validating the boundary element method for forward and inverse EEG computations in the presence of a hole in the skull. *Hum Brain Mapp*, 2002; 17: 179–192.

[31] Chaumon M, Bishop DVM, Busch NA, A practical guide to the selection of independent components of the electroencephalogram for artifact correction. *J Neurosci Methods* 2015; 250: 47–63. https://doi.org/10.1016/j.jneumeth.2015.02.025

[32] Vimal AK, Verma V, Khanna N, Joshi D, Investigating the effect of vibrotactile feedback in transfemoral amputee with and without movable ankle joint. *IEEE Trans Neural Syst Rehabilitation Eng* 2020; 28: 2890–2900. https://doi.org/10.1109/TNSRE.2020.3035833.

[33] Chen TL, Babiloni C, Ferretti A, Perrucci MG, Romani GL, Rossini PM, Tartaro A, Gratta CD, Human secondary somatosensory cortex is involved in the processing of somatosensory rare stimuli: an fMRI study. *NeuroImage* 2008; 40: 1765–1771. https://doi.org/10.1016/j.neuroimage.2008.01.020.

[34] Hagiwara K, Ogata K, Hironaga N, Tobimatsu S, Secondary somatosensory area is involved in vibrotactile temporal-structure processing: MEG analysis of slow cortical potential shifts in humans. *Somatosensory Mot Res* 2020; 37: 222–232. https://doi.org/10.1080/08990220.2020.1784127

[35] Hardwick RM, Rottschy C, Miall RC, Eickhoff SB, A quantitative meta-analysis and review of motor learning in the human brain. *Neuroimage* 2013; 67: 283–297.

[36] Pleger B, Ruff CC, Blankenburg F, Bestmann S, Wiech K, Stephan KE, Capilla A, Friston KJ, Dolan RJ, Neural coding of tactile decisions in the human prefrontal cortex. *J Neurosci* 2006; 26: 12596–12601. https://doi.org/10.1523/JNEUROSCI.4275-06.2006

[37] He W, Fong PY, Hong Leung TW, Huang TZ, Protocols of non-invasive brain stimulation for neuroplasticity induction. *Neurosci Lett* 2020; 719: 133437. https://doi.org/10.1016/j.neulet.2018.02.045.

12 Assessing the Impact of Body Mass Index on Gait Symmetry of Able-Bodied Adults Using Pressure-Sensitive Insole

Maria Rashid, Asim Waris,
Syed Omer Gilani, Faddy Al-Najjar,
Amit N. Pujari, and Imran Khan Niazi

12.1 INTRODUCTION

Human locomotion is a primary component of an individual's functional independence in performing daily life activities. It involves a complex yet coordinated interplay of musculoskeletal system and nervous system. Normal walking, synonymously called gait symmetry, is the notion that bilateral aspects of the lower limbs are identical. The lack of limb coordination causes an asymmetric gait pattern which serves as an indicator of cognitive decline [1] and life quality [2].

The incidence of gait disorders commonly increases with age. Gait parameters of the elderly change visibly as compared to those of young adults [3], [4]. Elderly people face significant gait impairments with the prevalence increasing from 10% in people aged 60–69 years to more than 60% in people aged above 80 years [5]. However, gait related problems and falls are largely underdiagnosed and receive minimal attention [6].

Asymmetry of gait is usually a consequence of: a) distorted neural input such as motor cortex disorders [7]–[10]; b) morphological disparity such as limb-length inequality and amputation [11]–[13]; and c) functional disability such as task-difference between dominant and non-dominant limb [14], [15].

Apart from disorders, psychological and physiological factors also play an etiologic role in this asymmetry. Depression and anxiety negatively affect symmetry of gait with delayed rehabilitative measures leading to slow prognosis and long-term disability [16]. Moreover, abnormal body mass distribution as indicated by a body mass index (BMI) seemingly affects bilateral coordination of gait. Body mass index is a ratio used to calculate a person's body weight relative to height (weight/height^2) and traditionally used to characterize obesity [17]. Although balance and gait are two distinct clinical entities, they are often intertwined. An excessive body mass

DOI: 10.1201/9781003201137-15

disrupts body geometry and shifts balance by influencing biomechanics thus causing functional gait limitations [18]. Similarly, atypical lack of body mass is more likely to inflict changes on gait and postural control [19].

Several studies have established the effect of adult obesity on spatiotemporal gait parameters as markers of gait pathology [20]–[22]. It is noteworthy that gait pathologies associated with obesity have mainly been reported in adults, but adequate evidence also exists for children [23]–[25] and adolescents [26]. Specifically, walking of underweight children showed significantly reduced symmetry when compared to those of their obese counterparts [27]. Additionally, significant spatiotemporal asymmetry has also been observed in obese children [28]. Furthermore, overweight adolescents, in addition to obese ones, also exhibited a significant asymmetry in spatiotemporal gait parameters [29]. Putting obesity-related research aside, up till now no earlier studies investigated the influence of BMI (all four classes) and gait symmetry of the able-bodied adult population.

In lieu of sparse research on the aforementioned topic, we quantified temporal gait parameters of the able-bodied adult population. The recent development of the force-sensitive insoles allowed us to analyze gait parameters in limited space as opposed to commercial force platforms which are more appropriate for clinical usage because of their cost and requirement of trained personnel.

The first aim of this study was to develop an inexpensive and user-friendly system capable of extracting temporal features of the gait. The second aim was to quantify gait asymmetry across all four classes of varying body mass distribution.

12.2   MATERIALS AND METHODS

12.2.1   DEVICE DEVELOPMENT

A thin sheet of pellite was used to fabricate a 2 mm thick insole since it was cost-effective, biocompatible, and readily available with required flexibility. Force sensitive resistors (model: IMS C-20, type: piezoresistive, diameter: 26 mm) were used to map pattern of loading under the foot subject to their accurate and linear response. Foot sole is physiologically divided into four areas such as heel, lateral mid-foot, metatarsal, and hallux based on weight-bearing capabilities which adjust the body balance by supporting its weight. Plantar pressure measured at these four points can be used to extract anatomical and physiological components of the lower limbs [30]. Therefore, four sensors were placed at these points on each insole as illustrated in Figure 12.1.

12.2.2   PARTICIPANTS

A total of 50 participants from students and the neighborhood of the university took part in this study. All of them were screened against the inclusion and exclusion criteria. Seven of them were found ineligible because of recent fractures and muscle injuries whereas four were left-leg dominant and therefore excluded to avoid any interference in results. The remaining thirty-nine participants were healthy with no history of musculoskeletal injuries, neurological disorders, leg-length discrepancies,

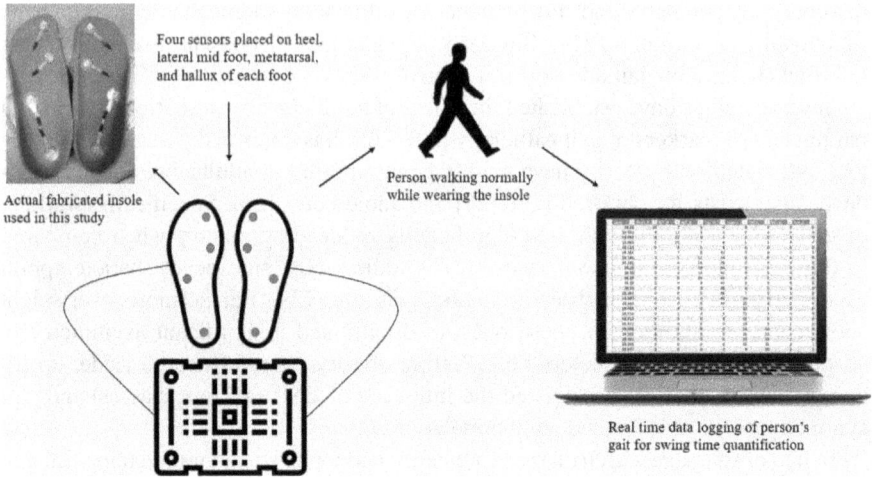

FIGURE 12.1   Schematic of the protocol used in this study.

TABLE 12.1
Demographics of the Study Population.

Class (BMI range)	Mean BMI (kg/m²)	Mean Age (Yrs.)	Gender (M, F)
I—Underweight (< 18.5)	16.8 ± 0.87	24.6 ± 0.5	(0,5)
II—Normal weight (18.5–24.9)	21.41 ± 1.71	24.6 ± 0.6	(10,5)
III—Overweight (25.0–29.9)	27.49 ± 1.55	24.6 ± 1.2	(7,8)
IV—Obese (30.0 +)	32.33 ± 1.55	36.5 ± 7.6	(1,3)

amputation, and idiopathic falling as verified by a registered medical officer (see demographics in Table 12.1). All the participants were recruited via convenience sampling but were retrospectively assigned numerical identifiers after being sorted into their respective BMI groups. The participants were provided with an informed written consent. The study was approved by the local ethical committee of the institute (NUST/ECLR/EXP/00567) and was carried out in accordance with rules of Declaration of Helsinki, 1975.

12.2.3 EXPERIMENTAL PROTOCOL

All the participants wore the apparatus weighing approximately 2 kg and walked at self-selected pace for 25 meters to conduct three consecutive trials. None of the subjects reported any discomfort by wearing the apparatus while normal walking. Real-time data logging was done by using Bluetooth module HC-05 at a baud rate of 115200 paired with Arduino Mega 2560 at a sampling frequency of 100 Hz and data were transferred to MATLAB R2013a (MathWorks®, Natick, MA, USA). By this system, we accurately mapped plantar pressures with reference to time corresponding to heel-strike and toe-off phase of subjects. Our aim was to measure asymmetry, so we calculated following previously described temporal variables as described by [31] separately for each foot.

1. Right swing time (SW_R): The time of right foot in the air averaged over three trials.
2. Left swing time (SW_L): The time of left foot in the air averaged over three trials.
3. Right swing variability: Co-efficient of variation of the right swing time.
4. Left swing variability: Co-efficient of variation of the left swing time.
5. Short (SSWT) and long (LSWT) swing time: Determined which foot has mean shorter and longer swing durations.
6. Gait asymmetry by formula described by [32], [33].

$$GA = 100 \times |\ln(SSWT / LSWT)| \qquad (12.1)$$

12.2.4 STATISTICAL ANALYSIS

One-way analysis of variance (ANOVA) was conducted to investigate the changes in swing time of all four classes. A probability value was set at $p < 0.05$ was used to infer the significance of this analysis. Levene's tests and normality checks were carried out and assumptions were met. Significant results were, then, followed by Hochberg's GT2 post-hoc test to cater for unequal sample size across all classes. All the results are reported as mean ± standard deviation from mean unless stated otherwise. Statistical analyses were performed using IBM®, SPSS®, Chicago, IL, USA.

12.3 RESULTS

12.3.1 DATA VISUALIZATION

We visualized the raw data of all eight sensors placed under both feet of subjects using a python algorithm. Comparative analysis of force patterns is evident from Figure 12.2. Exploratory data analysis shows the pressure shifting from different points under the foot following the pattern as Heel à Lateral midfoot (LM) à Metatarsal (MT) à Toe of right (R) and left foot (L), respectively. The outliers in 60 seconds of data occurred due to the dynamic nature of experimental protocol. These outliers were removed by using thresholding in MATLAB R2013a algorithm before applying statistics.

12.3.2 POTENTIAL PREDICTORS

We quantified the swing phase (foot not in contact with the ground) and calculated all the pre-defined variables to assess asymmetric walking of four classes. Table 12.2 summarizes the resulting variables with their mean values. Asymmetry, i.e.

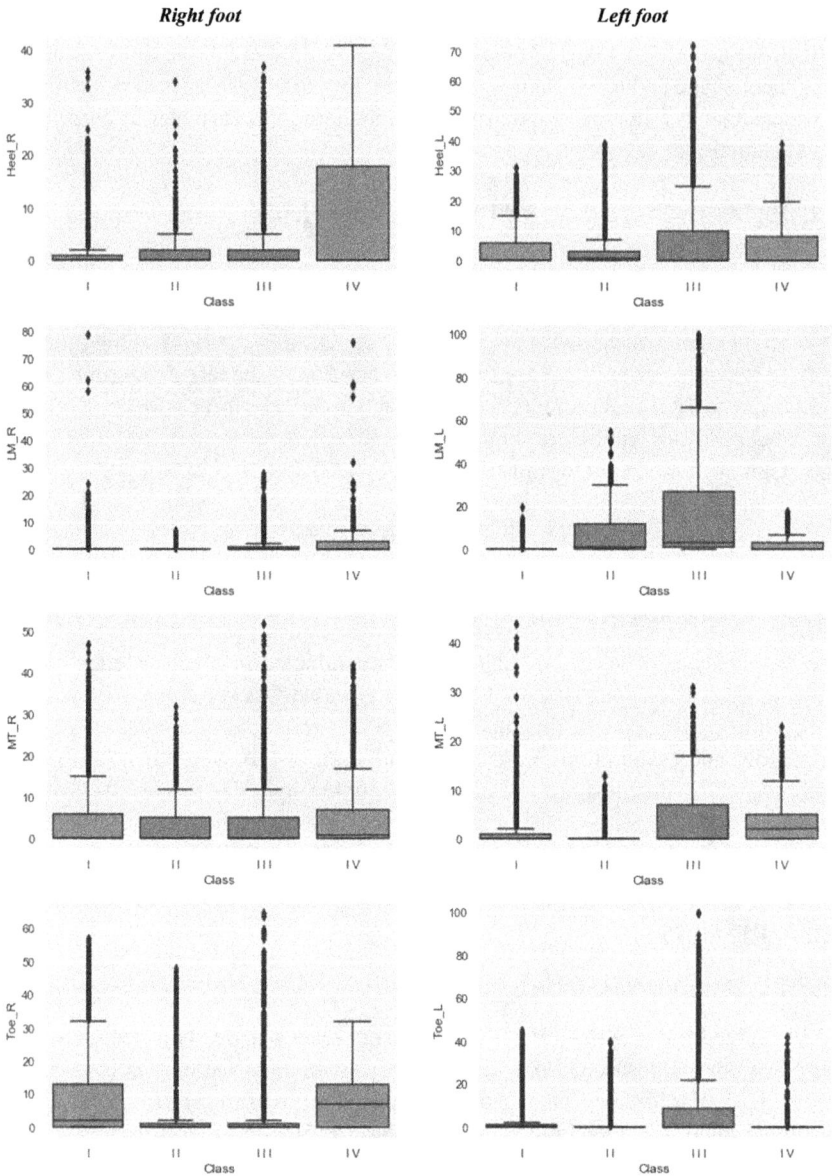

FIGURE 12.2 Data visualization of eight sensors at various points of right and left foot. X-axis represents participant classes and y-axis represents raw ADC values corresponding to change in pressure.

right-left limb coordination in gait increased as the difference between swing time of subjects increased. The values moved upwards and away from zero reflecting higher degrees of asymmetry (Figure 12.3). There exists no standard scale to assess asymmetry of adult able-bodied gait, but comparative analysis can be done. Earlier studies

TABLE 12.2

Average Temporal Gait of the Study Population. Values Are Expressed as Mean ± Standard Deviation.

Swing Time(sec)	Class I	Class II	Class III	Class IV
Right foot (SW_R)	0.44 ± 0.008	0.45 ± 0.03	0.45 ± 0.03	0.42 ± 0.06
Left foot (SW_L)	0.44 ± 0.005	0.45 ± 0.03	0.45 ± 0.04	0.41 ± 0.04
Short swing time (SSWT)	0.43 ± 0.04	0.45 ± 0.03	0.44 ± 0.04	0.39 ± 0.05
Long swing time (LSWT)	0.44 ± 0.03	0.46 ± 0.03	0.46 ± 0.03	0.45 ± 0.009
Gait asymmetry	1.82 ± 1.30	1.83 ± 1.28	4.53 ± 3.70	16.10 ± 12.51

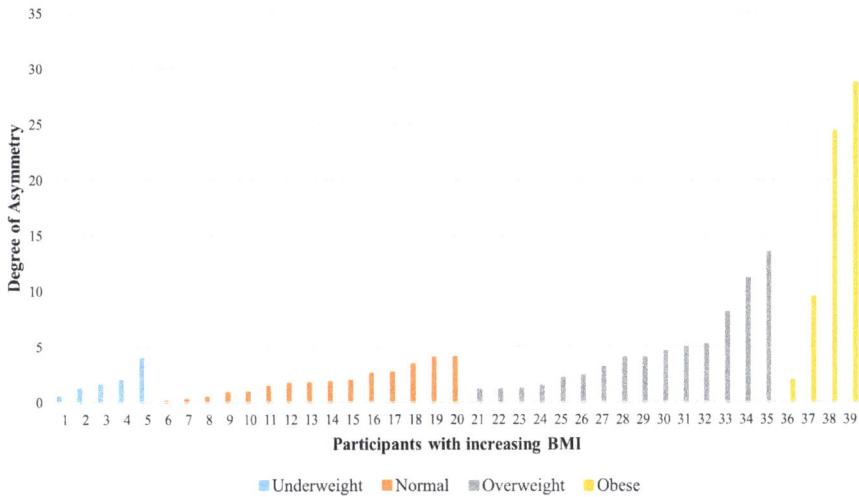

FIGURE 12.3   A column graph representing extent of asymmetry of the study population.

quantified the degree of asymmetries in the range of 55–60 for the patients suffering from severe neuro-degenerative diseases and visibly pathologic gait [31].

12.3.3   INTER-CLASS DIFFERENCES

Swing times of underweight and normal weight classes were quite consistent across both the limbs depicting symmetric walking patterns. Overweight class showed characteristic differences of right-left coordination among some subjects with varying BMIs, but the differences were not quantifiable at this time. However, marked changes were observed in the right-left swing times of obese class. Additionally, obese population showed distinct short and long swing time differences between both the limbs reflecting asymmetric gait (see Figure 12.4).

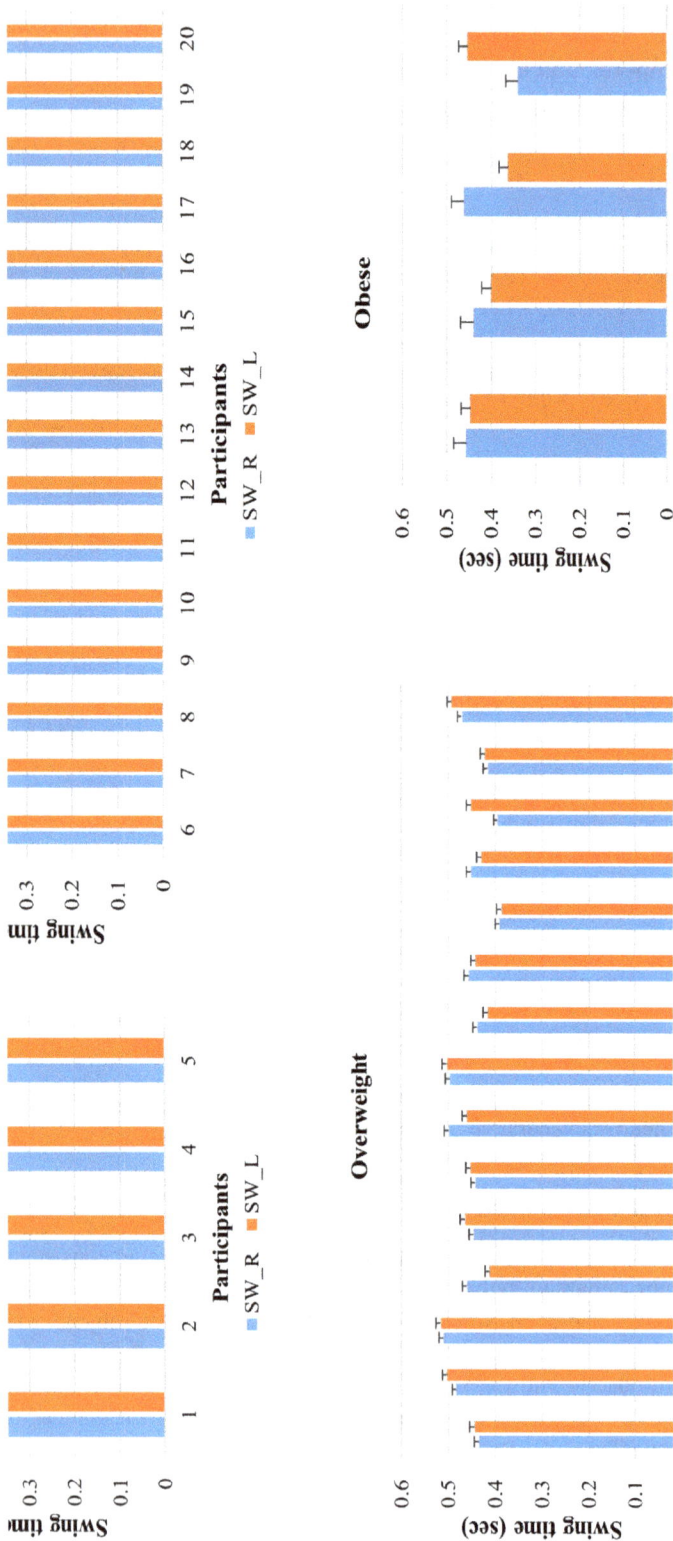

FIGURE 12.4   Column chart comparing mean swing time ± standard deviation of right and left foot of four classes.

12.3.4 STATISTICS

Since our sample pool consisted of a variable male to female ratio, we report the results without taking gender into the account. A one-way analysis of variance (ANOVA) returned the significant result as $F(3, 35) = 6.141$, $p = 0.002$ depicting that swing time was markedly changed with abnormal body masses. The post hoc analysis with Hochberg's GT2 test revealed that obese class showed highly asymmetric gait pattern as compared to that of normal and underweight class. The degree of asymmetry as calculated by employing swing time in obese class was approximately 13.5 and 13.7 times higher as compared to that of the underweight ($p = 0.023$) and normal weight class ($p = 0.004$) respectively. However, no significant asymmetry values were observed between overweight and obese classes in post-hoc comparison despite their initial higher swing time differences. As evident from Figure 12.3, the underweight and normal class represented index values below 5 as calculated by using equation 12.1 whereas the other two classes showed varying degrees of asymmetry with values lying farther away from zero.

12.4 DISCUSSION

The purpose of this study was to estimate gait asymmetry of able-bodied subjects in relation to their body mass index by using swing time. In this study, we were able to determine the right-left walking asymmetry of subjects with a significance of $p < 0.05$ among four classes of BMI.

The results further revealed that swing phase of the underweight and normal weight class were quite consistent reflecting a symmetric gait as compared to that of the overweight and obese class (see Figure 12.4). In overweight class, subjects with weights lying on the borderline of obesity showed higher degrees of asymmetry as compared to those who are relatively close to normal weight category. However, no significant asymmetric gait differences were observed in overweight class when compared to all other groups. We believe that a non-significant result as seen in post-hoc analysis of overweight class could be due to the limited sample size of classes.

Obese class showed highly asymmetric gait with higher ratios of swing time differences as consistent with the evidence of published literature [20]. However, a marked difference of coordination was seen for one subject where the gait was near-to-perfectly symmetric as shown in Figure 12.3. The subject's body mass distribution was mildly disturbed since the BMI was on the borderline of obesity. The intermittent differences might have occurred due to the physical changes such as hip circumference, muscle mass distribution, or adoption of certain behavioral practices like dynamic control while walking [20].

In the current study, we investigated the possible use of BMI as a key factor for gait trainings. Certain physiological and anatomical factors are considered such as leg-length discrepancy [34]–[36], electromyographical activity [37]–[39], and neurological input [40] depending upon the need and expected outcome while body mass distribution given the least importance. Since trauma and amputation results in greater imbalance and altered body geometry, BMI also tends to change and so does the gait patterns [41], [42]. Therefore, muscle mass distribution must be counted while forming rehabilitation regimes as coherent with the findings of current study. However, we suggest further exploration of other spatiotemporal variables to establish this positive association.

For this analysis, we used a ratio described earlier in equation 12.1 to calculate the temporal right-left asymmetry and have seen values within a range of 0–30 [31]. Some variations in the aforementioned ratio also exist which are beyond the scope of this study [43]. Previous studies used an index to calculate this right-left coordination of the lower limbs, but their values were compared against an average value [44]. However, no standardized scale has been established to-date which could be used to assess or compare able-bodied gait asymmetry regardless of the calculation procedure used. This study was a step forward towards formation of a standard scale on which asymmetries can be compared.

Perfect symmetry is unlikely in normal as well as pathologic gait evaluation and a certain level of asymmetry must be considered normal [15]. In our opinion, the lack of a standard scale or comparison criteria of gait asymmetry in able-bodied population is the root cause of sparse research in this area. Therefore, it is necessary to classify able-bodied gait asymmetries based on the phenomenological features to provide more objective evaluations of gait.

Clinically, certain high-frequency and high-resolution non-wearable sensors are mostly used which provide reliable measures of spatiotemporal gait parameters [45]. However, their advanced methodology and operation in a controlled research facility and cost makes them limited. Hence, a user-friendly, portable, and cost-effective system was fabricated for the purpose of this study requiring limited operational space.

We would like to acknowledge some of the limitations of this research. The influence of gender has not been considered since the number of males and females differed slightly among classes. Earlier studies emphasized and signified gender-based analysis of this right-left asymmetry [46], [47]. Moreover, the contrasting effects of dominant vs. non-dominant limb and their role in carrying out forward propulsion was also overlooked since all the participants were right-leg dominant [15], [48]. Further, mean age-difference between classes could be reduced further to describe any potential influence of age along with BMI on gait asymmetry.

We also observed some marked temporal changes in low-weight and normal-weight categories, otherwise invisible to the naked eye, which might have occurred due to an inherent factor. We suggest an increase in sample size and a broader inclusion criterion to yield significant and quantifiable results. We also propose that another factor such as body fat percentage might be considered to further validate the effects of excessively high or low body mass distribution on walking symmetry.

Since the primary focus of all healthcare research has been shifted to the etiology and treatment of coronavirus disease (COVID-19), we must, also, consider the importance of these handy protocols which can operate with limited technology and personnel. Seemingly, COVID-19 disproportionately affects overweight and obese individuals meanwhile long-COVID patients experience fatigue as a common symptom [49]. The said device could be improved in some form to cater for physical rehabilitation along with pulmonary rehabilitation of COVID-19 patients. Given the ongoing restrictions due to global pandemic, these systems must be validated and made commercially available to limit social interaction and ensure a more effective approach of exercising standard operating procedures.

12.5 CONCLUSION

The symmetry in gait serves as an important measure in guiding clinical decisions and rehabilitation outcomes. As reported earlier, gait becomes pathologic with increasing weight in the obese population due to unequal body mass distribution. Our study established that the overweight and obese population face significant changes in temporal gait parameters. Such findings suggest that body mass index has an influence, to some degree, over gait quality of able-bodied population and thus holds key-factorial importance in able-bodied gait pathology.

REFERENCES

[1] J. Verghese, C. Wang, R. B. Lipton, R. Holtzer, and X. Xue, "Quantitative gait dysfunction and risk of cognitive decline and dementia," *J. Neurol. Neurosurg. Psychiatry*, vol. 78, no. 9, pp. 929–935, 2007.

[2] M. Hirvensalo, T. Rantanen, and E. Heikkinen, "Mobility difficulties and physical activity as predictors of mortality and loss of independence in the community-living older population," *J. Am. Geriatr. Soc.*, vol. 48, no. 5, pp. 493–498, 2000.

[3] M. Saito, et al., "An in-shoe device to measure plantar pressure during daily human activity," *Med. Eng. Phys.*, vol. 33, no. 5, pp. 638–645, 2011.

[4] J. Woo, S. C. Ho, J. Lau, S. G. Chan, and Y. K. Yuen, "Age-associated gait changes in the elderly: pathological or physiological?," *Neuroepidemiology*, vol. 14, no. 2, pp. 65–71, 1995.

[5] P. Mahlknecht, et al., "Prevalence and burden of gait disorders in elderly men and women aged 60–97 years: a population-based study," *PLoS ONE*, vol. 8, no. 7, p. e69627, 2013.

[6] L. Z. Rubenstein, et al., "Detection and management of falls and instability in vulnerable elders by community physicians," *J. Am. Geriatr. Soc.*, vol. 52, no. 9, pp. 1527–1531, 2004.

[7] A. L. Hsu, P. F. Tang, and M. H. Jan, "Analysis of impairments influencing gait velocity and asymmetry of hemiplegic patients after mild to moderate stroke," *Arch. Phys. Med. Rehabil.*, vol. 84, no. 8, pp. 1185–1193, 2003.

[8] M. Plotnik, Y. Dagan, T. Gurevich, N. Giladi, and J. M. Hausdorff, "Effects of cognitive function on gait and dual tasking abilities in patients with Parkinson's disease suffering from motor response fluctuations," *Exp. Brain Res.*, vol. 208, no. 2, pp. 169–179, 2011.

[9] J. W. Chow, S. A. Yablon, T. S. Horn, and D. S. Stokic, "Temporospatial characteristics of gait in patients with lower limb muscle hypertonia after traumatic brain injury," *Brain Inj.*, vol. 24, no. 13–14, pp. 1575–1584, 2010.

[10] M. A. van Loo, A. M. Moseley, J. M. Bosman, R. A. De Bie, and L. Hassett, "Test-retest reliability of walking speed, step length and step width measurement after traumatic brain injury: a pilot study," *Brain Inj.*, vol. 18, no. 10, pp. 1041–1048, 2004.

[11] J. R. Perttunen, E. Anttila, J. Sodergard, J. Merikanto, and P. V. Komi, "Gait asymmetry in patients with limb length discrepancy," *Scand. J. Med. Sci. Sport.*, vol. 14, no. 1, pp. 49–56, 2004.

[12] M. Aiona, K. P. Do, K. Emara, R. Dorociak, and R. Pierce, "Gait patterns in children with limb length discrepancy," *J. Pediatr. Orthop.*, vol. 35, no. 3, pp. 280–284, 2015.

[13] M. Schaarschmidt, S. W. Lipfert, C. Meier-Gratz, H. C. Scholle, and A. Seyfarth, "Functional gait asymmetry of unilateral transfemoral amputees," *Hum. Mov. Sci.*, vol. 31, no. 4, pp. 907–917, 2012.

[14] H. Sadeghi, P. Allard, and M. Duhaime, "Functional gait asymmetry in able-bodied subjects," *Hum. Mov. Sci.*, vol. 16, no. 2–3, pp. 243–258, 1997.

[15] H. Sadeghi, P. Allard, F. Prince, and H. Labelle, "Symmetry and limb dominance in able-bodied gait: a review," *Gait Posture*, vol. 12, no. 1, pp. 34–45, 2000.

[16] L. Sudarsky, "Psychogenic gait disorders," *Seminars in Neurology*, vol. 26, no. 3, pp. 351–356, 2006.

[17] T. J. Cole, J. V. Freeman, and M. A. Preece, "Body mass index reference curves for the UK, 1990," *Arch. Dis. Child.*, vol. 73, no. 1, pp. 25–29, 1995.

[18] S. A. F. De Souza, et al., "Gait cinematic analysis in morbidly obese patients," *Obes. Surg.*, vol. 15, no. 9, pp. 1238–1242, 2005.

[19] V. Cimolin, et al., "Gait analysis in anorexia and bulimia nervosa," *J. Appl. Biomater. Funct. Mater.*, vol. 11, no. 2, pp. 122–128, 2013.

[20] C. Ling, T. Kelechi, M. Mueller, S. Brotherton, and S. Smith, "Gait and function in class III obesity," *J. Obes.*, vol. 2012, 2012.

[21] N. Cau, et al., "Center of pressure displacements during gait initiation in individuals with obesity," *J. Neuroeng. Rehabil.*, vol. 11, no. 1, p. 82, 2014.

[22] V. Cimolin, L. Vismara, M. Galli, G. Grugni, N. Cau, and P. Capodaglio, "Gait strategy in genetically obese patients: a 7-year follow up," *Res. Dev. Disabil.*, vol. 35, no. 7, pp. 1501–1506, 2014.

[23] C. C. T. Clark, C. M. Barnes, M. Holton, H. D. Summers, and G. Stratton, "Profiling movement quality and gait characteristics according to body-mass index in children (9–11 y)," *Hum. Mov. Sci.*, vol. 49, pp. 291–300, 2016.

[24] V. J. Blakemore, P. W. Fink, S. D. Lark, and S. P. Shultz, "Mass affects lower extremity muscle activity patterns in children's gait," *Gait Posture*, vol. 38, no. 4, pp. 609–613, 2013.

[25] B. McGraw, B. A. McClenaghan, H. G. Williams, J. Dickerson, and D. S. Ward, "Gait and postural stability in obese and nonobese prepubertal boys," *Arch. Phys. Med. Rehabil.*, vol. 81, no. 4, pp. 484–489, 2000.

[26] V. Cimolin, et al., "Computation of spatio-temporal parameters in level walking using a single inertial system in lean and obese adolescents," *Biomed. Tech.*, vol. 62, no. 5, pp. 505–511, 2017.

[27] V. Cimolin, et al., "Symmetry of gait in underweight, normal and overweight children and adolescents," *Sensors (Switzerland)*, vol. 19, no. 9, 2019.

[28] A. Phills and A. W. Parker, "Gait characteristics of obese pre-pubertal children," *Int. J. Rehabil. Res.*, vol. 14, no. 4, pp. 348–349, 1991.

[29] J. S. Dufek, et al., "Effects of overweight and obesity on walking characteristics in adolescents," *Hum. Mov. Sci.*, vol. 31, no. 4, pp. 897–906, 2012.

[30] L. Shu, T. Hua, Y. Wang, Q. Li, D. D. Feng, and X. Tao, "In-shoe plantar pressure measurement and analysis system based on fabric pressure sensing array," *IEEE Trans. Inf. Technol. Biomed.*, vol. 14, no. 3, pp. 767–775, 2010.

[31] G. Yogev, M. Plotnik, C. Peretz, N. Giladi, and J. M. Hausdorff, "Gait asymmetry in patients with Parkinson's disease and elderly fallers: when does the bilateral coordination of gait require attention?," *Exp. Brain Res.*, vol. 177, no. 3, pp. 336–346, 2007.

[32] Y. R. Yang, Y. C. Chen, C. S. Lee, S. J. Cheng, and R. Y. Wang, "Dual-task-related gait changes in individuals with stroke," *Gait Posture*, vol. 25, no. 2, pp. 185–190, 2007.

[33] M. Plotnik, N. Giladi, Y. Balash, C. Peretz, and J. M. Hausdorff, "Is freezing of gait in Parkinson's disease related to asymmetric motor function?," *Ann. Neurol.*, vol. 57, no. 5, pp. 656–663, 2005.

[34] M. K. Seeley, B. R. Umberger, J. L. Clasey, and R. Shapiro, "The relation between mild leg-length inequality and able-bodied gait asymmetry," *J. Sport. Sci. Med.*, vol. 9, no. 4, pp. 572–579, 2010.

[35] J. R. Perttunen, E. Anttila, J. Sodergard, J. Merikanto, and P. V. Komi, "Gait asymmetry in patients with limb length discrepancy," *Scand. J. Med. Sci. Sport.*, vol. 14, no. 1, pp. 49–56, 2004.

[36] S. Khamis and E. Carmeli, "Relationship and significance of gait deviations associated with limb length discrepancy: a systematic review," *Gait Posture*, vol. 57, pp. 115–123, 2017.

[37] J. Jonsdottir, et al., "Concepts of motor learning applied to a rehabilitation protocol using biofeedback to improve gait in a chronic stroke patient: an A-B system study with multiple gait analyses," *Neurorehabil. Neural Repair*, vol. 21, no. 2, pp. 190–194, 2007.

[38] R. A. Miller, M. H. Thaut, G. C. McIntosh, and R. R. Rice, "Components of EMG symmetry and variability in parkinsonian and healthy elderly gait," *Electroencephalogr. Clin. Neurophysiol.—Electromyogr. Mot. Control*, vol. 101, no. 1, pp. 1–7, 1996.

[39] H. N. Abdul Halim, A. Azaman, H. Manaf, S. Saidin, I. Zulkapri, and A. Yahya, "Gait asymmetry assessment using muscle activity signal: a review of current methods," *J. Phys. Conf. Ser.*, vol. 1372, no. 1, 2019.

[40] G. M. Rozanski, A. H. Huntley, L. D. Crosby, A. Schinkel-Ivy, A. Mansfield, and K. K. Patterson, "Lower limb muscle activity underlying temporal gait asymmetry post-stroke," *Clin. Neurophysiol.*, vol. 131, no. 8, pp. 1848–1858, 2020.

[41] D. T. Burke, S. Al-Adawi, N. B. Jain, and D. P. Burke, "The effect of body mass index on rehabilitation of patients with amputation," *JPO J. Prosthetics Orthot.*, vol. 30, no. 4, pp. 202–206, 2018.

[42] D. E. Rosenberg, et al., "Body mass index patterns following dysvascular lower extremity amputation," *Disabil. Rehabil.*, vol. 35, no. 15, pp. 1269–1275, 2013.

[43] R. A. Zifchock, I. Davis, J. Higginson, and T. Royer, "The symmetry angle: a novel, robust method of quantifying asymmetry," *Gait Posture*, vol. 27, no. 4, pp. 622–627, 2008.

[44] R. O. Robinson, W. Herzog, and B. M. Nigg, "Use of force platform variables to quantify the effects of chiropractic manipulation on gait symmetry," *J. Manipulative Physiol. Ther.*, vol. 10, no. 4, pp. 172–176, 1987.

[45] S. L. Chien, et al., "The efficacy of quantitative gait analysis by the GAITRite system in evaluation of parkinsonian bradykinesia," *Park. Relat. Disord.*, vol. 12, no. 7, pp. 438–442, 2006.

[46] J. Heredia-Jimenez and E. Orantes-Gonzalez, "Gender differences in patients with fibromyalgia: a gait analysis," *Clin. Rheumatol.*, vol. 38, no. 2, pp. 513–522, 2019.

[47] J. Heredia-Jimenez, E. Orantes-Gonzalez, and V. M. Soto-Hermoso, "Variability of gait, bilateral coordination, and asymmetry in women with fibromyalgia," *Gait Posture*, vol. 45, pp. 41–44, 2016.

[48] J. Suk Seo and S. Kim, "Prevention of potential falls of elderly healthy women: gait asymmetry," *Educ. Gerontol.*, vol. 40, no. 2, pp. 123–137, 2014.

[49] L. Vimercati, et al., "Association between long COVID and overweight/obesity," *J. Clin. Med.*, vol. 10, no. 18, pp. 1–8, 2021.

13 Analysis of Lower Limb Muscle Activities during Walking and Jogging at Different Speeds

Ganesh R. Naik

13.1 INTRODUCTION

From daily activities to professional training, walking, jogging, and running are widely witnessed. The public takes part in jogging or running for health reasons or as a hobby and the professionals use jogging as exercise or for recovery from injury. However, only a few studies analyzed the suitability of jogging for recovery from injury or as a proper activity for routine exercise. On the contrary, Tsuji et al. find that jogging might induce greater strain on certain muscles than running [1]. Many studies research the EMG pattern of walking, jogging, or running. Nonetheless, no comparison can be found between these tasks. Jogging, therefore, can by no means be considered as a proper measure for both recovery and routine exercise [1–5].

With the variation of tasks, different muscles can contribute differently. Mellor and Hodges [6] suggest that quadriceps femoris supports the patellofemoral joint to maintain dynamic stability. They further found that the extension angle of knee joint requires the coordination between vastus medialis and vastus lateralis. Nene et al. [7] suggested that rectus femoris is responsible for the flexion of hip and knee. It is claimed in [8] that the ground reaction force can affect the muscles differently with speed of walking, as the foot loading move to the lateral side of the foot with decline of the walking speed. Unfortunately, comparison of lower limb muscle activities when performing walking, jogging, and running cannot be found in existing research articles.

This study aims to provide an initial understanding of muscle activity at different speeds, which consist of slow walking, fast walking, slow jogging, and fast jogging (running). Initially, EMG signals were quantified by using root mean square (RMS) and median frequency. Later, an inter-day comparison is also investigated in our study.

13.2 METHODS

13.2.1 Subjects

This study was carried out on six healthy male volunteers (age: 25.8 ± 2.1 years, weight: 72.3 ± 4.3 Kg, height: 1.79 ± 3.6 m, body mass index (BMI): 22.85 ± 1.0

 DOI: 10.1201/9781003201137-16

Kg/m², and leg length: 44.75 ± 1.5 cm). None of them had a history of lower limb surgery, or neurological or musculoskeletal disorders. The research ethics committee of the University of Technology Sydney approved the experimental protocol.

13.2.2   STUDY DESIGN

Participants were required to perform four tasks in this study, which are: slow walking, fast walking, slow jogging, and fast jogging. Detailed explanation of trials was given to participants prior to the EMG recording. The comfortable speed of four tasks for each individual was determined during the trials. For each task, the EMG signal was recorded for 6 minutes, and 5 minutes break was given between each task so that sudden muscle fatigue can be prevented. All participants were asked to perform the four tasks twice on two separate sessions. The first session was on one day and the second session was on the next day. Therefore, the inter-day difference can be investigated.

The order of the walking and jogging tasks was different between participants and the participants were randomly assigned to different combinations of walking and jogging. All tasks were performed at laboratory on a normal surface. All participants wore shorts and sport shoes during the experiment.

13.2.3   DATA ACQUISITION

Six lower-limb muscles were investigated in this study, which include: rectus femoris (RF), vastus medialis (VM), tibialis anterior (TA), lateralis gastrocnemius (LG), biceps femoris (BF), and semitendinosus (ST). The electrodes were placed only on the right leg and the configuration of electrodes was decided according to SENIAM guidelines (Figure 13.1) [9]. The skin of the electrode placement site was shaved and cleaned with alcohol to reduce impedance. Six bipolar silver-silver triode electrodes (diameter: 10 mm, inter-electrode distance: 20 mm, Thought Technology, Montreal, Quebec, Canada) were connected on the selected sites. The Flexcomp Infiniti encoder and Biograph Infiniti were used to convert and display sEMG signals.

13.2.4   DATA PROCESSING

Data were exported from Biograph Infiniti and then processed in MATLAB R2012a. A Butterworth filter with 20 to 450 Hz band pass was applied to reject any frequency outside this range, which is likely to be noise. Median frequency, mean frequency, and RMS of the EMG signal of each participant were calculated using MATLAB software. The average of RMS was also obtained to quantify EMG signals. Finally, the median frequency, mean frequency, and RMS were averaged separately over all four participants (pooled results) for different muscle and different task.

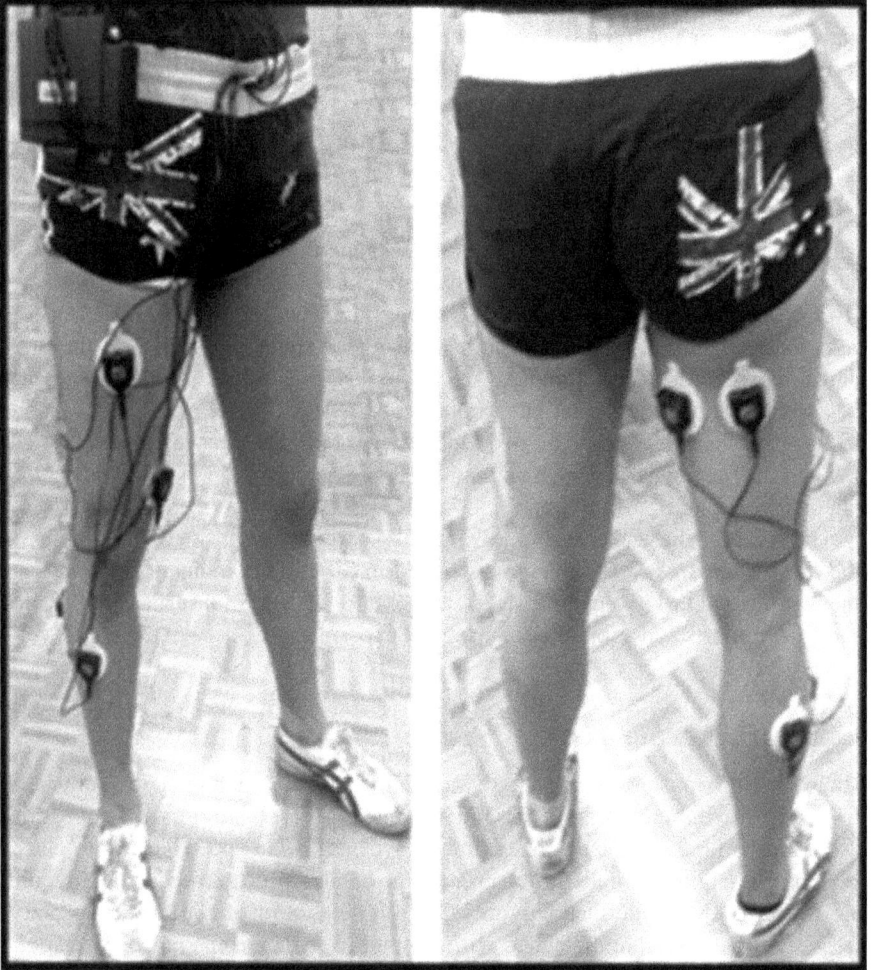

FIGURE 13.1   The configuration of electrodes.

13.3   RESULTS

13.3.1   ROOT MEAN SQUARE

For slow walking, the comparison of all data can be seen in Figure 13.2. All muscles, apart from BF, had the lowest RMS value than other three muscles (RF: 0.0135, VM: 0.03, TA: 0.037, LG: 0.049 and ST: 0.0355). BF had the highest averaged RMS (0.079) among investigated muscles. The second highest muscle was LG, followed by TA, ST, and VM. For fast walking, BF had an extremely high RMS (0.22), which was the highest among all four tasks. All other muscles had similar activities as slow walking, where RF was 0.029, VM was 0.033, TA was 0.044, LG was 0.054, and ST

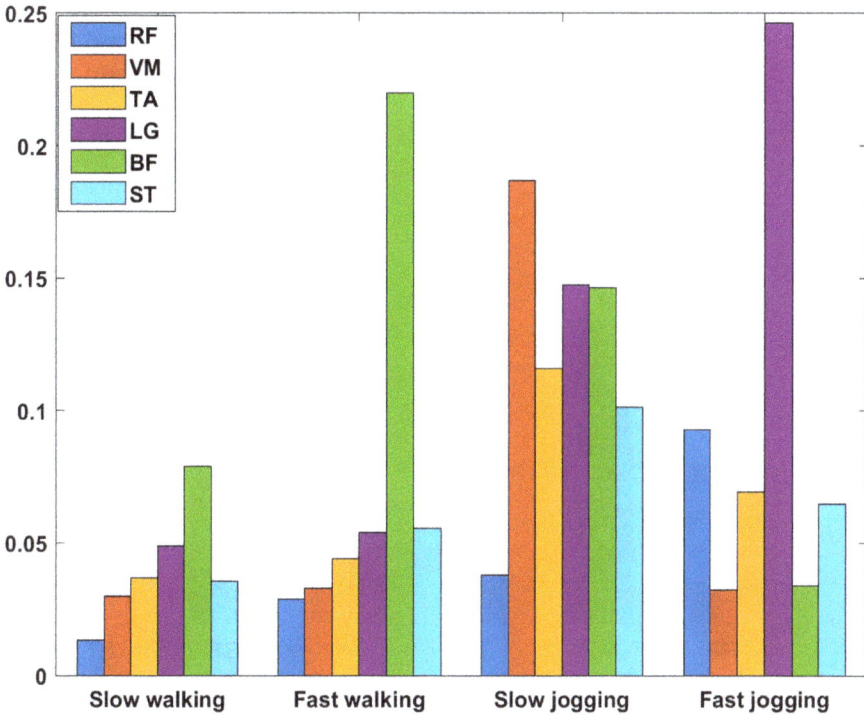

FIGURE 13.2   The averaged RMS over two sessions: 1) slow walking, 2) fast walking, 3) slow jogging, and 4) fast jogging.

was 0.0555. These RMS values were slightly higher than the RMS values of slow walking (0.02 compared to 0.003).

For slow jogging, interestingly, most of the muscles were highly activated compared to the other tasks. The VM had 0.187, which was the highest value among the four tasks and the most activated muscle for the slow jogging task. TA (0.116) and ST (0.1015) were also the highest among the four tasks. LG (0.1475) and BF (0.1465) were the second highest among all tasks. In contrast, RF (0.038) was just slightly higher than slow walking and fast walking. For fast jogging, the most activated muscle was LG with an RMS value of 0.2465. Surprisingly, the second highest RMS was seen in RF (0.093), which was also the highest value of RF among all four tasks. The RMS of TA (0.0695) and ST (0.065) were relatively high for fast jogging. However, the VM (0.0325) and BF (0.034) were inactivated in this task.

13.3.2 MEDIAN FREQUENCY

Considering median frequency, all four tasks had the same trend, where VM and LG were the most activated muscles (Figure 13.3–Figure 13.6). The RM,

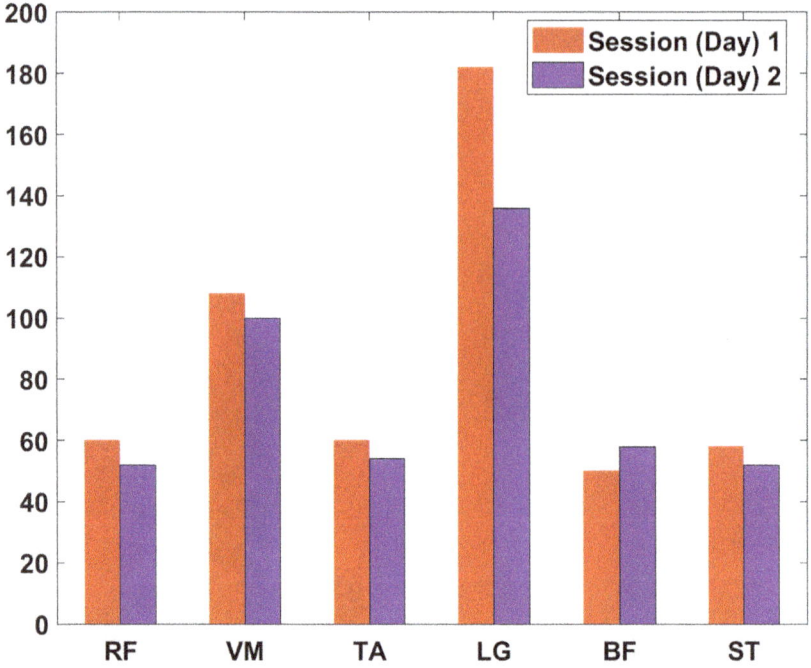

FIGURE 13.3   Median frequency of slow walking for two different sessions.

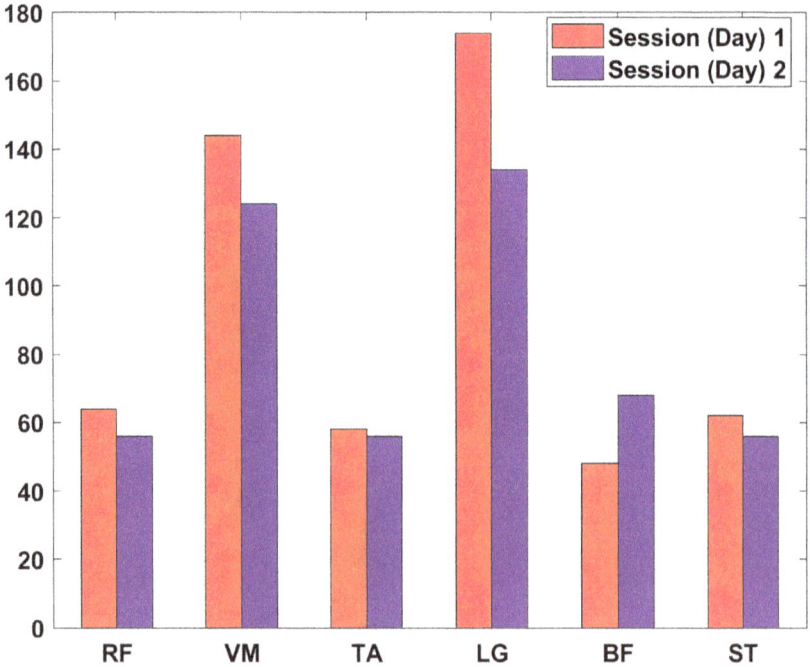

FIGURE 13.4   Median frequency of fast walking for two different sessions.

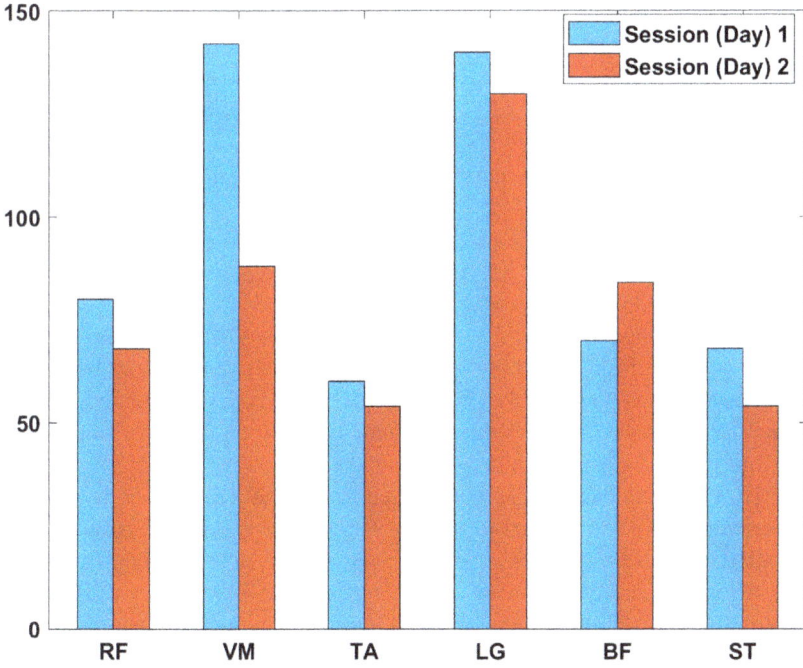

FIGURE 13.5 Median frequency of slow jogging for two different sessions.

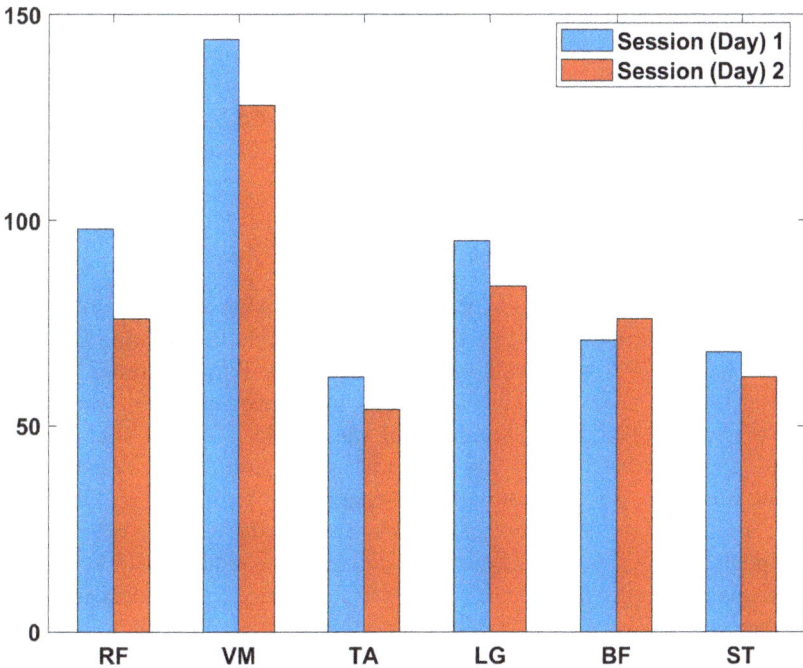

FIGURE 13.6 Median frequency of fast jogging for two different sessions.

TA, BM, and ST had relatively similar median frequency. With the increase of speed (from slow walking to fast jogging), most of the median frequency values increased as well, except LG, which decreased when the speed increased. More interestingly, apart from BF, the median frequency of all muscles showed a minor decline on the second session for all tasks. However, BF was the only muscle that showed a slight increase in median frequency on the second day session for all tasks.

For RF, although the day-two session had a lower median frequency than day-one session in all tasks, a constant rise of median frequency with speed was seen in both sessions. From slow walking to fast jogging, the median frequency was ranged from 60 to 98 on day one and 52 to 76 on day two. For VM, the median frequency was sharply grown from 109 (slow walking) to around (142–144) for the rest of the tasks on day one. On day two, there was no clear trend in the VM median frequency between the four tasks, as the values fluctuate between 88 (flow jogging) and 124 (fast jogging).

For TA, it had the most stable median frequency in all tasks for both session one and session two. All the median frequency values were around 60. After averaging the median frequency over two sessions, 57 was seen in slow walking, fast walking, and slow jogging and 58 was seen in fast jogging. For LG, slow walking had the highest frequency, 182 in session one and 136 in session two. It showed an opposite trend to all other muscles, where the median frequency steadily dropped from slow walking to fast jogging in both sessions.

For BF, the unique feature of its median frequency was that the second session was higher than the first session for all tasks. The median frequencies of slow walking and fast walking in the first session were 50 and 48. The slow jogging and fast jogging were 70 and 71, respectively. In the second session, the median frequency constantly increased from 58 to 84 (slow walking to slow jogging) but dropped to 76 in fast-jogging task. For semitendinosus, the median frequency was stable as well. In the first session, the median frequency ranged between 58 and 68. In the second session, the median frequency rose from 52 (slow walking) to 62 (fast jogging), apart from a slight drop in slow jogging.

13.3.3 MUSCLE UTILIZATION FOR WALKING AND JOGGING

Overall muscle utilization results for all four tasks (walking and jogging) are shown in Figure 13.7. The results indicate that for both slow and fast walking BF muscles are more active (32%—slow walking and 51%—fast walking) followed by LG and ST muscles. These findings are inconsistent with Stevenson et al. [10], who proposed that to maintain dynamic equilibrium during walking the BF muscles act to slow the forward progression of the body. On the other hand, for fast jogging LG muscle contribution was major (46%), whereas for slow jogging, VM, BF, and LG contributed almost similar fashion. A study that is conducted by Otter et al. [4] raised an interesting point, where they claimed that a burst of EMG amplitude could be found in a specific speed. This can explain the reason why some muscles showed extremely high amplitude in a certain task; for example, BF in fast walking, VM in slow jogging, and LG in fast jogging.

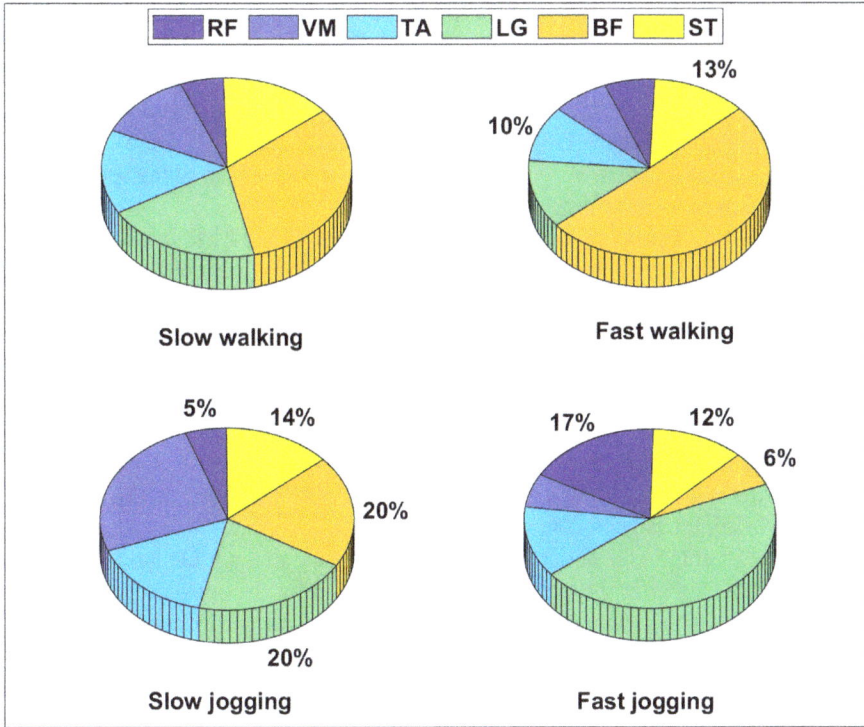

FIGURE 13.7   Muscle utilization for walking and jogging tasks.

13.4   DISCUSSION

In this present study, six muscles have been selected for investigating the locomotion of slow walking, fast walking, slow jogging, and fast jogging. By and large, the muscle activities had an increasing trend when speed increased. However, the RMS analysis indicates a huge variation of muscle activities between different tasks. Hardly can a common trend be found from RMS, even for muscles from the same muscular group, such as RF and VM which are innervated by the crural nerve (L2-L4) and BF and ST which are innervated by the sciatic nerve (L4-S2) [11]. This result agrees with the results from Gazendam and Hof [3]. Three factors that have been suggested by Gavilanes-Miranda et al. [12] may lead to the significant differences of muscle activities. The first is related to the inherent feature of different muscles such as mass, fibre length, transversal section area, and pennation angle. The second is that muscles from the same muscular group can have different functions. The third is that different muscles connect to a different number of joints.

The slow jogging task activated more muscles than the other three tasks investigated in our study. According to Tsuji et al. [1] and Gazendam & Hof [3], walking, jogging, and running are defined differently. Running is considered when there is an airborne phase, which means both legs are not in contact with the floor.

Consequently, the vertical ground reaction force will affect the muscle activities less than another task. Although our study did not specifically record whether there was an airborne phase in slow jogging and fast jogging, it is reasonable to consider that the fast jogging can be defined as running, while slow jogging was still walking with a faster speed than fast walking. Therefore, the muscle activities were maximized during slow jogging, as the vertical ground reaction force and the swing speed of legs affected the muscles the most.

Median power frequency shows a more reliable result than RMS in our study. This finding can be supported by the study of Smoliga et al. [13]. They concluded that MPF is the most reliable and precise measure to examine muscle activities in their study. In the present study, a more similar trend can be found with MPF among four tasks. In addition, MPF presented an inter-day difference, where all muscles had a higher MPF on the first day than the second day, except BF. MPF is widely used for examining muscle fatigue. The EMG signals are first converted from time domain to frequency domain by using fast Fourier transformation. The frequency that is located at the middle of the whole density spectrum is the median power frequency [13]. In the past studies [14, 15], it has been commonly accepted that decreasing MPF indicates muscle fatigue. Therefore, for our study, it can be concluded that muscle fatigue will still be present one day after the exercise. One explanation for the opposite pattern of BF is that BF is not a reliable muscle to examine muscle fatigue due to its inherent nature. Perttunen and Komi [8] observed that only BF muscle showed an inconsistent pattern among all muscles in their study. A large variation was presented in the EMG signal of BF, so that they concluded that BF is not suitable for analyzing muscle activities.

One limitation of our study is that the experiment was carried out in a laboratory setting, which is unlikely to imitate daily-life muscle activities that are affected by surface irregularities and ups and downs of the road. Furthermore, the speed of each task was not precisely instructed. Participants might perform the same task with different speeds. These factors can affect the accuracy of the result. In our future study, we plan to take the effect of speed into consideration. Different features of EMG signal will be applied as well. In addition, female participants would be highly desirable in our future study, as females are more vulnerable to injury than males during walking, jogging, or running.

13.5 CONCLUSION

This study has investigated the muscle activities during slow walking, fast walking, slow jogging, and fast jogging. It presents evidence that slow jogging is not an appropriate method for recovery from injury or daily exercise. Slow jogging can cause more muscle activities than other tasks. Moreover, from the median frequency, we can conclude that fatigue still affects muscle activities even after one day of exercise.

REFERENCES

[1] K. Tsuji, N. Soda, N. Okada, T. Ueki, K. Oba, Y. Ikedo, and H. Imaizumi, "A comparison of the lower limb muscles activities between walking and jogging performed at the same speed," *Journal of Physical Therapy Science*, vol. 24, pp. 23–26, 2012.

[2] E. S. Chumanov, C. Wall-Scheffler, and B. C. Heiderscheit, "Gender differences in walking and running on level and inclined surfaces," *Clinical Biomechanics (Bristol, Avon)*, vol. 23, pp. 1260–1268, 2008.

[3] M. G. Gazendam and A. L. Hof, "Averaged EMG profiles in jogging and running at different speeds," *Gait & Posture*, vol. 25, pp. 604–614, 2007.

[4] A. R. Otter, A. C. H. Geurts, T. Mulder, and J. Duysens, "Speed related changes in muscle activity from normal to very slow walking speeds," *Gait & Posture*, vol. 19, pp. 270–278, 2004.

[5] A. L. Hof, H. Elzinga, W. Grimmius, and J. P. K. Halbertsma, "Detection of non-standard EMG profiles in walking," *Gait & Posture*, vol. 21, pp. 171–177, 2005.

[6] R. Mellor and P. W. Hodges, "Motor unit synchronization of the vasti muscles in closed and open chain tasks," *Archives of Physical Medicine and Rehabilitation*, vol. 86, pp. 716–721, 2005.

[7] A. Nene, R. Mayagoitia, and P. Veltink, "Assessment of rectus femoris function during initial swing phase," *Gait & Posture*, vol. 9, pp. 1–9, 1999.

[8] J. R. Perttunen and P. V. Komi, "Effects of walking speed on foot loading patterns," *Journal of Human Movement Studies*, vol. 40, pp. 291–305, 2001.

[9] D. Stegeman and H. Hermens, *European Recommendations for Surface ElectroMyoGraphy*. Enschede, the Netherlands: Roessingh Research and Development, 1999.

[10] A. J. Stevenson, S. S. Geertsen, T. Sinkjær, J. B. Nielsen, and N. Mrachacz-Kersting, "Functionality of the contralateral biceps femoris reflex response during human walking," in *Replace, Repair, Restore, Relieve: Bridging Clinical and Engineering Solutions in Neurorehabilitation*. Berlin: Springer, 2014, pp. 765–773.

[11] M. Barbero, R. Merletti, A. Rainoldi, and SpringerLink (Online service). *Atlas of Muscle Innervation Zones Understanding Surface Electromyography and its Applications*. Berlin: Springer, 2012.

[12] B. Gavilanes-Miranda, J. Goiriena De Gandarias, and G. Garcia, "Walking and jogging: quantification of muscle activity of the lower extrmities," in *Advances in Applied Electromyography*, ed. J. Mizrahi. London: In Tech, 2011.

[13] J. M. Smoliga, J. B. Myers, M. S. Redfern, and S. M. Lephart, "Reliability and precision of EMG in leg, torso, and arm muscles during running," *Journal of Electromyography and Kinesiology*, vol. 20, pp. e1–9, 2010.

[14] S. Thongpanja, A. Phinyomark, P. Phukpattaranont, and C. Limsakul, "Mean and median frequency of EMG signal to determine muscle force based on time-dependent power spectrum," *Elektronika Ir Elektrotechnika*, vol. 19, pp. 51–56, 2013.

[15] G. T. Allison and T. Fujiwara, "The relationship between EMG median frequency and low frequency band amplitude changes at different levels of muscle capacity," *Clinical Biomechanics*, vol. 17, pp. 464–469, 2002.

Section 4

Wearables—Sensors
Signal Processing

14 Biosensors in Optical Devices for Sensing and Signal Processing Applications

Shwetha M. and Ganesh R. Naik

14.1 INTRODUCTION

Biosensors are considered a special sub classification of biomedical sensors: a biological recognition element, such as a purified enzyme, antibody, or receptor, that acts as a mediator and provides the selectivity required to sense the chemical component (usually referred to as the analyte) of interest, and a supporting structure that also acts as a transducer and is in close contact with the biological sensing sensed by the biological recognition element into a quantifiable measurement technique. The transducer's function is to translate the biological reaction into an optical, electrical, or physical signal proportionate to the concentration of a certain chemical. A blood pH sensor, for example, is not a biosensor according to this definition, despite the fact that it monitors a physiologically significant variable. It's nothing more than a chemical sensor that can be used to measure biological quantities [1].

In optical devices like ring resonator, mach zehender interferometer, and many other devices, the light passes through the waveguide and these devices can be used for various biomedical applications. These devices are helpful in point-of-care applications. Because of their sensitivity, specificity, speed, simplicity, and cost-effectiveness, nanomaterial-based biosensors have found widespread application in environmental and medical applications. Through the rapid and exact examination of numerous chemicals, silicon-based photonic biosensors integrated into semiconductor chip technology can lead to considerable breakthroughs in point-of-care applications, food diagnostics, and environmental monitoring. The ability to create sensor arrays is another advantage of SOI-based biosensors [2]. This enables for the simultaneous detection of many drugs. Interferometric or resonant structures can be used to build the SOI-based photonic biosensor. Enzyme-linked immunosorbent assay, electrochemical sensor, bilayer lipid membranes, high performance liquid chromatography, micro-electrode immunoassay, field immunoassay, and surface plasmon resonance spectroscopy are examples of competitive traditional biosensing techniques that are already commercially used. The key advantages of SOI-based ring resonators over standard benchtop electrochemical instruments are their small size, scalable mass-production, and quick readout [3].

DOI: 10.1201/9781003201137-18

The use of an optical microring resonator to improve biosensing efficiency is discussed [4]. Medical diagnostics are important in healthcare because they try to detect diseases early. To improve healthcare technologies, it is necessary to build efficient point-of-care diagnostics. Microring resonator-based dielectric waveguide integrated biosensors have the best performance of all the available biosensors. The evanescent fields produced by light propagation extend beyond the structure and are the sole source of field sensing. To improve the efficiency of biosensing, a couple of ways for designing high RI contrast silicon-on-insulator (SOI) microring resonators are addressed [5].

14.2 BIOSENSORS FOR POINT OF CARE

Point-of-care testing (POCT) is one of the most essential applications of biosensors. POCT is the practice of doing a diagnostic or prognostic test near the patient in order to offer rapid results, which means that the test must be quick and simple to perform without the use of expensive or specialized equipment. It also means that samples do not require the attention of a professional technician because no laboratory analysis is required, and there is no waiting period for the results to be collected and analyzed. In clinical applications for disease diagnosis and treatment monitoring at the point-of-care settings, the requirement for sensitive, robust, portable, and affordable biosensor platforms is of great importance. Accurate and timely diagnoses play a critical role in identifying the actual cause [6].

Figure 14.1 depicts the basic structure of a biosensor. POCT's ability to produce quick results in non-laboratory settings encourages greater patient-centered healthcare delivery and offers hope in the field of early detection. Currently, the attention and major focus are being switched toward early detection of a disease in order to efficiently manage therapies and lower patient mortality rates. The capacity of any biosensor to discriminate the target analyte from interfering molecules is critical to its utility. As a result, researchers have used a variety of biorecognition elements over the years, including antibodies, nucleic acids, cells, bacteriophages, and proteins. The next section will go through each of these biorecognition aspects and their strengths [7].

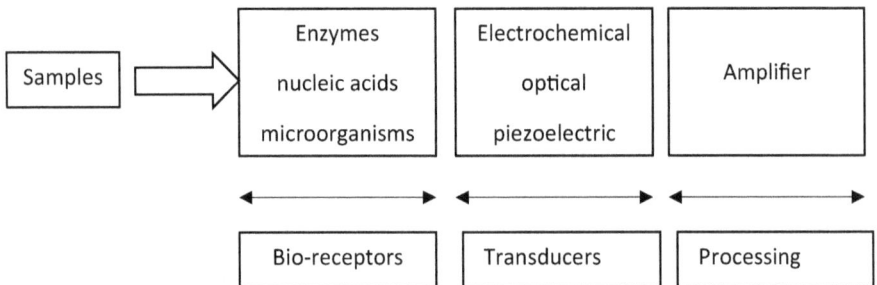

FIGURE 14.1   Basic structure of a biosensor.

14.2.1 Biorecognition Elements of Medical Biosensors

The biorecognition element is employed to encapsulate the target bioanalyte, and the transducer which subsequently generates a quantifiable signal that can be analyzed. A substantial quantifiable change in biosensor signal as a function of modest changes in bioanalyte concentration is referred to as high sensitivity. A biosensor is comprised of a biorecognition element coupled to a transducer. A few biorecognition elements are discussed here.

14.2.2 Antibodies

Antibodies are Y-shaped proteins that are naturally created by biological systems as a defense mechanism against invading germs or viruses. They attach to certain species (antigens) with a high degree of specificity through a combination of hydrogen bonds and other noncovalent interactions in the cleft of the protein molecule. The advantage is that they may be 'grown' by injecting the target antigen into laboratory animals; the animal's natural defensive systems are to generate antibodies to the antigen. The antibody can then be mounted onto a transducer to create a biosensor. One issue is that no easily detectable byproducts such as electrons or redoxactive species are created when antibodies attach to their antigens to form a complex. Another concern is that because antibody and antigen attach so tightly, there is no substrate turnover; the binding is basically irreversible. Sensors in this scenario are prone to saturation and can only be used once. Although extremes of pH or extremely ionic solutions can reverse the antibody–antigen reaction, they can destroy the antibody, resulting in permanent loss of function[8]. There has been a lot of research towards label-less detection systems that can merely identify the binding event directly without need for labeling.

14.2.3 Nucleic Acid Probes

Nucleic acid study and detection continue to pay close attention to a variety of genetic diseases and disorders because of their relationship with cancer. Different infections have their own nucleic acid sequences, which have been used in biosensors to detect them. DNA biosensors have been constructed using nucleic acids, including single-strand DNA, peptide nucleic acids (PNA), locked nucleic acids, G-quadraplexes, and DNAzymes as biorecognition elements.

14.2.4 Aptasensors

Aptamers are oligonucleotides that have the ability to bind to a wide range of targets like proteins, drugs, peptides, and cells. When aptamers connect to their targets, they frequently undergo conformational changes, such as folding around a tiny molecule. Aptamers are excellent candidates for sensing since these structural changes are generally easy to detect. Aptamers have several advantages over other types of recognition elements, such as enzymes. Figure 14.2 outlines the components of a biosensor.

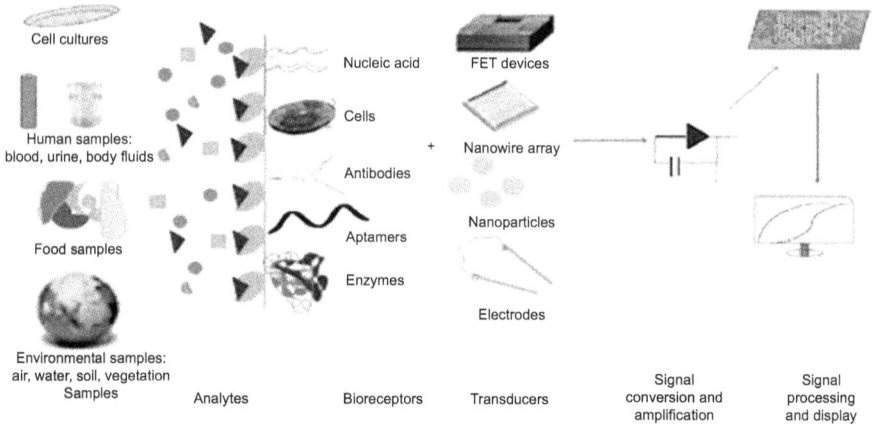

FIGURE 14.2 Outline displaying the components of a biosensor.

14.3 RING RESONATOR-BASED BIOSENSING

Photonic sensors based on SOI are commonly realized as interferometers or resonators, with the latter benefiting from higher sensitivities and smaller sizes, making them better suited for integrated sensing systems. Ring resonator technology, as originally proposed, is based on the looped propagation of light in the form of whispering gallery modes, which creates a resonance at frequencies that satisfy the resonance criteria. Due to total internal reflection, light going through the waveguides remains within the waveguides and forms an evanescent field near the waveguide surface, which interacts with the environment and enables the detection of analytes. A looped waveguide with a loop diameter of a few 10 to a few 100 m is positioned adjacent to a linear waveguide in the simplest devices [9]. Figure 14.3 shows the schematic of the principle of a ring resonator and its transmission spectrum.

Wavelengths (resonance wavelengths) with a harmonic relation to the ring's circumference of $2\pi Rn = m\lambda r$ (m being the mode number, R the ring radius, λr the resonance wavelengths, and n the effective refractive index) are trapped in the ring as light propagates through the straight waveguide from the in-coupling to the out-coupling fiber. Organic molecules immobilized on the ring vary n around the resonator, which is represented as Δn, trapping longer wavelengths of light. Photodetectors are used to monitor transmission spectra with resonance wavelengths and the sensing-induced resonance shift. To improve sensing ability, reduce the impact of production tolerances, and improve the detection limit, ring resonators with various topologies and geometries have been designed. A micro-ring resonator is used to identify proteins in most cases.

14.3.1 OPTICAL RING RESONATOR-BASED HB SENSOR

The interaction between the resonating light's evanescent field within the resonator and bioparticles (i.e. HB) present in the ambient environment such as blood causes

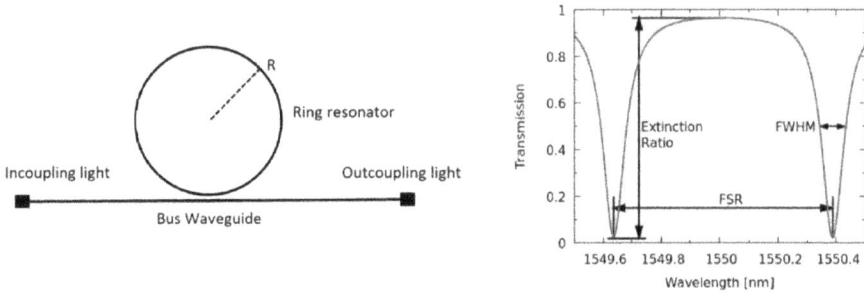

FIGURE 14.3   Schematic of the principle of ring resonators and its transmission spectrum.

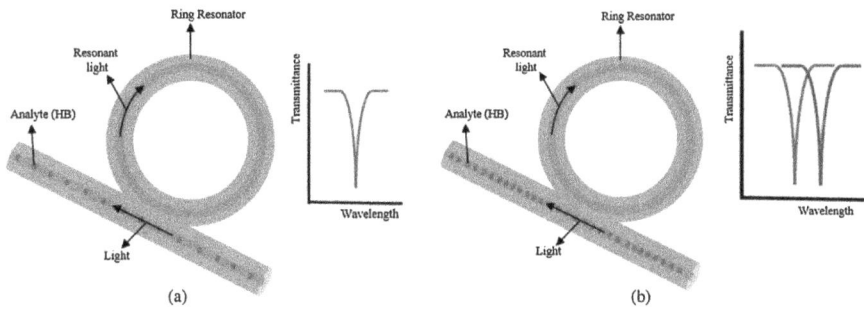

FIGURE 14.4   Conceptual illustration of an optical ring resonator HB sensor: (a) the solution contains a low concentration of HB, and (b) the solution contains a high concentration of HB. The HB amount has modified the interaction with the evanescent light field, resulting in a shift in the resonance wavelength (red color) of the resonator.

the input light's behavior to change in a ring resonator. The effective refractive index of the inner medium is affected by the presence of target bioparticles in the solution, resulting in a deviation of the resonator resonance conditions. The number of bioparticles (such as HB) in the medium affects the resonator's resonance wavelength shifting, which is the outcome of this interaction. By injecting the solution into the ring resonator, the generalized sensing mechanism of HB in a blood sample is accomplished [5].

By injecting the solution into the active layer or inner medium (i.e. core) of the straight waveguide of a ring resonator, the optical characteristics of the active layer or inner medium (i.e. core) can be changed. Depending on the amount of HB in the blood, resonant light interacts with target HB molecules, changing the effective RI of the resonating mode. A change in HB concentration can be shown by the spectral response shift.

Figure 14.4 shows the illustration of an optical ring resonator HB sensor: (a) the solution contains a low concentration of HB, and (b) the solution contains a high concentration of HB. The HB amount has modified the interaction with the evanescent light field, resulting in a shift in the resonance wavelength (red color) of the resonator. Photonic sensors based on optical ring resonators in SOI-technology are the subject of this research. A silicon-based waveguide stands on top of a buried oxide substrate in a ring resonator. A grating coupler or butt coupling is used to couple the light from a tunable laser or a super lumineszenz diode to the waveguide. If the resonance condition is met, the light is partially linked to the ring resonator, resulting in resonance peaks in the output spectrum. Depending on the light source, the light is linked to a photodetector or an optical spectrum analyzer at the output. Laser and photodiodes, as well as photonic sensors, can now be implemented on the same chip following recent improvements in heterogeneous and monolithic integration.

However, evanescent field sensing is the overall sensing process that underpins their operation. The resonance state of the ring resonator is changed if the evanescent field is changed due to an immobilization of analytes on the silicon waveguide, resulting in a resonance wavelength shift. By detecting either the resonance peak shift Dl or the intensity change DI, antibodies that only attach to their matching antigens can be detected with high specificity. The silicon waveguide is the most important component of all integrated photonic biosensors. The guided mode's evanescent field is partially penetrating the cladding material, where the analyte is found. Each waveguide configuration has a variable amount of light penetrating through the cladding, which correlates with undesired optical losses; i.e. the more light is penetrating into the cladding the higher the optical losses due to absorption and scattering [10].

Six important parameters determine a biosensor's performance: sensitivity, dynamic range, accuracy, sample-to-answer time, robustness, and amenability. The lowest concentration change that can be observed is referred to as sensitivity. It is sometimes, but not always, equivalent to the limit of detection, or the lowest concentration detectable.

14.3.2 GRAPHENE-BASED SPR BIOSENSORS

Surface plasmon polariton waves are used in SPR Biosensors to analyze the interaction between the biorecognition element and the sensor surface. In one experiment, graphene was adsorbed onto the surface of a gold sheet as a biorecognition element. This combination demonstrated that biomolecules with carbon-based ring structures, such as ssDNA, securely adhered to the surface and enhanced adsorption efficiency. This resulted in a local increase in refractive index (RI) along the metal's surface causing a deviation in the propagation constant of the surface plasmon polariton, hence increasing the sensitivity of the optical measurement [11].

14.3.3 FIBER BRAGG GRATING SENSORS

Now a days, most fiber optic sensor systems make use of fiber Bragg grating technology. Researchers have gained enormous attention in the field of fiber Bragg grating (FBG)-based sensing due to its inherent advantages, such as small size, fast response, distributed sensing, and immunity to the electromagnetic field. Fiber Bragg grating technology is popularly used in measurements of various physical parameters, such as pressure, temperature, and strain for civil engineering, industrial engineering, military, maritime, and aerospace applications. Nowadays, strong emphasis is given to structure health monitoring of various engineering and civil structures, which can be easily achieved with FBG-based sensors. Depending on the type of grating, FBG can be uniform, long, chirped, tilted, or phase shifted having periodic perturbation of refractive index inside core of the optical fiber. Basic fundamentals of FBG and recent progress of fiber Bragg grating-based sensors used in various applications for temperature, pressure, liquid level, strain, and refractive index sensing have been reviewed [12]. A major problem of temperature cross sensitivity that occurs in FBG-based sensing requires temperature compensation technique. This is a future work for obtaining the results.

14.4   OPTOELECTRONIC DEVICE APPLICATIONS

Optoelectronic devices and components are those electronic devices that operate on both light and electrical currents. This can include electrically driven light sources such as laser diodes and light-emitting diodes, components for converting light to an electrical current such as solar and photovoltaic cells and devices that can electronically control the propagation of light. Optoelectronic devices can be divided into two—light sensitive devices and light generating devices.

A photodetector is a type of optoelectronic device that can convert light signals into electrical signals and is widely used in civil and military applications such as satellite communication, remote sensing, missile guidance, night vision, and motion detection.

Photovoltaic devices, which turn sunlight into electricity, are a viable alternative for addressing the energy crisis. Due to its unique optoelectronic features, simple device fabrication procedure, solution processable approach, and tunable band structure, two-dimensional materials-based PV devices have received a lot of attention [13].

When an electrical current is conducted through an LED, it produces light. New types of LEDs with high efficiency, low power consumption, high brightness, and solution-processed manufacturing will play an increasingly essential role in flexible and wearable electronics in the future.

14.5   SYNTHESIS, CHARACTERIZATION, AND
APPLICATIONS OF NANOMATERIALS

Nanomaterials have a lot of potential for increasing the performance of electrochemical POC devices. These nanostructures have been actively investigated for POC medical applications due to their unique qualities, such as enlarged surface area, exceptionally high electrical conductivity, small size sufficient to interact with biological molecules, and stability [14]. Due to their unique features, metallic nanoparticles, carbon nanotubes, and graphene have been intensively explored, and the primary methods for production, characterization, and application of these nanomaterials in POC medical devices are reviewed.

14.5.1 Synthesis

Nanomaterials' electrical, optical, and catalytic capabilities are highly influenced by their size, composition, form, and purity. The method of synthesis and the conditions under which it is carried out can have a substantial impact on these parameters, which has been the subject of extensive research.

14.5.2 Chemical Reduction

Chemical reduction of ions by inorganic or organic chemicals, such as sodium citrate, sodium borohydride, or ascorbate, is one of the most frequent methods for generating metallic nanoparticles. The reaction is usually started using an aqueous solution containing ions of the desired metal (such as Ag+ or Au3+). After that, the reducing agent is introduced, and the ions are reduced to their metallic form, followed by agglomeration and cluster formation. The inclusion of a stabilizer agent, such as polymers, citrate, or any other surface agent, can prevent these clusters from growing and forming nanoparticles.

14.5.3 Characterization

Nanotechnology has advanced in recent years due to the availability of powerful characterization tools. A key component of nanotechnology research has been the ability to examine nanomaterial size, shape, surface qualities, composition, purity, stability, and dispersion. UV–Visible (UV–Vis) spectroscopy has become one of the most widely used techniques in nanomaterial characterization due to its capacity to analyze nanoparticle size, concentration, aggregation state, and bioconjugation in a simple, quick, and low-cost manner. Transmission electron microscopy (TEM) is a technique that uses high voltage to accelerate electrons toward a thin material, resulting in images with a spatial resolution of less than 0.1 nm. This method can provide valuable information regarding the shape, orientation, and structures of nanomaterials' narrow areas.

14.5.4 Application of Nanomaterials in Point-of-Care Medical Biosensors

Because of their physical, electrical, electrochemical, and chemical capabilities, carbon-based nanomaterials such as CNTs and graphene have drawn interest in the development of POC biosensors. Because of their high surface-to-volume ratio and electron-transport properties, one-dimensional nanostructures are particularly appealing for building high-density arrays for ultrasensitive protein detection [14]. Such nanostructures hold potential for the creation of various biomarker assays in ultra-small sample volumes, which is particularly appealing in the medical field. The ability to detect entire viruses and bacteria is critical for determining infection quickly and precisely, as well as identifying potentially dangerous places and contaminated equipment in health centers to avoid hospital-acquired diseases. Because of their ease of use, POC devices have been viewed as excellent candidates for whole-organism detection mainly due to their simple preparation, avoiding the typical steps of virus isolation, extraction, purification, and amplification of biomolecules [15].

14.6 CHALLENGES FACED IN SIGNAL PROCESSING OF THE SIGNALS

Various noise sources, including those introduced by the optical measuring setup, typically limit the final detection limit of optical biosensors. While expensive instrumentation changes may reduce noise, a simpler solution that can help all sensor platforms is the use of signal processing to minimize noise's negative impacts. Calculating an average phase difference between the filtered analyte and reference signals allows for a significant reduction in the detection limit, bringing it closer to current state-of-the-art approaches. It is possible to gather experimental data sets of thin film porous silicon sensors in buffered solution and complex media [16].

Real-time acquisition of data directly from the source by direct electrical connections to instruments avoids the need for people to measure, encode, and enter the data manually. Sensors attached to a patient convert biological signals, like blood pressure, pulse rate, mechanical movement, and electrical activity, e.g., of heart, muscle, and brain, into electrical signals, which are transmitted to the computer. The signals are sampled periodically and are converted to digital representation for storage and processing. Automated data-acquisition and signal-processing techniques are particularly important in patient monitoring settings [17].

Optical signal processing unites different fields of optics and signal processing. To be specific, it unites simple computerized signal, and propelled information to accomplish fast signal processing capacities that can conceivably work at the line rate of fiber optic interchanges. Data can be encoded in abundancy, stage, wavelength, polarization, and spatial highlights of an optical wave to accomplish high-limit transmission.

Biosensors are typically made up of three basic components, as shown in Figure 14.5. A biological sensing element, a physicochemical detector or transducer, and a signal

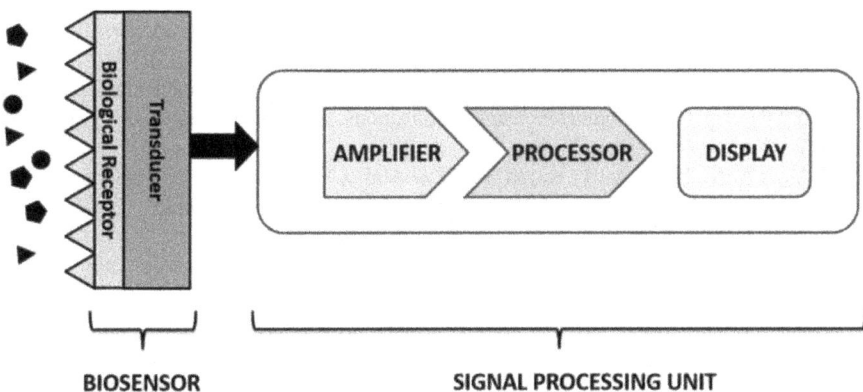

FIGURE 14.5 Biosensor design showing the various components necessary for generating a signal.

processing system are instances of biosensors. To generate a signal, biological sensing elements interact with the analyte of interest. Tissues, microbes, organelles, cell receptors, enzymes, antibodies, and nucleic acids are examples of sensing elements [18]. The transducer converts the signal created by the interaction of the sensing element and the analyte of interest into a measurable and quantifiable electrical signal. As a result, the signal processing system amplifies the electrical signal and sends it to a data processor, which converts it into a measurable signal such as a digital display, printout, or color change.

14.7   CONCLUSION

In this chapter, the review of biosensors in optical devices has been done, as we know optical biosensors are projected to have the greatest impact in the healthcare, biomedical, and biopharmaceutical industries. They can provide novel small-scale analytical tools as well as large-scale high-throughput sensitivity screening of a wide range of samples for a variety of characteristics. Optical biosensors have been successfully tested in many fields such as medicine, pharmacy, food safety, environment, biotechnology, defense, and security, and various sensitivity-enhancement techniques for optical biosensors can be developed to improve the signal-to-noise ratio and reduce the detection limit. Optical biosensors, on the other hand, are still in the early stages of development and are largely used in academic and pharmaceutical settings. The only exceptions are lateral flow assay biosensors marketed as test strips for home, point-of-care, or laboratory testing. In addition, the challenges faced in biosensor signal processing are outlined for future work.

REFERENCES

1. Wang, J., et al. Silicon-Based Integrated Label-Free Optofluidic Biosensors: Latest Advances and Roadmap. *Adv. Mater. Technol.* **5** (2020).
2. Bogaerts, W., et al. Silicon Microring Resonators. *Laser Photonics Rev.* **6**, 47–73 (2012).
3. Steglich, P., Hülsemann, M., Dietzel, B. & Mai, A. Optical Biosensors Based on Silicon-On-Insulator Ring Resonators: A Review. *Molecules* **24** (2019).
4. Sarkaleh, A. K., Lahijani, B. V., Saberkari, H. & Esmaeeli, A. Optical Ring Resonators: A Platform for Biological Sensing Applications. *J. Med. Signals Sens.* **7**, 185–191 (2017).
5. Ajad, A. K., Islam, M. J., Kaysir, M. R. & Atai, J. Highly Sensitive Bio Sensor Based on WGM Ring Resonator for Hemoglobin Detection in Blood Samples. *Optik (Stuttg).* **226**, 166009 (2021).
6. Chen, Y. T., et al. Review of Integrated Optical Biosensors for Point-of-Care Applications. *Biosensors* **10**, 1–22 (2020).
7. Narayan, R. J. *Medical Biosensors for Point of Care (POC) Applications. Medical Biosensors for Point of Care (POC) Applications* (2016). https://www.sciencedirect.com/book/9780081000724/medical-biosensors-for-point-of-care-poc-applications.
8. Pereira, A., Sales, M. & Rodrigues, L. *Biosensors for Rapid Detection of Breast Cancer Biomarkers. Advanced Biosensors for Health Care Applications* (USA Philadelphia: Elsevier Inc., 2019). doi:10.1016/b978-0-12-815743-5.00003-2.

9. Malthesh, S. & Krishnaswamy, N. Improvement in Quality Factor of Double Microring Resonator for Sensing Applications. *J. Nanophotonics* **13**, 1 (2019).

10. Steglich, P., Mai, C., Villringer, C. & Mai, A. Direct Observation and Simultaneous Use of Linear and Quadratic Electro-optical Effects. *J. Phys. D. Appl. Phys.* **53** (2020).

11. Nurrohman, D. T. A Review of Graphene-Based Surface Plasmon Resonance and Surface-Enhanced Raman Scattering Biosensors: Current Status and Future Prospects. *Nanomaterials* **11**, 1 (2021).

12. Sahota, J. K., Gupta, N. & Dhawan, D. Fiber Bragg Grating Sensors for Monitoring of Physical Parameters: A Comprehensive Review. *Opt. Eng.* **59**, 1 (2020).

13. Shwetha, M., Prajwal, P., Praveen, N. M. & Narayan, K. Modeling and Analysis of Microring Resonator for Bio-sensing Applications. In *2018 3rd IEEE International Conference on Recent Trends in Electronics, Information and Communication Technology, RTEICT 2018—Proceedings* (2018), IEEE. doi:10.1109/RTEICT42901.2018.9012287.

14. Shwetha, M., Reddy, N. K., Pattnaik, P. K. & Narayan, K. Design and Analysis of Silicon Ring Resonator for Bio-sensing Application. In *Proceedings of SPIE—The International Society for Optical Engineering* 10690 (2018). https://www.spiedigitallibrary.org/.

15. Steglich, P. Silicon-on-Insulator Slot Waveguides: Theory and Applications in Electro-Optics and Optical Sensing. In *Emerging Waveguide Technology* (London: InTech, 2018). doi:10.5772/intechopen.75539.

16. Pan, Q., Brulin, D. & Campo, E. Current Status and Future Challenges of Sleep Monitoring Systems: Systematic Review. *JMIR Biomed. Eng.* **5**, e20921 (2020).

17. Rajeswari, J. & Jagannath, M. Advances in Biomedical Signal and Image Processing—A Systematic Review. *Informatics Med. Unlocked* **8**, 13–19 (2017).

18. Henriksson, A., Kasper, L., Jäger, M., Neubauer, P. & Birkholz, M. An Approach to Ring Resonator Biosensing Assisted by Dielectrophoresis: Design, Simulation and Fabrication. *Micromachines* **11** (2020).

Index

A

adaptive filter, 122
adaptive inverse filtering method, 95
adjustable artifact removal, 157
arousal prediction, 213
artefacts, 120
artificial intelligence, 200

B

band power, 13
bayesian, 160
bidirectional long-short term memory, 46
biosensors for point of care, 272

C

cardiovascular diseases, 139
cepstral peak prominence, 94
channel selection, 26
chemical reduction, 278
classification, 183
classification and regression trees, 36
common spatial patterns, 8
confidence interval, 95
continuos wavelet transform, 18
convolution neural network, 139

D

decision trees, 158
directed acyclic graph, 160
discrete wavelet transform, 20, 125

E

electrical control activity, 62
electroencephalography, 115, 119, 149
electrogastrography, 62
electromyography, 225
empirical mode decomposition, 125
entropy, 37
epilepsy, 126
extreme learning machine, 183

F

fast fourier transform, 14
feature classification, 28
feature transformation, 26
fourier analysis, 13

G

galvanic skin response, 173

H

harmonics to noise ratio, 94
histogram, 11
human analysis of transcripts, 44

I

independent component analysis, 7, 122
intrinsic mode functions, 143

L

linear discriminant analysis, 29
linear regression, 174
logistic model trees, 38
low pass filter, 22, 63

M

mean absolute error, 210
median frequency, 261
medical biosensors, 278
mel-frequency cepstral coefficients, 202
multilayer perceptron, 174, 207
muscle utilisation, 264
mutual information, 174

N

neural network, 31

O

optoelectronic devices, 277

P

Parkinson's disease, 91
particle swarm optimization, 25
power spectral density, 73, 76
principal component analysis, 6
pseudo wigner ville distribution, 125

Q

quadratic discriminant analysis, 30, 176

R

radial basis function, 208
relative absolute error, 210
ring reonator biosensing, 274
root mean square, 10, 260
root mean square energy, 95
running spectral analysis, 71

S

segmentation, 4
short time fourier transform, 17, 23
signal processing, 5

signal transformation, 124
smote, 161
spatial filtering, 5
Spearman's correlation, 210
support vector machine, 183, 207

V

valence prediction, 212
vibrotile feedback, 236

W

wavelet denoising, 123
wavelet transformation, 156
Wilcoxon rank test, 106
windowing, 4

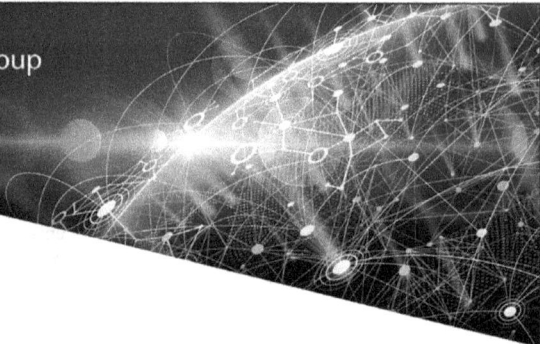

For Product Safety Concerns and Information please contact our EU
representative GPSR@taylorandfrancis.com
Taylor & Francis Verlag GmbH, Kaufingerstraße 24, 80331 München, Germany